METHODS IN MOLECULAR BIOLOGY™

Series Editor
John M. Walker
School of Life Sciences
University of Hertfordshire
Hatfield, Hertfordshire, AL10 9AB, UK

For other titles published in this series, go to
www.springer.com/series/7651

Regulatory T Cells

Methods and Protocols

Edited by

George Kassiotis

Division of Immunoregulation, MRC National Institute for Medical Research, London, UK

Adrian Liston

VIB Autoimmune Genetics Laboratory, K.U. Leuven, Leuven, Belgium

Humana Press

Editors
George Kassiotis
Division of Immunoregulation
MRC National Institute
for Medical Research
London
UK
gkassio@nimr.mrc.ac.uk

Adrian Liston
VIB Autoimmune Genetics Laboratory
K.U. Leuven, Leuven
Belgium
adrian.liston@vib.be

ISSN 1064-3745 e-ISSN 1940-6029
ISBN 978-1-4939-6158-0 ISBN 978-1-61737-979-6 (eBook)
DOI 10.1007/978-1-61737-979-6
Springer New York Dordrecht Heidelberg London

Printed on acid-free paper

Humana Press is part of Springer Science+Business Media (www.springer.com)

Preface

Given the fundamental importance of immune regulation for control over effective immunity and avoidance of autoimmunity and immune pathology, the existence of multiple immune regulators with overlapping fields of function is expected. The presence of a regulatory subset of T cells with naturally-endowed immune suppressive activity has been postulated for more than three decades. We now recognize regulatory T cells as the most numerous subset of immune regulators in the body, with critical functions in a wide array of immune responses. Despite this current acceptance, mechanisms of regulatory T cell immune modulation, and indeed their very existence, remained contentious for many years. A significant contribution to this uncertainty was due to methodological limitations, whereby the presence of regulatory T cells was usually assessed indirectly, by the reduction they caused on the more readily-measurable immune response of effector cells. The collapse of the suppressor T cell edifice built without the foundations of robust lineage markers in the 1980s (Fig. 1) added further to the skepticism.

The recent revival of regulatory T cells has been driven by methodological success in identifying reliable lineage markers, first with the use of the IL-2 receptor α chain and other markers, and more recently using the transcription factor FoxP3. This capacity to directly identify regulatory T cells has driven the exponential growth in publications on regulatory T cells since 2000 (Fig. 1). Further, methodological innovations outlined in this book have lead to insights on the suppressive mechanisms and biology of regulatory T cells. Although many of these assays still remain complex and, furthermore, they may not always assay a property unique to regulatory T cells, they have firmly established this subset in the immunological center stage and have been instrumental in the dissemination of both the expertise and interest in regulatory T cells, reflected in the wealth of scientific

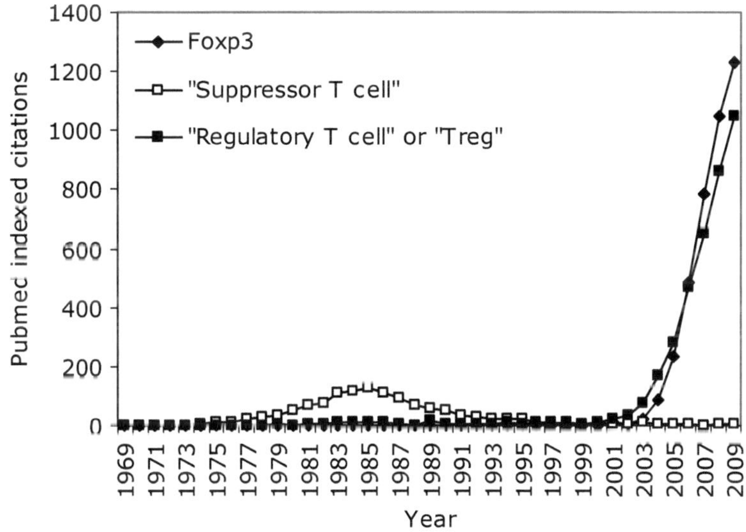

Fig. 1. Annual publication rates of papers indexed in Pubmed under "Suppressor T cell", "Regulatory T cell" (or "Treg") and "Foxp3".

publications in this field, to the point where ~4% of all immune-related papers in 2007 were related to regulatory T cells. The aim of this volume is to offer a collection of current methods and protocols for the study of regulatory T cells. These are distilled through several years of optimization and standardization to allow reliable and reproducible use by both the young and experienced cellular and molecular immunologists.

London, UK *George Kassiotis*
Leuven, Belgium *Adrian Liston*

Contents

PART IV HUMAN

Contributors

ARNE N. AKBAR • *Department of Immunology, Infection, and Immunity, University College London, London, UK*
SUZANNE ASAD • *T Cell Biology Research Program, Centenary Institute and Faculty of Medicine, University of Sydney, Sydney, NSW, Australia*
MARIA BETTINI • *Department of Immunology, St. Jude Children's Research Hospital, Memphis, TN, USA*
ALEXANDER G. BETZ • *Laboratory of Molecular Biology, Medical Research Council, Cambridge, UK*
JANE H. BUCKNER • *Benaroya Research Institute at Virginia Mason, Seattle, WA, USA*
HARALD VON BOEHMER • *Dana-Farber Cancer Institute, Harvard Medical School, Boston, MA, USA*
ANDREW J. CATON • *The Wistar Institute, Philadelphia, PA, USA*
TIMOTHY M. CLAY • *Departments of Surgery and Immunology, Duke University Medical Center, Durham, NC, USA*
LAUREN W. COLLISON • *Department of Immunology, St. Jude Children's Research Hospital, Memphis, TN, USA*
CAROLIN DANIEL • *Dana-Farber Cancer Institute, Harvard Medical School, Boston, MA, USA*
BARBARA FAZEKAS DE ST GROTH • *T Cell Biology Research Program, Centenary Institute and Faculty of Medicine, University of Sydney, Sydney, NSW, Australia*
EVA D'HENNEZEL • *Center for the Study of Host Resistance, Montreal QC, Canada*
KIM J. HASENKRUG • *Laboratory of Persistent Viral Diseases, Rocky Mountain Laboratories, National Institute of Allergy and Infectious Diseases, National Institutes of Health, Hamilton, MT, USA*
AMY C. HOBEIKA • *Department of Surgery, Duke University Medical Center, Durham, NC, USA*
LESZEK IGNATOWICZ • *Medical College of Georgia, Augusta, GA, USA*
MAKIO IWASHIMA • *Department of Microbiology and Immunology, Stritch School of Medicine, Loyola University Chicago, Maywood, IL, USA*
NADIA M. JEREMIAH • *VIB Autoimmune Genetics Laboratory, K.U. Leuven, Leuven, Belgium*
KATHARINA LAHL • *School of Medicine, Stanford University, Stanford, CA, USA*
LORETTA LEE • *T Cell Biology Research Program, Centenary Institute and Faculty of Medicine, University of Sydney, Sydney, NSW, Australia*
ADRIAN LISTON • *VIB Autoimmune Genetics Laboratory, K.U. Leuven, Leuven, Belgium*
H. KIM LYERLY • *Department of Surgery and the Duke Comprehensive Cancer Center, Duke University Medical Center, Durham, NC, USA*
DEREK C. MACALLAN • *Centre for Infection, Cellular, and Molecular Medicine, St George's, University of London, London, UK*

JOOST P.M. VAN MEERWIJK • *Tolerance and Autoimmunity Section, Institut National de la Santé et de la Recherche Médicale, U563, Toulouse, France; Université Toulouse III Paul Sabatier, Toulouse, France; Institut Universitaire de France, Paris, France*
MICHAEL A. MORSE • *Department of Medicine, Duke University Medical Center, Durham, NC, USA*
LARA M. MYERS • *Laboratory of Persistent Viral Diseases, Rocky Mountain Laboratories, National Institute of Allergy and Infectious Diseases, National Institutes of Health, Hamilton, MT, USA*
CLÉMENCE NOUZÉ • *Tolerance and Autoimmunity Section, Institut National de la Santé et de la Recherche Médicale, U563, Toulouse, France; Université Toulouse III Paul Sabatier, Toulouse, France*
TAKUYA OSADA • *Department of Surgery, Duke University Medical Center, Durham, NC, USA*
RAFAL PACHOLCZYK • *Medical College of Georgia, Augusta, GA, USA*
LISE PASQUET • *Tolerance and Autoimmunity Section, Institut National de la Santé et de la Recherche Médicale, U563, Toulouse, France; Université Toulouse III Paul Sabatier, Toulouse, France*
SHARON PEPLINSKI • *Department of Surgery, Duke University Medical Center, Durham, NC, USA*
CIRIACO A. PICCIRILLO • *Center for the Study of Host Resistance, Montreal, QC, Canada*
MEENU R. PILLAI • *Department of Immunology, St. Jude Children's Research Hospital, Memphis, TN, USA*
HIDDE PLOEGH • *Department of Biology, Whitehead Institute for Biomedical Research, Massachusetts Institute of Technology, Cambridge, MA, USA*
JEROLD E. REHG • *Department of Pathology, St. Jude Children's Research Hospital, Memphis, TN, USA*
SHIMON SAKAGUCHI • *Department of Experimental Pathology, Institute for Frontier Medical Sciences, Kyoto University, Kyoto, Japan; WPI Immunology Frontier Research Center, Osaka University, Suita, Japan*
MILKA SARRIS • *Laboratory of Molecular Biology, Medical Research Council, Cambridge, UK*
ANYA SCHNEIDER • *Benaroya Research Institute at Virginia Mason, Seattle, WA, USA*
DONALD M. SIMONS • *The Wistar Institute, Philadelphia, PA, USA*
NAGENDRA SINGH • *Medical College of Georgia, Augusta, GA, USA*
TIM SPARWASSER • *Institute of Infection Immunology, TWINCORE, Center for Experimental and Clinical Infection Research, Hannover, Germany*
DARIO A.A. VIGNALI • *Department of Immunology, St. Jude Children's Research Hospital, Memphis, TN, USA*
MILICA VUKMANOVIC-STEJIC • *Department of Immunology, Infection, and Immunity, University College London, London, UK*
CREG J. WORKMAN • *Department of Immunology, St. Jude Children's Research Hospital, Memphis, TN, USA*
YAN ZHANG • *Centre for Infection, Cellular, and Molecular Medicine, St George's, University of London, London, UK*

YE ZHENG • *Nomis Center for Immunobiology and Microbial Pathogenesis, The Salk Institute for Biological Studies, La Jolla, CA, USA*
ERHUA ZHU • *T Cell Biology Research Program, Centenary Institute and Faculty of Medicine, University of Sydney, Sydney, NSW, Australia*

Part I

Introduction

Chapter 1

Regulatory T Cells: History and Perspective

Shimon Sakaguchi

Abstract

Despite the skepticism that once prevailed among immunologists, it is now widely accepted that the normal immune system harbors a T-cell population, called regulatory T cells (Treg cells), specialized for immune suppression. It was first shown that depletion of a T-cell subpopulation from normal rodents produced autoimmune disease. Search for a molecular marker specific for such autoimmune-preventive Treg cells has revealed that the majority, if not all, of them constitutively express the CD25 molecule as depletion of $CD25^+CD4^+$ T cells spontaneously evokes autoimmune disease in otherwise normal rodents. The expression of CD25 by Treg cells has made it possible to delineate their developmental pathways, in particular their thymic development, and establish simple in vitro assay for assessing their suppressive activity. The marker and the in vitro assay have helped to identify human Treg cells with similar functional and phenotypic characteristics. Recent efforts have shown that natural Treg cells specifically express the transcription factor Foxp3 and that mutations of the Foxp3 gene produce a variety of immunological diseases in humans and rodents. Specific expression of Foxp3 in natural Treg cells has enabled their functional and developmental characterization by genetic approach. These studies altogether have provided firm evidence for $Foxp3^+CD25^+CD4^+$ Treg cells as an indispensable cellular constituent of the normal immune system for establishing and maintaining immunologic self-tolerance and immune homeostasis. Treg cells are now within the scope of clinical use to treat immunological diseases and control physiological and pathological immune responses.

Key words: Regulatory T cells, Suppressor T cells, Immunological self-tolerance, CD25, Il-2, Foxp3, IPEX

Abbreviations

APC	Antigen-presenting cell
ATx	Adult thymectomy
IBD	Inflammatory bowel disease
IPEX	Immune dysfunction, polyendocrinopathy, enteropathy, X linked syndrome
NTx	Neonatal thymectomy
T1D	Type 1 diabetes mellitus
TCR	T-cell receptor
Treg cells	Regulatory T cells

George Kassiotis and Adrian Liston (eds.), *Regulatory T Cells: Methods and Protocols*, Methods in Molecular Biology, vol. 707, DOI 10.1007/978-1-61737-979-6_1,

1. Introduction

Among various mechanisms for establishing and sustaining immunological self-tolerance and immune homeostasis, T-cell-mediated suppression of immune responses toward self and nonself antigens has recently attracted enormous interest (1). The idea of suppressor T cells, now called regulatory T cells (Treg cells), is not a new one for immunologists since early 1970s. In 1970, Gershon and Kondo made the seminal finding that T cells not only augmented but also dampened immune responses and that this down-regulation was mediated by T cells that were different from helper T cells (2). This T-cell population, called suppressor T cells, was intensively studied over the following years in various fields of immunology. However, active research of suppressor T cells, involving many immunologists, abruptly collapsed in the mid-1980s when scrutiny of the mouse MHC gene by molecular biology techniques showed no existence of the I-J region, which was assumed to encode a putative molecule intimately associated with their suppressive function (3, 4). With this bewildering I-J episode as a turning point, immunologists' interest in suppressor T cells rapidly waned, forming, in the late 1980s and early 1990s, an atmosphere in which they even shied away from using the word "suppressor T cells" in interpreting suppressive or inhibitory immunological phenomena (5). In retrospect, there are several other reasons for this decline in the study, e.g., failure in finding reliable markers for distinguishing suppressor T cells from other T cells, ambiguity in the molecular basis of suppression, and difficulty in preparing antigen-specific suppressor T-cell clones amenable to fine cellular and molecular analyses. Clinical immunologists failed to obtain definitive evidence for anomaly of suppressor T cells as a primary cause of any immunological disease. In contrast with the stagnation in suppressor T-cell research, molecular characterization of various cytokines, including the newly found immunosuppressive IL-10, in the 1980s revealed their pleiotropism and cross-regulation in function (6). These findings altogether generated a climate in which T-cell-involving suppressive phenomena were attributed to T cells secreting immunosuppressive or cross-regulatory cytokines, with little meaningful part played by suppressor T cells. In this atmosphere in the 1990s, it is quite understandable that IL-10-secreting Treg cells, called Tr1 cells, produced in vitro by antigenic stimulation of naïve T cells in the presence of IL-10, or TGF-β-secreting Treg cells, called Th3 cells, propagated from animals via oral tolerance encountered little resistance to be accepted (7, 8).

In parallel with the study of suppressor T cells briefly depicted above, there has been a different stream of endeavor to investigate T-cell suppression. A notable feature of the latter is that it examined from the beginning how autoimmune disease can be produced by breaching natural self-tolerance and how it can be inhibited to develop, rather than analyzing tolerance or suppression induced experimentally toward a particular exogenous antigen. This approach led to the finding that the normal immune system naturally harbors T cells and thymocytes with autoimmune-suppressive activity, later called regulatory T cells (1). This article reviews how Treg cells, in particular naturally arising $CD4^+$ Treg cells engaged in the sustenance of self-tolerance, have been investigated for years. It also discusses a perspective on future Treg cell research and their application in clinic.

2. $CD4^+$ T Cells with Autoimmune-Suppressive Activity

Two important findings made nearly 40 years ago have contributed to the identification and characterization of naturally occurring Treg cells currently investigated by many researchers. Nishizuka and Sakakura showed in 1969 that neonatal thymectomy (NTx) of normal mice between day 2 and 4 after birth resulted in the destruction of ovaries, which they first supposed to be due to deficiency of a certain ovary-tropic hormone secreted by the thymus, hence was called "ovarian dysgenesis" (9). This ovarian lesion later turned out to be of autoimmune nature because subsequent investigation demonstrated that NTx produced inflammatory tissue damage in other organs. Further, it was connected with the appearance of tissue-specific autoantibodies in the circulation. Depending on the mouse strain used, NTx, which is also called 3dTx because it is most efficient if the thymus is removed 3 days after birth, leads to the development of thyroiditis, gastritis, orchitis, prostatitis, and sialadenitis (10). In 1973, Penhale et al. reported that adult thymectomy (ATx) of normal rats (e.g., PVG rats) followed by four rounds of biweekly sublethal X-irradiation (2–2.5 Gray) produced autoimmune thyroiditis accompanied by antithyroglobulin autoantibody production (11). They and others later showed that the same protocol was able to elicit type 1 diabetes (T1D) in other strains of rats (12, 13). Importantly, inoculation of normal T cells from normal syngeneic animals inhibited disease development in both systems (14, 15). $CD4^+$ T cells and $CD4^+CD8^-$ mature thymocytes in particular inhibited NTx-induced murine autoimmune disease (14). On the other hand, once autoimmunity has developed, $CD4^+$ T cells were able to adoptively transfer the disease to syngeneic T-cell-deficient mice as helper T cells for autoantibody formation and effectors of cell-mediated immune destruction (16).

These results altogether indicated the following scenario of autoimmune disease. The normal thymus continuously produces a population of $CD4^+$ T cells with an autoimmune-suppressive activity; NTx of mice shortly after birth abrogates developmentally determined thymic production of autoimmune-suppressive $CD4^+$ T cells, allowing those self-reactive $CD4^+$ T cells that have been produced before NTx to become spontaneously activated and cause autoimmune disease because of the paucity of suppressive $CD4^+$ T cells in the periphery. Likewise, ATx and X-irradiations abrogates thymic supply of such T cells and reduce them in the periphery presumably because they are relatively radiosensitive. The results also suggested that there might coexist two types of $CD4^+$ T cells in the periphery of normal untreated mice and rats, one potentially capable of mediating autoimmune diseases and the other dominantly suppressing them (17).

3. Naturally Arising $CD25^+CD4^+$ Treg Cells and Their Crucial Role in Self-Tolerance

A next obvious question from above findings was how the two populations of $CD4^+$ T cells can be distinguished in normal animals and whether specific and direct removal of the autoimmune-suppressive population can break self-tolerance and cause autoimmune disease similar to the one produced by NTx in mice or ATx and X-irradiation in rats. Attempts were made to separate the two putative $CD4^+$ populations in normal naïve mice by the expression of cell surface molecules (17–23). Our experiments in 1985 showed that when splenic $CD4^+$ T-cell suspensions from normal BALB/c mice were depleted of $CD5^{high}CD4^+$ T-cells ex vivo and the remaining $CD5^{low}CD4^+$ T cells were transferred to congenitally T-cell-deficient BALB/c athymic nude mice, the nude mice spontaneously developed autoimmune disease in multiple organs (stomach, thyroid, ovaries, or testes) in a few months after the cell transfer (17). Cotransfer of normal untreated $CD4^+$ T cells with $CD5^{low}CD4^+$ T cells inhibited autoimmunity. Likewise, transfer of $CD5^{low}CD4^+$ T cells from normal C3H mice to T-cell-depleted C3H mice produced autoimmune thyroiditis at a high incidence (18). In 1990, Powrie and Mason reconstituted PVG athymic nude rats with splenic T-cell suspensions that were depleted of $CD45RC^{low}CD4^+$ T cells, thereby showing that the transferred $CD45RC^{high}CD4^+$ T cells elicited a systemic disease resembling graft-versus-host disease and autoimmune tissue damage in multiple organs including thyroid and Langerhans' islets (20). McKeever et al. conducted a similar experiment and showed that transfer of splenic cell suspensions depleted of $RT6.1^+$ T cells was able to produce T1D and thyroiditis in histocompatible athymic nude rats (21). Powrie et al. and Morrissey et al. then independently showed that transfer of BALB/c CD45RB^{high}CD4^{+} T cells to T/B-cell-deficient BALB/c SCID mice induced inflammatory bowel disease (IBD) (24, 25).

These findings prompted us to search for a cell surface molecule that would be more specific than CD5 or CD45RB (or CD45RC) in defining such autoimmunity- and inflammation-suppressive CD4^{+} T cells. In 1995, we identified the CD25 molecule (the IL-2 receptor α-chain) as a candidate because CD25^{+} T cells, which constituted 5–10% of peripheral CD4^{+} T cells (and less than 1% of peripheral CD8^{+} T cells) in normal naive mice, were confined in the CD5^{high} and CD45RB^{low} fraction of CD4^{+} T cells (22, 23). Transfer of BALB/c splenic cell suspensions depleted of CD25^{+}CD4^{+} T cells to BALB/c athymic nude mice indeed produced histologically and serologically evident autoimmune diseases at higher incidences and in a wider spectrum of organs (including stomach, thyroid, ovaries, adrenal glands, and Langerhans' islets) than the transfer of CD5^{low} or CD45RB^{high} T cells prepared from the same number of splenic cell suspensions. Cotransfer of a small number of CD25^{+}CD4^{+} T cells with the depleted cell suspensions clearly inhibited the autoimmunity. Removal of CD25^{+}CD4^{+} T cells not only elicited autoimmune disease but also enhanced immune responses to nonself antigens including soluble xenogeneic proteins and allografts; reconstitution with CD25^{+}CD4^{+} T cells normalized the responses (22). Transfer of steroid-resistant CD4^{+}CD8^{-} mature thymocyte suspensions depleted of CD25^{+} thymocytes also produced similar autoimmune diseases in syngeneic nude mice (26). Furthermore, the appearance of CD25^{+}CD4^{+} T cells in the spleen correlated well with the findings in NTx system. CD25^{+}CD4^{+} T cells became detectable in the periphery of normal mice from around day 3 after birth, rapidly increasing to the adult level (i.e., 5–10% of CD4^{+} T cells) in 3 weeks, though some CD25^{+}CD4^{+} T cells can already be detected in the lymph nodes of 2-day-old mice (27). Further, inoculation of CD25^{+}CD4^{+} T cells from normal mice within a limited period after NTx prevented autoimmune development (23). Interestingly, they were also able to suppress autoimmune disease induced by already active antigen-specific effector T cells (27).

Thus, the attempts to delineate autoimmune-suppressive CD4^{+} T cells, which are present in the normal immune system, by utilizing cell surface markers revealed thymus-produced CD25^{+}CD4^{+} T cells that engage in the maintenance of natural self-tolerance and also the control of immune responses to nonself antigens. The thymus is at any time producing functionally mature CD25^{+}CD4^{+} suppressive T cells and also some potentially pathogenic self-reactive T cells. With these results that defined a specific small subset of T cells with suppressive activity, the suppressive cells were then called Treg cells.

4. Regulatory T Cells for Transplantation Tolerance

Besides the investigations on Treg cells for maintaining natural self-tolerance discussed above, there have been other important studies that have contributed to our current conceptualization of Treg cells. For example, studies from the early 1990s have demonstrated that dominant transplantation tolerance can be established by administration of anti-CD4 or other monoclonal antibodies, the immunosuppressant cyclosporine A, or transplanting allogeneic or xenogeneic thymic epithelial cells into embryos (28–30). There is recent evidence that these types of graft tolerance are maintained by suppressive $CD4^+$ T cells, which are, at least in part, similar to $CD25^+CD4^+$ Treg cells functionally and phenotypically (31).

5. The Functional Role of IL-2 and CD25 for Natural Treg Cells

Following the discovery of CD25 as a useful marker for operationally distinguishing endogenous Treg cells from other T cells in normal naïve animals, several studies revealed that the molecule was not a mere marker for natural Treg cells but essential for their function. IL-2-deficient mice, which spontaneously develop severe autoimmunity/inflammation, were found to have a substantially reduced number of $CD25^+CD4^+$ T cells despite a normal number of T cells and a normal composition of CD4/CD8 subsets (32, 33). Bone marrow chimera of IL-2-deficient and IL-2-intact T cells failed to develop autoimmunity or inflammation and had normal generation of $CD25^+CD4^+$ Treg cells (33). CD25-deficient or CD122 (the IL-2Rβ-chain)-deficient mice were afflicted with similar autoimmunity and inflammation, which was prevented by inoculation of normal $CD25^+CD4^+$ T cells (34–36). Besides, neutralization of circulating IL-2 by administration of anti-IL-2 monoclonal antibody selectively reduced $CD25^+CD4^+$ T cells in normal mice and consequently provoked autoimmune disease (37). These findings collectively indicate that IL-2 is a key growth and survival factor for natural Treg cells and that CD25 as a component of the high affinity IL-2R is therefore not a mere marker for Treg cells but also an indispensable molecule for their maintenance.

6. Establishment of In Vitro Functional Assay for Natural Treg Cells

The discovery of CD25 as a highly Treg-specific cell surface marker enabled easy isolation of natural Treg cells from normal rodents and encouraged to establish in vitro assay for their suppressive function. In 1998, two groups showed that CD25$^+$CD4$^+$ T cells potently suppressed in vitro proliferation of other CD4$^+$ and CD8$^+$ T cells when both populations were cocultured and stimulated with specific antigen (or polyclonal T-cell receptor [TCR] stimulator such as anti-CD3 mAb) in the presence of antigen-presenting cells (APCs) (38, 39). The studies also revealed Treg cells' inability to produce IL-2 upon stimulation, their in vitro hypo-proliferative response to antigenic stimulation, and their proliferation upon TCR stimulation in the presence of high dose IL-2. Further, although the mechanism of in vitro suppression is still contentious, this assay has shown that Treg cells directly suppress CD4$^+$ T cells via cell contact with no need for soluble factors (38, 39). Notably, this simple and reliable in vitro assay, together with the CD25 marker, made it possible to identify human CD25$^+$CD4$^+$ Treg cells with similar phenotype and function as those in rodents (reviewed in (40)).

7. The Transcription Factor Foxp3 as a Key Control Molecule of Treg Cell Development and Function

A recent mile stone in Treg cell research is the discovery of the function of Foxp3. The *Foxp3* gene was identified in 2001 as the disease-causative gene in Scurfy mice, which spontaneously develop severe autoimmunity/inflammation as a result of a single gene mutation on X chromosome (41). Mutations of the human *FOXP3* gene, the ortholog of murine *Foxp3*, were immediately found to be the cause of a similar human disease called IPEX (Immune dysregulation, polyendocrinopathy, enteropathy, X-linked syndrome), which is characterized by autoimmune disease in multiple endocrine organs (such as T1D and thyroiditis), IBD, and severe allergy (42–44). Similarities in autoimmune disease and IBD in IPEX to those produced in mice by Treg cell depletion prompted several groups to investigate possible roles of Foxp3 in natural Treg cells. In 2003, they reported that Foxp3 was indeed a key molecule essential for Treg cell development and function. CD25$^+$CD4$^+$ peripheral T cells and CD25$^+$CD4$^+$CD8$^-$ thymocytes specifically expressed *Foxp3* mRNA, and activation of CD25$^-$CD4$^+$ T cells was unable to induce *Foxp3* expression (45–47). Retroviral transduction of *Foxp3* to normal CD25$^-$CD4$^+$ T cells converted them into phenotypically and functionally Treg-like cells. Such transduced cells displayed in vivo and in vitro suppressive activity, in vitro hypo-proliferation and hypo-production of IL-2, and up-regulation of CD25 and other Treg cell-associated molecules (such as CTLA-4 and GITR) (45, 46). In BM chimera with a mixture of BM cells from wild type and *Foxp3*-deficient mice, *Foxp3*-deficient BM cells failed to give rise to CD25$^+$CD4$^+$ T cells, while *Foxp3*-intact BM cells generated them and suppressed disease development (46). These findings collectively indicated that the transcription factor Foxp3 could be a master controller of the development and function of natural $CD25^+CD4^+$ Treg cells.

With the specific expression of Foxp3 in natural Treg cells, genetically engineered mice have been prepared that express the reporter GFP or diphtheria toxin receptor under the control of the Foxp3 promoter (48–50). The use of these Foxp3-reporter mice confirmed the previous findings that were made by the use of CD25 as a specific Treg cell marker, e.g., the ontogeny of Treg cells, the requirement of IL-2 and CD25 for Treg cell maintenance, and induction of autoimmunity by depletion of Treg cells. Raising monoclonal antibody to the Foxp3 protein and its use for intracellular staining of Foxp3 have also showed that Foxp3 is abundantly expressed in natural Treg cells and so far the most reliable molecular marker for them (51). This has enabled more reliable analyses than before on the dynamics of Treg cells in physiological and pathological immune responses in humans and rodents.

8. Perspective and Current Key Issues of Treg Research

A historical sketch of Treg research depicted above shows that $Foxp3^+CD25^+CD4^+$ Treg cells are an indispensable cellular constituent of the normal immune system. There are several key issues pertinent to further understanding of the function and development of Treg cells.

1. Given that Foxp3 expression suffices to confer suppressive activity to naïve T cells, how does Foxp3 control the activity? Foxp3 appears to activate or repress hundreds of genes directly or indirectly through forming a transcription complex with other key transcription factors such as NFAT and AML1/Runx1 (52–55). It has been shown that multiple suppressive mechanisms are mediated by $Foxp3^+$ Treg cells, e.g., cell-contact-dependent inhibition of the activation and proliferation of T cells, killing or inactivation of APCs and/or T cells, and suppression via cytokines such as IL-10, IL-35, and TGF-β (56–59). A central question is then whether there is a single core suppressive mechanism shared by every Treg cell and several complementary mechanisms; whether a particular mechanism may play a dominant role under a particular condition, with different mechanisms operating in various situations; alternatively, whether multiple suppressive mechanisms operate simultaneously and synergistically; and whether dysfunction of any of them is not sufficient to seriously impair

suppression. In other words, one can ask whether defect of any molecule that is controlled by Foxp3 and associated with suppressive function should impair in vitro and in vivo suppressive activity of Foxp3⁺ Treg cells and cause autoimmune/inflammatory disease as observed in Foxp3 deficiency.

2. How are the cell fate of Treg cells and their TCR repertoire determined in the thymus? It has been shown that, in developing T cells, TCR engagement by a high-affinity self-ligand initiates signaling cascades that induce Foxp3 expression, which further drives thymocytes to the Treg cell lineage (reviewed in ref. (60)). The precise mechanism of this selection and differentiation of Foxp3⁺ Treg cells and stable maintenance of Foxp3 expression in Treg cells remain to be elucidated. Recent studies suggest that Foxp3 is not required for the initial commitment of the Treg cell lineage: without Foxp3, some developing thymocytes are able to acquire partial Treg cell phenotype (such as the expression of CD25, CTLA-4, and GITR) without having suppressive activity (61, 62). A cross-sectional analysis of the Treg cell signature in Treg-like cells generated under a number of conditions with or without Foxp3 has also revealed that much of the Treg cell signature is not ascribable to Foxp3 (63). These findings indicate that a higher level of regulation, which is independent of Foxp3, might exist in the Treg lineage commitment. Thus, it is an intriguing issue to determine what mechanisms define the Treg differentiation program and turn on Foxp3 gene expression in developing thymocytes. One of the key elements for initiating the program may be TCR signal. Given that the TCR repertoire of Foxp3⁺ Treg cells is as broad as conventional T cells and characteristically skewed to higher self-reactivity than the latter, one can ask whether self-reactivity of a TCR expressed by a developing thymocyte can determine its commitment to the Treg cell lineage, hence the formation of self-reactive TCR repertoire of Treg cells. At the same time, one can ask how Treg cells with specificity for conventional antigens (e.g., microbial antigen) can be produced by TCR-dependent initiation of Treg cell lineage commitment.

3. To what extent do induced Foxp3⁺ Treg cells contribute to self-tolerance and immune homeostasis? Besides thymic production of natural Treg cells, naïve T cells in the periphery can acquire Foxp3 expression and Treg cell function in several experimental settings, such as in vitro antigenic stimulation in the presence of TGF-β, in vivo chronic suboptimal antigenic stimulation, and targeting antigen to immature dendritic cells (DCs) (64, 65). Physiologically, the induction of Foxp3⁺ Treg cells from naïve T cells takes place, at least, in

the intestine. Yet it remains to be determined whether the induced Treg cells are functionally stable, survive long, and circulate systemicly to other lymphoid organs to maintain self-tolerance and immune homeostasis.

4. How are the activation, expansion, and differentiation of Treg cells controlled systemically and locally? Mature DCs expand Foxp3⁺ Treg cells in a CD80/86 dependent fashion (66, 67). Activated DCs secrete IL-6, which renders antigen-responding non-Treg cells resistant to Treg-mediated suppression in vitro (68). Naïve CD4⁺ T cells may differentiate into Foxp3⁺ Treg cells in the presence of TGF-β or into IL-17-secreting Th17 cells in the presence of TGF-β and IL-6 (69, 70). IL-2 facilitates this differentiation of naïve CD4⁺ T cells into Foxp3⁺ Treg cells but inhibits their differentiation into Th17 cells (71). Thus, costimulatory molecules expressed by APCs and cytokines secreted by APCs and other T cells crucially contribute to the control of various aspects of Treg cell development, differentiation, and function in a complex manner. Precise mechanisms of local and systemic control of Treg cell number, activation, and differentiation need to be elucidated for effective control of immune responses.

9. Clinical Perspective

Human Treg cells have been investigated for a decade since the demonstration of the existence of Treg cells functionally and phenotypically similar to the mouse counterpart. A typical illustration of the role of Foxp3⁺ natural Treg cells for self-tolerance and immune homeostasis is IPEX syndrome as discussed above.

In contrast to IPEX, in which genetic anomaly of Treg cells is primarily causative, it is obscure whether any Treg cell anomaly, genetically determined or environmentally induced, should play a substantial role for the development of common immunological diseases, such as T1D in particular, which are apparently polygenic (72). It has been well documented that polymorphisms of the CTLA-4, IL-2, and CD25 genes significantly contribute to genetic susceptibility to T1D in humans and also in NOD mice with spontaneous T1D (72, 73). Given that total genetic deficiency of these genes, particularly the IL-2 and CD25 genes, produces severe autoimmunity mainly through affecting Treg cell development and function (see above), it is possible that the polymorphisms of these genes may alter Treg cell development or function and thereby render the host susceptible to autoimmune disease. Whether known polymorphisms of other autoimmune susceptibility genes, especially the CTLA-4 gene, might affect

Treg cells needs to be examined (74). In addition, given that $Foxp3^+$ Treg cells play crucial roles in allergy and immunopathology (such as IBD) as observed in IPEX syndrome, it remains to be determined whether anomaly of Treg cells is conducive to common immunological diseases such as allergy and IBD (74).

There is also accumulating evidence that Foxp3+ Treg cells hamper effective immunity against tumor cells. They abundantly infiltrate into tumor tissues, and high ratios of $Foxp3^+$ cells to $CD8^+$ T cells indicate poor prognosis of cancer patients (75). On the other hand, abundant infiltration of Foxp3+ cells into transplanted organs correlates with the state of operational graft tolerance, indicating possible contribution of Foxp3+ Treg cells to the maintenance of stable transplantation tolerance (76).

For clinical use of Treg cells, natural $Foxp3^+$ Treg cells bear unique immunological properties that make them a suitable therapeutic target. They are naturally present in the circulation and can be phenotypically distinguished from other T cells, although cell surface markers specific for Treg cells still need to be found for their reliably pure isolation. They can recognize a broad repertoire of self and nonself antigens. They can be stimulated to proliferate by in vivo antigenic stimulation and are functionally stable, retaining their suppressive activity after clonal expansion in vivo and in vitro. By exploiting these characteristics, in vivo and in vitro strategies that clonally expand antigen-specific natural Treg cells are useful to strengthen or reestablish self-tolerance in autoimmune disease, induce tolerance to nonself-antigens in organ transplantation, allergy and IBD, or augment feto-maternal tolerance in pregnancy. As a reciprocal approach, selective reductions in the number or function of natural Treg cells while retaining or enhancing effector T cells may be a strategy for provoking and augmenting tumor immunity in cancer patients or microbial immunity in chronic infection.

10. Conclusion

Research for years has established that the normal immune system harbors Treg cells specialized for immune suppression. In addition to $Foxp3^+CD25^+CD4^+$ natural Treg cells, on which this review focuses, other types of Treg cells, such as IL-10-secreting Tr1 cells, also contribute to peripheral immune homeostasis. Antigen induced suppressor T cells that were intensively studied in the 1970s and early 1980s remain to be reinvestigated from a vantage point of the present. Further investigation of these various types of Treg cells, especially their common cellular and molecular basis, will enable better control of physiological and pathological immune responses in humans.

Acknowledgements

The author thanks Atsushi Tanaka for the critical reading of the manuscript. The author's research is supported by grants-in-aid from the Ministry of Education, Science, Sports and Culture, and the Ministry of Human Welfare.

References

1. Sakaguchi, S. (2000) Regulatory T cells: key controllers of immunologic self-tolerance. *Cell* **101**, 455–458.
2. Gershon, R. K. and Kondo, K. (1970) Cell interactions in the induction of tolerance: the role of thymic lymphocytes. *Immunology* **18**, 723–737.
3. Green, D. R., Flood, P. M. and Gershon, R. K. (1983) Immunoregulatory T-cell pathways. *Annu. Rev. Immunol.* **1**,439–463.
4. Kronenberg, M., Steinmetz, M., Kobori, J., Kraig, E., Kapp, J. A., Pierce, C. W., et al. (1983) RNA transcripts for I-J polypeptides are apparently not encoded between the I-A and I-E subregions of the murine major histocompatibility complex. *Proc. Natl. Acad. Sci. U.S.A.* **80**, 5704–5708.
5. Bloom B. R., Salgame, P. and Diamond, B. (1992) Revisiting and revising suppressor T cells. *Immunol. Today* **13**, 131–136.
6. O'Garra, A. and Murphy, K. (1994) Role of cytokines in determining T-lymphocyte function. *Curr. Opin. Immunol.* **6**, 458–466.
7. Chen, Y., Kuchroo, V. K., Inobe, J., Hafler, D. A. and Weiner, H. L. (1994) Regulatory T cell clones induced by oral tolerance: suppression of autoimmune encephalitis. *Science* **265**, 1237–1240.
8. Groux, H., O'Garra, A., Bigler, M., Rouleau, M., Antonenko, S., de Vries, J. E. and Roncarolo, M. G. (1997) A $CD4^+$ T-cell subset inhibits antigen-specific T-cell responses and prevents colitis. *Nature* **389**, 737–742.
9. Nishizuka, Y. and Sakakura, T. (1969) Thymus and reproduction: sex-linked dysgenesia of the gonad after neonatal thymectomy in mice. *Science* **166**, 753–755.
10. Kojima, A. and Prehn, R. T. (1981) Genetic susceptibility to post-thymectomy autoimmune diseases in mice. *Immunogenetics* **14**, 15–27.
11. Penhale, W. J., Farmer, A., McKenna, R. P. and Irvine, W. J. (1973) Spontaneous thyroiditis in thymectomized and irradiated Wistar rats. *Clin. Exp. Immunol.* **15**, 225–236.
12. Penhale, W. J., Stumbles, P. A., Huxtable, C. R., Sutherland, R. J. and Pethick, D. W. (1990) Induction of diabetes in PVG/c strain rats by manipulation of the immune system. *Autoimmunity* **7**, 169–179.
13. Fowell, D. and Mason, D. (1993) Evidence that the T cell repertoire of normal rats contains cells with the potential to cause diabetes. Characterization of the $CD4^+$ T cell subset that inhibits this autoimmune potential. *J. Exp. Med.* **177**, 627–636.
14. Sakaguchi, S., Takahashi, T. and Nishizuka, Y. (1982) Study on cellular events in post-thymectomy autoimmune oophoritis in mice. II. Requirement of Lyt-1 cells in normal female mice for the prevention of oophoritis. *J. Exp. Med.* **156**, 1577–1586.
15. Penhale, W. J., Irvine, W. J., Inglis, J. R. and Farmer, A. (1976) Thyroiditis in T cell-depleted rats: suppression of the autoallergic response by reconstitution with normal lymphoid cells. *Clin. Exp. Immunol.* **25**, 6–16.
16. Sakaguchi, S., Takahashi, T. and Nishizuka, Y. (1982) Study on cellular events in post-thymectomy autoimmune oophoritis in mice. I. Requirement of Lyt-1 effector cells for oocytes damage after adoptive transfer. *J. Exp. Med.* **156**, 1565–1576.
17. Sakaguchi, S., Fukuma, K., Kuribayashi, K. and Masuda, T. (1985) Organ-specific autoimmune diseases induced in mice by elimination of T-cell subset. I. Evidence for the active participation of T cells in natural self-tolerance: deficit of a T-cell subset as a possible cause of autoimmune disease. *J. Exp. Med.* **161**, 72–87.
18. Sugihara, S., Izumi, Y., Yoshioka, T., Yagi, H., Tsujimura, T., Tarutani, O., et al. (1988) Autoimmune thyroiditis induced in mice depleted of particular T-cell subset. I. Requirement of $Lyt\text{-}1^{dull}$ $L3T4^{bright}$ normal T cells for the induction of thyroiditis. *J. Immunol.* **141**, 105–113.

19. Smith, H., Lou, Y. H., lacy, P. and Tung, K. S. K. (1992) Tolerance mechanism in experimental ovarian and gastric autoimmune disease. *J. Immunol.* **149**, 2212–2218.
20. Powrie, F. and Mason, D. (1990) OX-22^{high} CD4^{+} T cells induce wasting disease with multiple organ pathology: prevention by OX-22^{low} subset. *J. Exp. Med.* **172**, 1701–1708.
21. McKeever, U., Mordes, J. P., Greiner, D. L., Appel, M. C., Rozing, J., Handler, E. S. and Rossini, A. A. (1990) Adoptive transfer of autoimmune diabetes and thyroiditis to athymic rats. *Proc. Natl. Acad. Sci. U.S.A.* **87**, 7618–7622.
22. Sakaguchi, S., Sakaguchi, N., Asano, M., Itoh, M. and Toda, M. (1995) Immunologic tolerance maintained by activated T cells expressing IL-2 receptor α-chains (CD25): breakdown of a single mechanism of self-tolerance causes various autoimmune diseases. *J. Immunol.* **155**, 1151–1164.
23. Asano, M., Toda, M., Sakaguchi, N. and Sakaguchi, S. (1996) Autoimmune disease as a consequence of developmental abnormality of a T cell subpopulation. *J. Exp. Med.* **184**, 387–396.
24. Powrie, F., Leach, M. W., Mauze, S., Caddle, L. B. and Coffman, R. L. (1993) Phenotypically distinct subsets of CD4^{+} T cells induce or protect from chronic intestinal inflammation in C. B-17 scid mice. *Int. Immunol.* **5**, 1461–1471.
25. Morrissey, P. J., Charrier, K., Braddy, S., Liggitt, D. and Watson, J. D. (1993) CD4^{+} T cells that express high levels of CD45RB induce wasting disease when transferred into congenic severe combined immunodeficient mice. Disease development is prevented by cotransfer of purified CD4^{+} T cells. *J. Exp. Med.* **178**, 237–244.
26. Itoh, M., Takahashi, T., Sakaguchi, N., Kuniyasu, Y., Shimizu, J., Otsuka, F. and Sakaguchi, S. (1999) Thymus and autoimmunity: production of CD25^{+}CD4^{+} naturally anergic and suppressive T cells as a key function of the thymus in maintaining immunologic self-tolerance. *J. Immunol.* **162**, 5317–5326.
27. Suri-Payer, E., Amar, A. Z., Thornton, A. M. and Shevach, E. M. (1998) CD4^{+}CD25^{+} T cells inhibit both the induction and effector function of autoreactive T cells and represent a unique lineage of immunoregulatory cells. *J. Immunol.* **160**, 1212–1218.
28. Qin, S., Cobbold, S. P., Pope, H., Elliott, J., Kioussis, D., Davies, J. and Waldmann, H. (1993) "Infectious" transplantation tolerance. *Science* **259**, 974–977.
29. Hall, B. M., Pearce, N. W., Gurley, K. E. and Dorsch, S. E. (1990) Specific unresponsiveness in rats with prolonged cardiac allograft survival after treatment with cyclosporine. III. Further characterization of the CD4^{+} suppressor cell and its mechanisms of action. *J. Exp. Med.* **171**, 141–157.
30. Le Douarin, N., Corbel, C., Bandeira, A., Thomas-Vaslin, V., Modigliani, Y., Coutinho, A. and Salaun, J. (1996) Evidence for a thymus-dependent form of tolerance that is not based on elimination or anergy of reactive T cells. *Immunol. Rev.* **149**, 35–53.
31. Waldmann, H., Adams, E., Fairchild, P. and Cobbold, S. (2006) Infectious tolerance and the long-term acceptance of transplanted tissue. *Immunol. Rev.* **212**, 301–313.
32. Schorle, H., Holtschke, T., Hunig, T., Schimpl, A. and Horak, I. (1991) Development and function of T cells in mice rendered interleukin-2 deficient by gene targeting. *Nature* **352**, 621–624.
33. Almeida, A. R., Legrand, N., Papiernik, M. and Freitas, A. A. (2002) Homeostasis of peripheral CD4^{+} T cells: IL-2R alpha and IL-2 shape a population of regulatory cells that controls CD4^{+} T cell numbers. *J. Immunol.* **169**, 4850–4860.
34. Willerford, D. M., Chen, J., Ferry, J. A., Davidson, L., Ma, A. and Alt, F. W. (1995) Interleukin-2 receptor alpha chain regulates the size and content of the peripheral lymphoid compartment. *Immunity* **3**, 521–530.
35. Suzuki, H., Zhou, Y. W., Kato, M., Mak, T. W. and Nakashima, I. (1999) Normal regulatory alpha/beta T cells effectively eliminate abnormally activated T cells lacking the interleukin 2 receptor beta in vivo. *J. Exp. Med.* **190**, 1561–1572.
36. Malek. T. R., Yu, A., Vincek, V., Scibelli, P. and Kong, L. (2002) CD4 regulatory T cells prevent lethal autoimmunity in IL-2Rbeta-deficient mice. Implications for the nonredundant function of IL-2. *Immunity* **17**, 167–178.
37. Setoguchi, R., Hori, S., Takahashi, T. and Sakaguchi, S. (2005) Homeostatic maintenance of natural Foxp3^{+}CD25^{+}CD4^{+} regulatory T cells by interleukin (IL)-2 and induction of autoimmune disease by IL-2 neutralization. *J. Exp. Med.* **201**, 723–735.
38. Takahashi, T., Kuniyasu, Y., Toda, M., Sakaguchi, N., Itoh, M., Iwata, M., Shimizu, J. and Sakaguchi, S. (1998) Immunologic self-tolerance maintained by CD25^{+} CD4^{+} naturally anergic and suppressive T cells: induction of autoimmune disease by breaking their anergic/suppressive state. *Int. Immunol.* **10**, 1969–1980.

39. Thornton, A. M. and Shevach, E. M. (1998) CD4+ CD25+ immunoregulatory T cells suppress polyclonal T cell activation in vitro by inhibiting interleukin 2 production. *J. Exp. Med.* **188**, 287–296.
40. Shevach, E. M. (2001) Certified professionals: CD4+CD25+ suppressor T cells. *J. Exp. Med.* **193**, F41–F46.
41. Brunkow, M. E., Jeffery, E. W., Hjerrild, K. A., Paeper, B., Clark, L. B., Yasayko, S. A., et al. (2001) Disruption of a new forkhead/winged-helix protein, scurfin, results in the fatal lymphoproliferative disorder of the scurfy mouse. *Nat. Genet.* **27**, 68–73.
42. Chatila, T. A., Blaeser, F., Ho, N., Lederman, H. M., Voulgaropoulos, C., Helms, C. and Bowcock, A. M. (2000) JM2, encoding a fork head-related protein, is mutated in X-linked autoimmunity-allergic disregulation syndrome. *J. Clin. Invest.* **106**, R75–R81.
43. Wildin, R. S., Ramsdell, F., Peake, J., Faravelli, F., Casanova, J. L., Buist, N., et al. (2001) X-linked neonatal diabetes mellitus, enteropathy and endocrinopathy syndrome is the human equivalent of mouse scurfy. *Nat. Genet.* **27**, 18–20.
44. Bennett, C. L., Christie, J., Ramsdell, F., Brunkow, M. E., Ferguson, P. J., Whitesell, L., et al. (2001) The immune dysregulation, polyendocrinopathy, enteropathy, X-linked syndrome (IPEX) is caused by mutations of FOXP3. *Nat. Genet.* **27**, 20–21.
45. Hori, S., Nomura, T. and Sakaguchi, S. (2003) Control of regulatory T cell development by the transcription factor Foxp3. *Science* **299**, 1057–1061.
46. Fontenot, J. D., Gavin, M. A. and Rudensky, A. Y. (2003) Foxp3 programs the development and function of CD4+CD25+ regulatory T cells. *Nat. Immunol.* **4**, 330–336.
47. Khattri, R., Cox, T., Yasayko, S. A. and Ramsdell, F. (2003) An essential role for Scurfin in CD4+CD25+ T regulatory cells. *Nat. Immunol.* **4**, 337–342.
48. Fontenot, J. D., Rasmussen, J. P., Williams, L. M., Dooley, J. L., Farr, A. G. and Rudensky, A. Y. (2005) Regulatory T cell lineage specification by the forkhead transcription factor foxp3. *Immunity* **22**, 329–341.
49. Kim, J. M., Rasmussen, J. P. and Rudensky, A. Y. (2007) Regulatory T cells prevent catastrophic autoimmunity throughout the lifespan of mice. *Nat. Immunol.* **8**, 191–197.
50. Lahl, K., Loddenkemper, C., Drouin, C., Freyer, J., Arnason, J., Eberl, G., et al. (2007) Selective depletion of Foxp3+ regulatory T cells induces a scurfy-like disease. *J. Exp. Med.* **204**, 57–63.
51. Roncador, G., Brown, P. J., Maestre, L., Hue, S., Martinez-Torrecuadrada, J. L., Ling, et al. (2005) Analysis of FOXP3 protein expression in human CD4+CD25+ regulatory T cells at the single-cell level. *Eur. J. Immunol.* **35**, 1681–1691.
52. Zheng, Y., Josefowicz, S. Z., Kas, A., Chu, T. T., Gavin, M. A. and Rudensky, A. Y. (2007) Genome-wide analysis of Foxp3 target genes in developing and mature regulatory T cells. *Nature* **445**, 936–940.
53. Marson, A., Kretschmer, K., Frampton, G. M., Jacobsen, E. S., Polansky, J. K., MacIsaac, K. D., et al. (2007) Foxp3 occupancy and regulation of key target genes during T-cell stimulation. *Nature* **445**, 931–935.
54. Wu, Y., Borde, M., Heissmeyer, V., Feuerer, M., Lapan, A. D., Stroud, J. C. et al. (2006) FOXP3 controls regulatory T cell function through cooperation with NFAT. *Cell* **126**, 375–387.
55. Ono, M., Yaguchi, H., Ohkura, N., Kitabayashi, I., Nagamura, Y., Nomura, T., et al. (2007) Foxp3 controls regulatory T-cell function by interacting with AML1/Runx1. *Nature* **446**, 685–689.
56. von Boehmer H. (2005) Mechanisms of suppression by suppressor T cells. *Nat. Immunol.* **6**, 338–344.
57. Tang, Q. and Bluestone, J. A. (2008) The Foxp3+ regulatory T cell: a jack of all trades, master of regulation. *Nat. Immunol.* **9**, 239–244.
58. Vignali, D. A., Collison, L. W. and Workman, C. J. (2008) How regulatory T cells work. *Nat. Rev. Immunol.* **8**, 523–532.
59. Shevach, E. M. (2009) Mechanisms of foxp3+ T regulatory cell-mediated suppression. *Immunity* **30**, 636–645.
60. Sakaguchi, S., Yamaguchi, T., Nomura, T. and Ono, M. (2008) Regulatory T cells and immune tolerance. *Cell* **133**, 775–787.
61. Gavin, M. A., Rasmussen, J. P., Fontenot, J. D., Vasta, V., Manganiello, V.C., Beavo, J. A. and Rudensky, A. Y. (2007) Foxp3-dependent programme of regulatory T-cell differentiation. *Nature* **445**, 771–775.
62. Lin, W., Haribhai, D., Relland, L. M., Truong, N., Carlson, M. R., Williams, C. B., and Chatila, T. A. (2007) Regulatory T cell development in the absence of functional Foxp3. *Nat. Immunol.* **8**, 359–368.
63. Hill, J. A., Feuerer, M., Tash, K., Haxhinasto, S., Perez, J., Melamed, R., et al. (2007) Foxp3 transcription-factor-dependent and -independent regulation of the regulatory T cell transcriptional signature. *Immunity* **27**, 786–800.

64. Chen, W., Jin, W., Hardegen, N., Lei, K. J., Li, L., Marinos, N., et al. (2003) Conversion of peripheral $CD4^+CD25^-$ naive T cells to $CD4^+CD25^+$ regulatory T cells by TGF-beta induction of transcription factor Foxp3. *J. Exp. Med.* **198**, 1875–1886.
65. Apostolou, I. and von Boehmer, H. (2004) In vivo instruction of suppressor commitment in naive T cells. *J. Exp. Med.* **199**, 1401–1408.
66. Yamazaki, S., Iyoda, T., Tarbell, K., Olson, K., Velinzon, K., Inaba, K. and Steinman, R. M. (2003) Direct expansion of functional $CD25^+$ $CD4^+$ regulatory T cells by antigen-processing dendritic cells. *J. Exp. Med.* **198**, 235–247.
67. Fehervari, Z. and Sakaguchi, S. (2004) Control of $Foxp3^+$ $CD25^+CD4^+$ regulatory cell activation and function by dendritic cells. *Int. Immunol.* **16**, 1769–1780.
68. Pasare, C. and Medzhitov, R. (2003) Toll pathway-dependent blockade of CD4+CD25+ T cell-mediated suppression by dendritic cells. *Science* **299**, 1033–1036.
69. Veldhoen, M., Hocking, R. J., Atkins, C. J., Locksley, R. M. and Stockinger, B. (2006) TGFbeta in the context of an inflammatory cytokine milieu supports de novo differentiation of IL-17-producing T cells. *Immunity* **24**, 179–189.
70. Bettelli, E., Carrier, Y., Gao, W., Korn, T., Strom, T. B., Oukka, M., et al. (2006) Reciprocal developmental pathways for the generation of pathogenic effector TH17 and regulatory T cells. *Nature* **441**, 235–238.
71. Laurence, A., Tato, C. M., Davidson, T. S., Kanno, Y., Chen, Z., Yao, Z., et al. (2007) Interleukin-2 signaling via STAT5 constrains T helper 17 cell generation. *Immunity* **26**, 371–381.
72. Wellcome Trust Case Control Consortium. (2007) Genome-wide association study of 14,000 cases of seven common diseases and 3,000 shared controls. *Nature* **447**, 661–678.
73. Encinas, J. A., Wicker, L. S., Peterson, L. B., Mukasa, A., Teuscher, C., Sobel, R., et al. (1999) QTL influencing autoimmune diabetes and encephalomyelitis map to a 0.15-cM region containing Il2. *Nat. Genet.* **21**, 158–160.
74. Sakaguchi, S., Ono, M., Setoguchi, R., Yagi, H., Hori, S., Fehervari, Z., et al. (2006) $Foxp3^+$ $CD25^+$ $CD4^+$ natural regulatory T cells in dominant self-tolerance and autoimmune disease. *Immunol. Rev.* **212**, 8–27.
75. Nishikawa, H. and Sakaguchi, S. (2010) Regulatory T cells in tumor immunity. *Int. J. Cancer* **127**, 759–777.
76. Li, Y., Zhao, X., Cheng, D., Haga, H., Tsuruyama, T., Wood, K., et al. (2008) The presence of FOXP3 expressing T cells within grafts of tolerance human liver transplant recipients. *Transplantation* **86**, 1837–1843.

Part II

In Vitro

Chapter 2

In Vitro Treg Suppression Assays

Lauren W. Collison and Dario A.A. Vignali

Abstract

Determining the activity of a regulatory T-cell population in vitro is often the first step in analyzing its function. To obtain reliable and reproducible results, it is critical to follow the protocol that is most applicable to your experimental question. We have outlined below a basic in vitro suppression assay as well as a variety of alternative/additional protocols that can be utilized alone or in combination as desired.

Key words: Treg, In vitro, Suppression, Foxp3

1. Introduction

The first in vitro assays to measure regulatory T-cell (T_{reg}) function were described by two groups over a decade ago (1, 2). The observation that a CD25$^+$ T-cell population possessed regulatory activity enabled isolation of natural T_{regs} cells from mice and humans. With this knowledge, it was shown that CD4$^+$CD25$^+$ T cells could potently suppress the proliferation of activated CD4$^+$CD25$^-$ and CD8$^+$ T cells when the populations were cocultured in vitro. In vitro suppression assays are now widely used to determine the suppressive capacity of T_{regs}. The benefits of this assay include ease and simplicity of setup and reliability. In addition, few reagents are needed to perform the basic protocol, making it an appropriate initial test of suppressive capacity. Given that conventional T cells (T_{conv}) and T_{regs} can be purified from genetically deficient mice, the role that individual molecules play in suppression can easily be determined. In addition, ex vivo suppressive capacity of T_{regs} obtained from normal or diseased patients can provide information regarding immunocompetance. Lastly, due

George Kassiotis and Adrian Liston (eds.), *Regulatory T Cells: Methods and Protocols*, Methods in Molecular Biology, vol. 707, DOI 10.1007/978-1-61737-979-6_2, © Springer Science+Business Media, LLC 2011

to the simplicity of assay setup, numerous variables including type of activation, cell number, and degree of proliferation can be manipulated within a single experiment. The primary weakness of in vitro T_{reg} suppression assays is that they do not necessarily recapitulate in vivo processes. In vivo, T_{regs} are strongly proliferative, yet in vitro T_{regs} are hypoproliferative in response to antigenic stimulation (1, 2). Another limitation is that antigen-specific assays are limited due to reduced numbers of antigen-specific T_{regs} that can be purified following immunologic response to a specific pathogen or disease state. For this reason, polyclonal T_{regs} are typically assayed for their ability to suppress T_{conv} cell proliferation. Finally, the use of in vitro suppression assays lead to the conclusion that T_{regs} suppress in a cytokine-dependent manner, yet the role of soluble factors in T_{reg}-mediated suppression in vivo is clear (3–9). Fortunately, however, a new variation of the standard in vitro T_{reg} suppression assay has been developed that demonstrates the importance of soluble factors in T_{reg}-mediated suppression (10).

2. Materials

2.1. Basic Protocol

1. Murine cell culture media: RPMI (Mediatech) supplemented with 10% FBS (optimal manufacturer and lot to be determined empirically), 2 mM L-glutamine (Mediatech), 1 mM sodium pyruvate (Mediatech), 100 mM non-essential amino acids (Mediatech), 5 mM HEPES free acid (Mediatech), 10 ml of 5.5×10^{-2} 2-mercaptoethanol (Invitrogen), and 100 U/ml Penicillin/Streptomycin (Mediatech) (see Note 1).
2. Human cell culture media: X-VIVO™ 15 Chemically Defined Medium, with gentamicin and phenol red (Lonza) supplemented with 15% male human serum (Lonza) and 10% L-glutamine (Mediatech) (see Note 1).
3. Gey's solution for red blood cell lysis: 12 mM potassium bicarbonate ($KHCO_3$), 156 mM ammonium chloride (NH_4Cl), diluted in water. Filter sterilize through a 0.2 μm filter to maintain sterility.
4. Murine anti-CD3 Ab, clone 2c11 (NALE/functional grade) and murine anti-CD28, clone 37.51 (NALE/functional grade).
5. Human anti-CD3 Ab, clone OKT3 (NALE/functional grade) and human anti-CD28, clone CD28.6 (NALE/functional grade).
6. Round bottom 96 well tissue culture plate (Nunc).
7. Purified, azide-free, endotoxin-free anti-CD3 and anti-CD28 antibodies.

8. 5 μM Sulfate latex beads (4% solid) (Molecular Probes).
9. [^{3}H]-Thymidine (Amersham Biosciences).
10. 70 μM Nylon cell strainer (BD).
11. 50 ml Conical tubes (BD).
12. Normal mouse serum (Gibco).
13. Phosphate buffered saline (PBS) (Mediatech).
14. Hanks Balanced Salt Solution (Mediatech).
15. 1 ml Syringes, use plunger for homogenization.
16. Fluorescently tagged antibodies (CD4, CD25, CD45RB).
17. 40 μM Nylon cell strainer (BD).
18. Recombinant human IL-2 (R&D Systems).
19. Ficoll Paque Plus (GE Healthcare).
20. Plasma transfer set (Charter Medical).
21. Phosphate buffer (4.82 g/l monohydrate, monosodium phosphate, pH 6.5).

2.2. Variations of Basic Protocol

1. Frosted glass microscope slides (Fisher).
2. 1,500 U/ml Collagenase Type III, High specific activity (Worthington).
3. 300 U/ml DNase I, 2,000 U/vial (Sigma).
4. Anti-CD11c antibody (eBioscience).
5. Peptides (e.g., $Ova_{3326-339}$, PCC_{88-104}, or $HA_{110-120}$ as desired).
6. PMA and Ionomycin (Calbiochem).
7. Bovine serum albumin (BSA) (Sigma).
8. CFSE (carboxyfluorescein succinimidyl ester) or SNARF-1 (Seminaphtharhodafluor) (Molecular Probes).
9. MTT cell proliferation assay kit (Cayman Chemical).
10. Transwell: Millicell 96 well receiver plate (Millipore).
11. Transwell: Millicell 96 cell culture insert plate (0.4 μM) (Millipore).
12. Foxp3 staining kit and fluorescently conjugated anti-Foxp3 antibody (eBioscience).

3. Methods

3.1. Basic Protocol

The following protocol describes a basic type of in vitro T_{reg} suppression assay where T_{reg} function is measured in the absence of antigen-presenting cells (APCs). In this protocol, activation is mediated by anti-CD3 + anti-CD28 coated beads and, therefore,

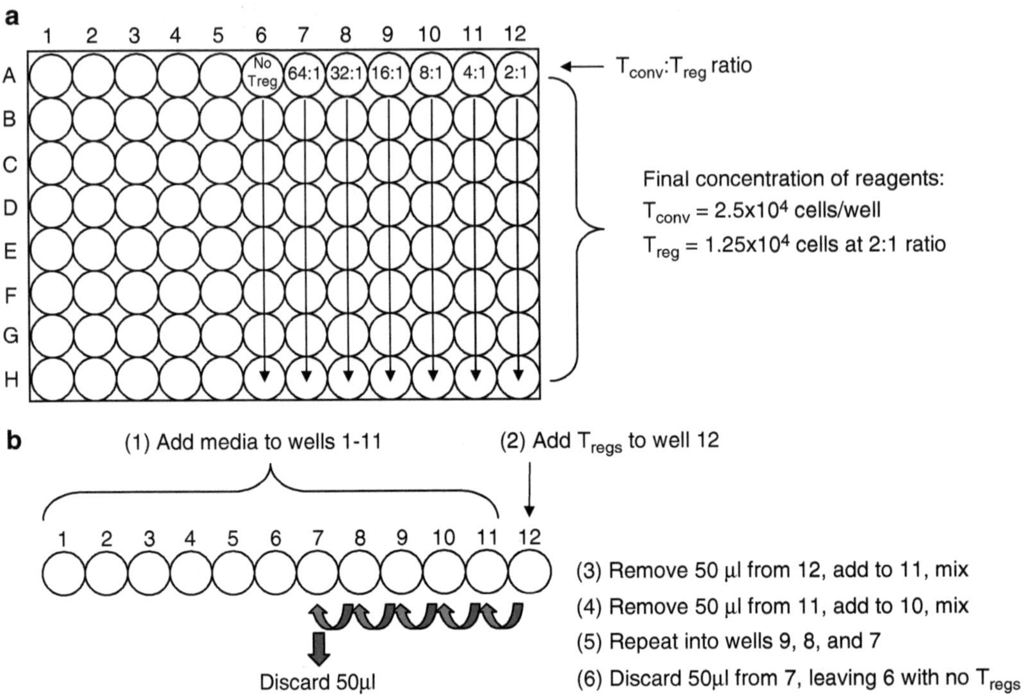

Fig. 1. Plate diagram for T_{reg} assay. T_{regs} are titrated into a T_{conv} cell proliferation assay starting at a 2:1 T_{conv}:T_{reg} ratio.

includes only two cell types, the target T_{conv} and test T_{regs}. In this protocol, the experiment is setup in a 96-well round-bottom plate in a total volume of 200 μl. All reagents are prepared at four times their desired final concentration and added to assay in 50 μl such that in the total volume of 200 μl, their concentration will be correct. See Fig. 1a for a 96-well plate layout (see Note 2).

1. Purify T_{regs} and T_{conv} from desired source (see Subheading 3.8).
2. Count T_{regs} and T_{conv} and adjust in T-cell culture medium (see Subheading 2.1) to 2.5×10^5/ml and 5×10^5/ml, respectively.
3. In round-bottom 96-well plate, add 50 μl culture media to wells 1–11 (see Fig. 1b).
4. Add 100 μl T_{reg} to well 12.
5. Mix T_{regs} thoroughly with a pipet and titrate 50 μl of T_{regs} into well 11 to generate a twofold dilution. For multiple Treg populations, use a multichannel pipet to titrate multiple wells at the same time.
6. Repeat mixing and titration into successive wells, 50 μl at a time, leaving the well 6 with no T_{reg} to determine maximum proliferation of T_{conv}.
7. Add 50 μl T_{conv} cells to all wells.

8. Add 100 μl anti-CD3/CD28-coated sulfate latex beads to all wells (see Subheading 3.9).
9. Incubate plate at 37°C, 5% CO_2 for 72 h.
10. Pulse plates with 0.1 μCi [^{3}H]-thymidine (<!> – Caution: Radioactive material. Institutional approval to handle radioactive materials is required) per well 8 h prior to completion of experiment.
11. Harvest cultures with a commercial cell harvester and determine counts per minute (cpm) with a direct beta counter (see Notes 3 and 4).

3.2. Variations of Basic Protocol: Antigen Presenting Cell Activation

1. *Murine APC activation of T_{conv} cell proliferation.* Irradiated splenocytes or purified dendritic cells combined with soluble anti-CD3 or peptide may be substituted for anti-CD3 + anti-CD28 coated beads to stimulate T_{conv} cell proliferation. The benefit of using APCs is the more physiological activation of T_{conv} cells. However, these cells add an additional variable to the assay in that APCs may also mediate/modulate both T_{conv} and T_{reg} cell function and must be considered when interpreting results. It is important to ensure that only T_{conv} cell proliferation is measured and that irradiated splenocytes and T_{regs} do not contribute to the proliferation observed. To this end, control wells containing (a) APCs alone + antigen (or antibody) and (b) T_{regs} + antigen (or antibody) must be included in all experiments (see Note 5).
   (a) Splenocytes as APCs: Make a single cell suspension of splenocytes by homogenizing spleen with a 1 ml syringe through a 0.7 μM filter into a 50 ml conical tube. Alternatively, splenocytes may be homogenized between two frosted glass microscope slides. Following homogenization, lyse red blood cells using commercial lysis solution or Gey's solution (see Subheading 2.1). Quench lysis reaction with 10 ml HBSS. Irradiate splenocytes using 3,000 rads (<!> – Caution: Institutional approval to irradiate materials is required).
   (b) Dendritic cells as APCs: Make 10× digestion mix by dissolving 2 vials of Collagenase and 5 vials of DNase in 32 ml PBS, filter sterilize, and freeze in 4 ml aliquots (−20°C). Cut spleen into small pieces using sterile scissors. Digest spleen with 4 ml/spleen of RPMI medium containing 5% Fetal Bovine Serum and 10% digestion mix. Incubate 1 h in 37°C shaking water bath. Homogenize through a 0.7 μM filter into a 50 ml conical tube. Lyse red blood cells with commercial lysis solution or Gey's solution (see Subheading 2.1) and stain cells with a fluorescently conjugated anti-CD11c antibody for purification by FACS.

(c) Resuspend splenocytes (for protocol 1) at 1×10^6/ml or DCs (for protocol 2) at 1×10^5/ml and add anti-CD3 antibody at 1 μg/ml.

(d) Omit anti-CD3 + anti-CD28 beads in basic protocol and replace with 50 μl each APCs and soluble anti-CD3 antibody in all wells.

(e) Add T_{conv} and titrations of T_{reg} cells to wells as described in Subheading 2.1 (see Note 6).

2. *Human APC activation of T_{conv} cell proliferation.* For assays with human cord blood or PBMC derived T_{conv}, irradiated PBMCs can be used as antigen-presenting cells in a standard mixed lymphocyte reaction. Assays are to be performed in a 96-well round-bottom plate in a final volume of 200 μl of complete medium.

   (a) Add T_{conv} and titrations of T_{reg} cells to wells as described in Subheading 2.1.

   (b) Irradiate allogeneic PBMCs or unmanipulated cord blood cells with 2,500 rads (<!> – Caution: Institutional approval to irradiate materials is required) and suspend at 1×10^6/ml.

   (c) Add 50 μl PBMCs or cord blood cells per well to serve as APCs.

   Alternatively, irradiated syngeneic PBMCs can be cultured with anti-CD3 (OKT3) peptide to activate T_{conv} cells.

3. *Murine antigen-specific suppression assays.* Suppression of antigen-specific responses can be determined by utilizing murine TCR transgenic T_{conv} and T_{reg} cells instead of a polyclonal T-cell population. The benefit of this variation to the basic protocol is that monoclonal or polyclonal T_{regs} as well as T_{conv} cells and T_{regs} with a variety of specificities can be utilized in suppression assays. However, many TCR transgenic mice have limited numbers of clonotype positive T_{regs}, which has to be considered when designing these experiments (e.g., on a *Rag1*$^{-/-}$ background, OTII transgenic mice have none, while AND transgenic mice have ~10% of normal Treg numbers (11). However, the use of endogenous TCR chains often endows T_{regs} from TCR transgenic mice with potent peptide specific regulatory capacity.

   (a) Prepare irradiated splenocytes for culture as described above.

   (b) Dilute cognate antigen in media at desired concentration (0.1–10 μg/ml). For example, T cells from OTII, AND, or 6.5 TCR transgenic mice are cultured with their cognate antigen: $Ova_{326-339}$, PCC_{88-104}, or $HA_{110-120}$, respectively (see Note 7).

(c) Omit anti-CD3 + anti-CD28 beads in basic protocol and replace with 50 μl each APCs and cognate antigen in all wells.

3.3. Variations of Basic Protocol: T_{reg} Activation State

Recent studies using pre-activated T_{reg} have contributed to our understanding of the characteristics and conditions required for T_{reg} to suppress T_{conv} proliferation (1, 12). Reports indicate that previously activated T_{regs} do not require restimulation through their TCR to suppress T_{conv} proliferation (12). Freshly isolated T_{regs} can be utilized for a number of protocols; however, activated or expanded T_{regs} are sometimes desired. Pre-activated murine T_{regs} have been shown to have superior suppressive capacity when compared to naïve, freshly purified T_{regs}. Moreover, human cord blood T_{regs} are naïve and require activation to suppress T_{conv} cell proliferation effectively. For this reason, it is sometimes advisable to activate T_{regs} prior to assaying. In addition, when T_{regs} numbers are limiting, they can be expanded in vitro to obtain greater numbers of cells.

1. Freshly isolated T_{regs} can be directly assayed for regulatory capacity as described in Subheading 2.1.
2. Alternatively: "Pre-Activated" T_{regs} can be generated and used in assays by activating purified T_{regs} for 24 h at 5×10^5/ml in a 96-well round-bottom plate containing anti-CD3 (1 μg/ml) and anti-CD28 (2 μg/ml). Following activation, T_{regs} should be washed and adjusted to 2.5×10^5/ml for use in suppression assays (as described in Subheading 2.1).
3. Murine T_{reg} expansion: Several murine T_{reg} in vitro expansion protocols have been described. This could be useful when the number of purified T_{regs} is very limited, such as when isolated from sites of infection, tumors, or autoimmune lesions (see Chapter 9).
4. Human T_{reg} expansion: Human T_{regs} are activated at a density of 5×10^5 cells/ml in a 24-well plate in complete X-VIVO 15 media supplemented with anti-CD3/anti-CD28 coated beads at a 3:1 (bead:cell) ratio and 500 IU of IL-2. Cells are passaged to maintain cell density of 5×10^5cells/ml. Following 10 days culture, T_{reg} expansion is approximately 20-fold. Expanded T_{regs} maintain FoxP3 expression and suppressive capacity.

3.4. Variations of Basic Protocol: MTT Assay as a Readout of Suppression

Suppression of proliferation can be monitored without the use of radioisotopes or fluorescence chemistries by using Cayman Chemical's MTT Cell Proliferation Assay Kit. This method utilizes the reduction of MTT reagent by intracellular NAD(P)H oxidoreductases as a measure of cellular proliferation.

Reagent Preparation: Dissolve the Cell Based Assay Buffer tablet in 100 ml of distilled water. Prepare MTT reagent by dissolving the 25 mg vial of reagent in 5 ml Assay Buffer. Store at 4°C.

1. Approximately 4 h prior to completion of assay: Add 20 μl MTT reagent to each well, mix gently, and return to incubator.
2. Allow cells to reduce MTT reagent for 4 h. Formazan produced by the cells will appear as purple/black dots in the wells.
3. Centrifuge the plate at 400 × *g* for 10 min to pellet the cells. Aspirate supernatant.
4. Add 100 μl of Crystal Dissolving Detergent Solution to the wells and pipet to mix.
5. Measure the absorbance of the samples at 570 nm using a microplate reader.

3.5. Variations of Basic Protocol: CFSE as a Readout of Suppression

Suppression of proliferation can be monitored without the use of radioisotopes by monitoring the dilution of a green fluorochrome ester CFSE, or the red alternative, SNARF-1 (Seminaphtharhodafluor) by flow cytometry. In addition, CFSE analysis allows for the determination of the number of cell divisions with or without T_{reg} suppression.

Reagent Preparation: Prepare solution of sterile PBS + 0.1% BSA to use as a diluent. Prepare single use aliquots of CFSE and store at −20°C.

1. Wash T_{conv} cells once with PBS.
2. Resuspend cells at 2–3 × 10^6/ml in PBS + 0.1% BSA and keep on ice.
3. Prepare 8 μM CFSE in PBS + 0.1% BSA. Discard unused CFSE solution.
4. *While vortexing cells*, add volume of CFSE solution equivalent to volume of cells (i.e., for 2 × 10^6 cells, resuspend in 1 ml PB and add 1 ml CFSE solution).
5. Incubate at room temperature without agitation for 10 min.
6. *While vortexing cells*, quench reaction as quickly as possible with three times the staining volume of ice-cold FBS (i.e., 2 ml staining volume, add 6 ml FBS).
7. Put on ice immediately for 2 min.
8. Wash cells two times with 10 ml T-cell culture medium, centrifuging at 300 × *g* for 10 min in between washes.
9. Count CFSE labeled T_{conv} cells, resuspend at 5 × 10^5/ml, and add to suppression assay as described in Subheading 2.1.
10. Analyze proliferation as determined by CFSE dilution on a cytometer. See Fig. 2 for a representative flow cytometric histogram of CFSE dilution of T_{conv} in the presence and absence of T_{regs} (see Notes 8 and 9).

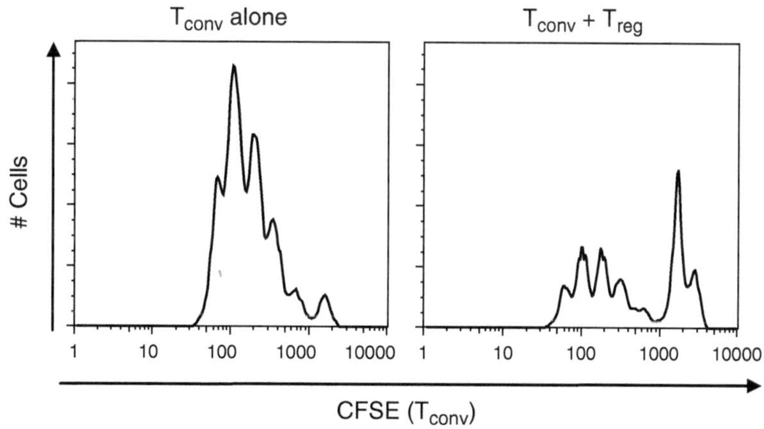

Fig. 2. T_{reg}-mediated suppression as measured by carboxyfluorescein succinimidyl ester (CFSE) dilution. T_{conv} were isolated from C57BL/6 mice and labeled with 5 μM CFSE. Cells were activated with anti-CD3 + anti-CD28 coated beads and cultured either alone or in the presence of T_{regs} at a 4:1 T_{conv}:T_{reg} ratio. After 72 h, proliferation was determined by CFSE dilution and flow cytometric analysis.

3.6. Reporting Data as cpm Versus Percent Suppression

The results of in vitro T_{reg} suppression assays are most commonly reported as cpm when [^{3}H]-thymidine is incorporated into proliferating cells. Wells containing both T_{conv} and T_{regs} will have lower cpm than wells that contain T_{conv} cells alone because coculture of T_{regs} with T_{conv} cells reduces the proliferative capacity of T_{conv} cells. In addition, as the ratio of T_{conv} cells to T_{reg} increases, the cpm values will increase proportionately. As T_{regs} proliferate very poorly in vitro, they do not contribute significantly to cpm values. Control wells containing activated T_{regs} and no T_{conv} cells should have cpm values of less than 1,000, similar to that seen in wells containing unstimulated $T_{conv.}$ Due to day to day or sample to sample variability, experimental replicates will often not result in identical cpm values. For this reason, a percent suppression (% supp) calculation assay can be graphed in order to depict many experiments with slightly (or significantly) different cpm values. Percent suppression can be calculated using the following formula: ((cpm of T_{conv} cells alone − cpm of T_{conv} cells treated with T_{reg})/cpm of T_{conv} cells alone)*100. Alternatively, a representative experiment can be depicted with cpm. The data graphed are the same; however, the graphs will appear differently (see Fig. 3 for examples).

Statistical Analysis of Results: To determine statistical significance between groups, a variety of different statistical methods can be used. For comparisons of two samples, an unpaired T test can be used. For this analysis, a two-tailed *p* value with a confidence interval of 95% is recommended. For analyses of three or more samples, one-way ANOVA with a confidence interval of 95% is recommended.

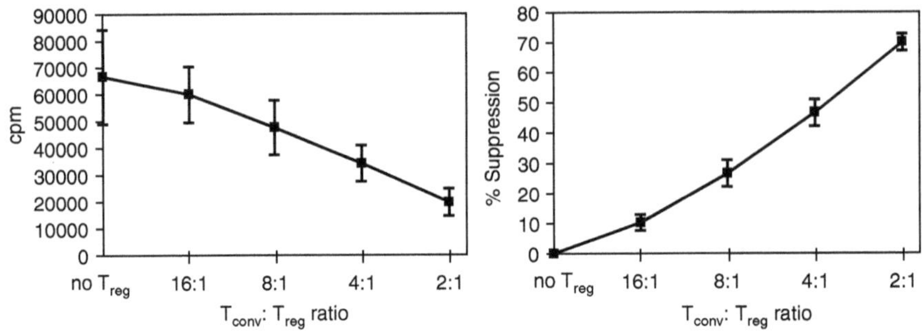

Fig. 3. T_{reg}-mediated suppression. T_{reg} cells were purified by FACS and mixed at different ratios with naïve wild type T_{conv} cells and anti-CD3+anti-CD28 coated beads for 72 h. Proliferation was determined by [^{3}H]-thymidine incorporation.

3.7. Variations of Basic Protocol: Transwell Suppression Assay

The importance of cytokines in mediating in vitro T_{reg} suppression has been controversial. Neutralizing IL-10 and TGFβ in a conventional in vitro T_{reg} assays does not inhibit suppression by T_{reg}, suggesting that these cytokines are not required for T_{reg}-mediated suppression in vitro (1, 2, 13, 14). However, cytokines are critical for T_{reg}-mediated suppression in vivo (3–9), making it difficult to reconcile these differential requirements. By using a specialized 96 well plate in which a permeable membrane called a Transwell membrane is inserted, cells can be separated from one another via a membrane that permits exchange of soluble molecules between cells but does not allow cell–cell contact. Addition of T_{conv} and T_{reg} alone or in combination on either side of the Transwell membrane allows one to permit cell contact between populations as desired (see Fig. 4 for a Transwell plate diagram).

1. Add freshly purified T_{conv} cells (5×10^4/well) in 50 µl media in the bottom chamber of a 96 well receiver plate.
2. Add 50 µl anti-CD3/CD28 coated sulfate latex beads to all bottom wells (see Subheading 3.9).
3. Add 50 µl T cell culture media to bring all wells to a final volume of 200 µl.
4. Gently insert 0.4 µM Transwell membrane into bottom chamber of receiver plate.
5. Add cells that are to be tested for regulatory capacity to the top chamber wells. (ex) T_{conv} and T_{reg} either alone at 1.25×10^4/ well or coculture at a ratio of 4:1 with a total of 2.5×10^4 cells in top chamber.
6. Add 50 µl anti-CD3/CD28 coated sulfate latex beads to all top wells.
7. Where necessary, add T-cell culture media to bring top wells to a final volume of 150 µl.

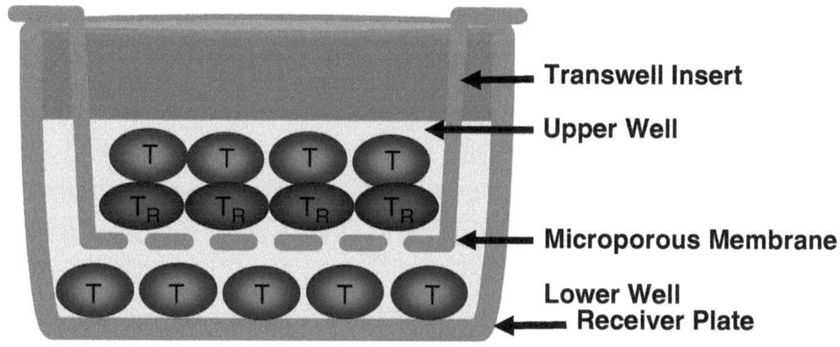

Fig. 4. Transwell plate setup. T_{conv} cell proliferation in the lower well of a Transwell plate can be suppressed T_{reg} cells in the top well of a Transwell plate when they are activated in the presence of T_{conv} cells. Proliferation of lower well T_{conv} cells is determined by [^{3}H]-thymidine incorporation.

8. After 64 h in culture, remove top chambers, and add [^{3}H]-thymidine directly to the responder T_{conv} cells in the bottom chambers of the original receiver plate.
9. Harvest as described above (see Note 10).

3.8. Purification of T_{conv} Cells and T_{regs} for Assay

An important difference between murine and human in vitro T_{reg} suppression assays is the source of cells. Murine T_{conv} and T_{reg} are predominantly purified from spleens and lymph nodes on the basis of CD25 expression. However, human T_{conv} and T_{reg} can be isolated from peripheral blood (from PBMCs or apheresis rings, depending upon availability) or umbilical cord blood. In human peripheral blood, suppressive capacity is not associated with all $CD25^+$ cells, as it is in the mouse, but instead with the brightest subset of $CD25^+$ cells (termed $CD25^{bright}$). Another complication with using peripheral blood T_{conv} and T_{regs} is that unlike in the mouse, Foxp3 can be expressed in both T_{reg} and activated T_{conv}, making classification and purity analysis difficult. For this reason, a number of additional cell surface markers have been used to help purify peripheral blood T_{conv} and T_{reg}, with relative degrees of success. For detailed information regarding purification, subsets of human T_{regs}, and alternative cell surface markers for identification of T_{regs}, see refs. (15–18). An alternative source of human T_{conv} and T_{reg} is umbilical cord blood. Unlike peripheral T cells, cord blood T_{conv} have not encountered any peripheral antigen; therefore, CD25 expression is a much better marker for $T_{regs.}$ In addition, both Foxp3 expression and suppressive capacity are exclusively within the $CD25^+$ population. The complications with using umbilical cord blood samples are (1) access to samples (2) both T_{conv} and T_{reg} are antigen inexperienced as they have never entered peripheral circulation. For this reason, additional manipulation is required; IL-2 supplementation to achieve strong proliferation of T_{conv} and pre-activation for maximum suppressive capacity of $T_{regs.}$

1. *Purification of murine T_{conv}/T_{reg} (CD4, CD45RB, CD25).* Murine T_{conv} and T_{reg} can be separated using only CD4 and CD25 markers. However, by also staining with CD45RB, naïve T_{conv} can be separated from memory T_{conv} and T_{reg}, resulting in better purity of both populations. A similar strategy can be utilized by staining cells with CD44 and CD62L, where $CD44^{low}$, $CD62L^{high}$ populations represent the naïve T_{conv} cells. The only disadvantages with this staining is that an additional fluorochrome-conjugated antibody is needed that adds to the expense of purification as well as utilizing another fluorescent channel (thus eliminating this flow cytometer channel for staining for downstream applications such as intracellular staining).
   (a) Harvest spleen and lymph nodes from mice.
   (b) Homogenize tissue with a 1-ml syringe through a 70-μm cell strainer into a 50-ml conical tube. Rinse strainer two times with HBSS to recover all cells.

   Alternatively, splenocytes may be homogenized between two frosted glass microscope slides.
   (c) Centrifuge homogenate at 300 × *g* for 10 min.
   (d) Resuspend homogenate in 1 ml Gey's solution (see Subheading 2.1) per spleen. Gently swirl for 2 min and then quench reaction by adding 12 ml of HBSS.
   (e) Centrifuge at 300 × *g* for 10 min.
   (f) Resuspend cells in blocking solution at 0.5 ml/spleen (10% mouse serum in PBS + 5% FBS).
   (g) Incubate cells for 10 min at 4°C.
   (h) Add 0.5 ml/spleen fluorescently conjugated antibodies at final concentration of 1:200 for 20–30 min at 4°C, for example, anti-CD4 Alexa 647 (or APC), anti-CD45RB (PE), and anti-CD25 FITC (see Note 11).
   (i) Wash cells with 5 ml PBS + 5% FBS. Centrifuge cells at 300 × *g* for 10 min.
   (j) Resuspend cells in PBS + 5% FBS and strain through 40 μm filter.
   (k) Purify cells by FACS according to the profile shown in Fig. 5.
2. Purification of human PBMC or cord blood $T_{conv}/T_{reg.}$
   (a) Obtain PBMCs or cord blood samples (see Note 12).
   (b) In hood, wipe down tip of ring or bag with 70% ethanol.
   (c) Ensure that the clamp that comes in the unit is secured tightly.
   (d) Attach the plasma transfer set to collect blood.

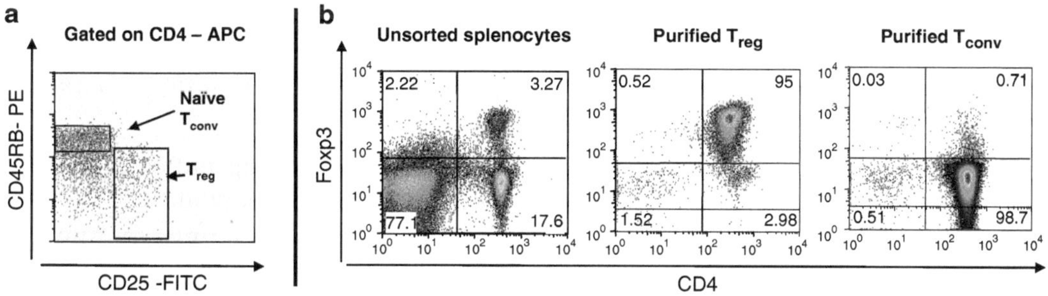

Fig. 5. T_{conv}/T_{reg} purification. (**a**) Murine splenocytes were processed and red blood cells lysed prior to staining with anti-CD4, anti-CD25 and anti-CD45RB antibodies. T_{conv} and T_{reg} were purified by FACS based on the profile shown. (**b**) Red blood cell depleted murine splenocytes were stained with anti-CD4 and anti-Foxp3 antibodies. In parallel, purified T_{conv} ($CD4^{+}CD25^{-}CD45RB^{hi}$) and T_{reg} ($CD4^{+}CD25^{+}CD45RB^{lo}$) were stained with anti-CD4 and anti-Foxp3 antibodies and % Foxp3+ cells were determined by flow cytometry.

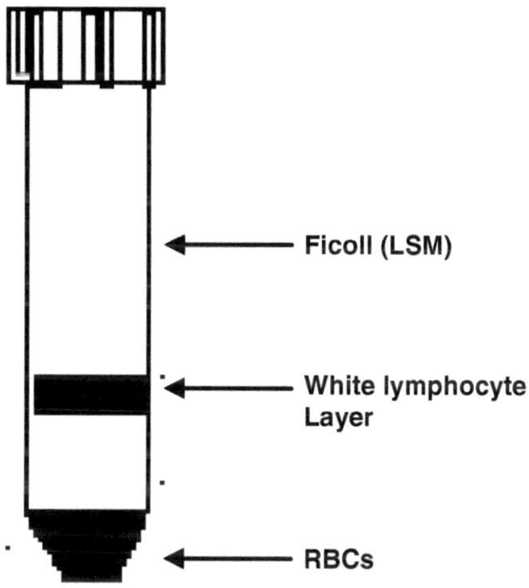

Fig. 6. Ficoll gradient for T-cell purification. Depiction of lymphocyte layer following Ficoll separation of cord blood or PBMCs.

(e) Clamp the set closed and remove the plastic piercing cover.

(f) Open new port of blood unit and insert piercing pin.

(g) Remove female adaptor, open up clamp and pour blood from female adaptor port into 50 ml conical tube(s).

(h) Pellet blood at 1,800 ×*g* for 15 min at room temperature. Discard supernatant (serum).

(i) Resuspend pellet at 1:2.5–3 ratio of pellet volume: PBS.

(j) Overlay 15 ml diluted blood onto 25–30 ml Ficoll. Centrifuge at 1,150 ×*g* for 20 min without brake at room temperature.

(k) After centrifugation, sample will separate into bands (shown in Fig. 6).

(l) Aspirate excess Ficoll into biohazard container.

(m) With 5 ml pipet, slowly remove white lymphocyte layer and put into new 50 ml conical tube.

(n) Fill tube to 50 ml with sterile PBS. Centrifuge at 480 × *g* for 10 min. Max brake.

(o) Resuspend cells in antibody staining buffer containing anti-CD4 and anti-CD25 at 1:20 dilution.

(p) Incubate on ice for 30 min. Add 5 ml PBS + 5% FBS and centrifuge 480 × *g* for 10 min with max brake.

(q) Resuspend in PBS + 5% FBS. Filter cells through a 40 μM strainer and purify by FACS (see Note 13).

3.9. Labeling of Anti-CD3 + Anti-CD28 Coated Latex Beads

1. Make antibody mix in Phosphate Buffer:
   Anti-CD3 Ab (NALE/functional grade) – murine 13.3 μg/ml, human 26.6 μg/ml
   Anti-CD28 (NALE/functional grade) – murine and human 26.6 μg/ml
   Add 750 μl sterile 5 mM phosphate buffer (4.82 g/l monohydrate, monosodium phosphate, pH 6.5).
2. Incubate 5 μM sulfate latex beads (4% solid) in a 1:4 dilution of antibody mix to make 1% solid. (ex) 250 μl beads + 750 μl antibody mix in a 1.5 ml tube.
3. Incubate overnight at room temperature either by vortexing or by rotation.
4. Wash beads three times with Phosphate Buffer, centrifuging at 200 × *g* to remove buffer between washes.
5. Count beads with a hemacytometer and resuspend beads at 5×10^7/ml in sterile Phosphate Buffer with 2 mM BSA.
6. Optimal bead concentration is typically between 3:1 and 10:1 (T cell:bead ratio); however, this must be determined empirically by titrating beads into a proliferation assay prior to suppression assays. Desired T_{conv} cell proliferation is 40,000–80,000 cpm following 8 h [^{3}H]-thymidine culture for the final 72 h of assay. Alternatively, at least four CFSE peaks should be visible by flow cytometry following 72 h assay (e.g., see Fig. 2).

3.10. Foxp3 Staining to Determine Purity

To ensure purity of isolated T_{conv} and T_{regs}, Foxp3 staining of cells before and after purification should be performed (e.g., see Fig. 5). The Foxp3 staining kit manufactured by eBioscience provides all of the reagents needed for optimal Foxp3 staining and is the recommended kit for this purpose.

1. Add 100 μl of prepared cells (2×10^5/well) to a v-bottom 96 well plate.
2. Stain surface molecules such as CD4, CD8, CD25, etc. in PBS. Incubate at 4°C for 20 min.

3. Wash in 50 μl cold PBS, centrifuging at 200 × *g* for 2 min.
4. Resuspend cell pellet with pulse vortex and add 100 μl of freshly prepared Fixation/Permeabilization working solution to each sample. Pulse vortex again.
5. Incubate at 4°C for 30–60 min in the dark.
6. Wash once by adding 100 μl 1× Permeabilization Buffer (made from 10× Permeabilization Buffer) followed by centrifugation and decanting of supernatant.
7. Add 100 μl fluorochrome conjugated anti-Foxp3 antibody or isotype control at 1:100 dilution in 1× Permeabilization Buffer and incubate at 4°C for 30 min in the dark.
8. Wash cells with 200 μl 1× Permeabilization Buffer. Centrifuge and decant supernatant.
9. Resuspend in appropriate volume of PBS and analyze on cytometer.

4. Notes

1. The optimal manufacturer and lot number of FBS can vary; therefore, this must be determined empirically. Prior to use in assays, FBS must be heat inactivated for 30 min at 56°C. Following heat inactivation, FBS can be stored at 4°C for up to 1 month.
2. Sterility during all steps of the protocols is essential. Sterile technique must be followed, and all reagents used including buffers and antibodies must be sterile filtered through a 0.2 μm filter.
3. Human cord blood T_{conv} are naïve and require IL-2 supplementation and longer stimulation to obtain optimal proliferation. Therefore, for assays with human cord blood T_{reg}, recombinant human IL-2 is added (10 U/ml) and cultures are harvested after 6 days. Human PBMC derived T_{conv} are fully capable of responding to stimulation without exogenous IL-2 within the 3 days assay; therefore, no alterations from the basic protocol are needed to perform assays with PBMC derived T_{conv}.
4. For large scale isolation of T_{regs}, or if purification by FACS is not possible, magnetic-based cell separations provide an alternative means of T_{reg} isolation. For a detailed protocol describing purification by MACS of human T_{regs}, see ref. (19).
5. When performing APC driven T_{reg} suppression assays, it is imperative to use mice of the same genetic background and sex.

6. With the addition of APCs and anti-CD3 or peptide, the volume will be 200 μl (50 μl T_{conv}, 50 μl T_{regs}, 50 μl APCs, and 50 μl anti-CD3 or peptide); therefore, no additional media should be added to culture wells.
7. TCR specific T cells are optimally stimulated by different concentrations of peptides. A titration must be done to determine optimal antigen concentration.
8. To obtain clear CFSE peaks, it is critical that CFSE is intercalated into all cells at the same time, hence the reason for vortexing cells while adding CFSE solution. Furthermore, CFSE quenching must occur quickly, completely, and while vortexing. Deviation from this protocol will yield less clear results.
9. T_{regs} are not labeled with CFSE and can, therefore, easily be distinguished from proliferating T_{conv} cells as a CFSE negative population. The use of T_{conv} and T_{regs} with different congenic markers (i.e., Thy1.1 T_{conv} and Thy1.2 T_{regs}) can help to distinguish T_{conv} and T_{regs} by flow cytometry.
10. If so desired, T_{conv} in the top well can be fixed with 4% formaldehyde (<!> Caution: Irritant and suspected carcinogen) in media for 10 min, at room temperature in order to eliminate any contribution of T_{conv}-derived soluble factors. T_{conv} should be fixed at 1×10^6 cells/ml and washed four times with 10 ml of fresh media prior to assay. Care must be taken to thoroughly wash cells to eliminate formaldehyde carryover into culture.
11. Antibodies used can be altered depending upon lasers available, and optimal antibody concentrations must be determined empirically.
12. Institutional Review Board (IRB) approval must be obtained prior to use unless samples are purchased from commercial sources.
13. Additional cell surface molecules such as CD127, HLA-DR, etc. may be used in addition to CD4 and CD25, as desired (15–18).

Acknowledgments

We wish to thank members of the Vignali lab for many discussions regarding these methods. We are particularly grateful to Andrea Szymczak-Workman (for advice on anti-CD3/CD28 bead conjugation), Creg Workman and Andrea Szymczak-Workman (set up of murine antigen specific suppression assays), Janice Riberdy (human suppression assay setup), and Sam Connell (CFSE labeling). LWC is supported by an Individual NIH NRSA (F32 AI072816). DAAV is supported by the National Institutes of Health (NIH) (AI39480, AI52199, AI072239), Juvenile

Diabetes Research Foundation International (1-2004-141 [The Robert and Janice Compton Research Grant, In Honor of Elizabeth S. Compton] and 1-2006-847), a Cancer Center Support CORE grant (CA21765), and the American Lebanese Syrian Associated Charities (ALSAC).

References

1. Takahashi, T., Kuniyasu, Y., Toda, M., Sakaguchi, N., Itoh, M., Iwata, M., Shimizu, J., and Sakaguchi, S. (1998) Immunologic self-tolerance maintained by CD25+CD4+ naturally anergic and suppressive T cells: induction of autoimmune disease by breaking their anergic/suppressive state. *Int Immunol* 10, 1969–1980.
2. Thornton, A. M., and Shevach, E. M. (1998) CD4+CD25+ immunoregulatory T cells suppress polyclonal T cell activation in vitro by inhibiting interleukin 2 production. *J Exp Med* 188, 287–296.
3. Annacker, O., Pimenta-Araujo, R., Burlen-Defranoux, O., and Bandeira, A. (2001) On the ontogeny and physiology of regulatory T cells. *Immunol Rev* 182, 5–17.
4. Asseman, C., Mauze, S., Leach, M. W., Coffman, R. L., and Powrie, F. (1999) An essential role for interleukin 10 in the function of regulatory T cells that inhibit intestinal inflammation. *J Exp Med* 190, 995–1004.
5. Belkaid, Y., Piccirillo, C. A., Mendez, S., Shevach, E. M., and Sacks, D. L. (2002) CD4+CD25+ regulatory T cells control Leishmania major persistence and immunity. *Nature* 420, 502–507.
6. Cavinato, R. A., Casiraghi, F., Azzollini, N., Mister, M., Pezzotta, A., Cassis, P., Cugini, D., Perico, N., Remuzzi, G., and Noris, M. (2007) Role of thymic- and graft-dependent mechanisms in tolerance induction to rat kidney transplant by donor PBMC infusion. *Kidney Int* 71, 1132–1141.
7. Kingsley, C. I., Karim, M., Bushell, A. R., and Wood, K. J. (2002) CD25+CD4+ regulatory T cells prevent graft rejection: CTLA-4- and IL-10-dependent immunoregulation of alloresponses. *J Immunol* 168, 1080–1086.
8. McGeachy, M. J., Stephens, L. A., and Anderton, S. M. (2005) Natural recovery and protection from autoimmune encephalomyelitis: contribution of CD4+CD25+ regulatory cells within the central nervous system. *J Immunol* 175, 3025–3032.
9. Read, S., Malmstrom, V., and Powrie, F. (2000) Cytotoxic T lymphocyte-associated antigen 4 plays an essential role in the function of CD25(+)CD4(+) regulatory cells that control intestinal inflammation. *J Exp Med* 192, 295–302.
10. Collison, L. W., Pillai, M. R., Chaturvedi, V., and Vignali, D. A. (2009) Regulatory T cell suppression is potentiated by target T cells in a cell contact, IL-35- and IL-10-dependent manner. *J Immunol* 182, 6121–6128.
11. Szymczak-Workman, A. L., Workman, C. J., and Vignali, D. A. (2009) Cutting edge: regulatory T cells do not require stimulation through their TCR to suppress. *J Immunol* 182, 5188–5192.
12. Liu, H., Komai-Koma, M., Xu, D., and Liew, F. Y. (2006) Toll-like receptor 2 signaling modulates the functions of CD4+ CD25+ regulatory T cells. *Proc Natl Acad Sci USA* 103, 7048–7053.
13. Dieckmann, D., Plottner, H., Berchtold, S., Berger, T., and Schuler, G. (2001) Ex vivo isolation and characterization of CD4(+) CD25(+) T cells with regulatory properties from human blood. *J Exp Med* 193, 1303–1310.
14. Jonuleit, H., Schmitt, E., Stassen, M., Tuettenberg, A., Knop, J., and Enk, A. H. (2001) Identification and functional characterization of human CD4(+)CD25(+) T cells with regulatory properties isolated from peripheral blood. *J Exp Med* 193, 1285–1294.
15. Baecher-Allan, C., Brown, J. A., Freeman, G. J., and Hafler, D. A. (2001) CD4+CD25high regulatory cells in human peripheral blood. *J Immunol* 167, 1245–1253.
16. Baecher-Allan, C., and Hafler, D. A. (2004) Suppressor T cells in human diseases. *J Exp Med* 200, 273–276.
17. Baecher-Allan, C., Viglietta, V., and Hafler, D. A. (2004) Human CD4+CD25+ regulatory T cells. *Semin Immunol* 16, 89–98.
18. Vignali, D. A., Collison, L. W., and Workman, C. J. (2008) How regulatory T cells work. *Nat Rev Immunol* 8, 523–532.
19. Wichlan, D. G., Roddam, P. L., Eldridge, P., Handgretinger, R., and Riberdy, J. M. (2006) Efficient and reproducible large-scale isolation of human CD4+ CD25+ regulatory T cells with potent suppressor activity. *J Immunol Methods* 315, 27–36.

Chapter 3

Generation of T Cell Hybridomas from Naturally Occurring FoxP3⁺ Regulatory T Cells

Nagendra Singh, Rafal Pacholczyk, Makio Iwashima, and Leszek Ignatowicz

Abstract

Generation of regulatory T cells (or Treg) derived hybridomas offers a tool to study their antigen specificity. T cells hybridomas are produced by fusing TCR α-β-thymoma BW5147 with highly dividing T cell population. In vitro anergy of Tregs is an obstacle in generation of highly dividing Treg population for their fusion. In this chapter, we describe a simple and efficient method to generate large number of blasting Treg and their successful fusion with thymoma BW5147. The resultant hybridomas lose Treg-specific transcription factor FoxP3, respond to antigenic stimulation by producing IL-2, and thus allow the evaluation of antigen specific, Tregs-derived TCRs.

Key words: CD4 T cells, Foxp3, Hybridomas

1. Introduction

Regulatory T cells or Tregs express transcription factor FoxP3 and suppress the immune responses against self and foreign antigens. Recognition of MHC-peptide complexes by Tregs TCR is required for Treg-mediated suppression. However, antigen-specificity of Treg-mediated suppression has been a matter of debate. Validation of Treg TCR specificities requires studying a large pool of Tregs-derived TCRs that is not possible by most of the current procedures (e.g., Treg clones). Generation of Tregs-derived T cell hybridomas offers a tool to test the functional specificity of a larger number of Tregs-derived TCRs.

One of the critical steps toward the production of T cell hybridomas is generation of activated and highly expanding T cell populations that will be fused with the growing BW5147 thymoma

George Kassiotis and Adrian Liston (eds.), *Regulatory T Cells: Methods and Protocols*, Methods in Molecular Biology, vol. 707, DOI 10.1007/978-1-61737-979-6_3, © Springer Science+Business Media, LLC 2011

lacking TCR α and β chains (1). For generation of Tregs hybridomas, Tregs need to be purified using conventional markers, e.g., $CD4^{+}CD25^{+}$/$CD4^{+}CD25^{+}CD62L^{high}$/$CD4^{+}CD25^{+}GITR^{+}$/$CD4^{+}CD25^{+}GITR^{+}CD127^{low}$/$CD4^{+}FoxP3GFP^{+}$ (however, the latter is possible only on a few genetic backgrounds) followed by their expansion. All the currently published Treg expansion techniques have two disadvantages: (1) they do not actively eliminate effector T cells and/or (2) they expand effector T cells better than Tregs. As a result, in Treg expansion cultures, contaminating effector T cells in the initial seed of sorted Tregs overwhelms the culture with the time, and fusion of expanded T cells to BW5147 will result in production of T cell hybridomas pool dominated by TCR derived from effector T cells. We have recently discovered that sustained plate bound CD3 and CD28 stimulation procedure expands Tregs vigorously, while inducing apoptosis in effector T cells (2), and described in Subheading 3. Under this condition, effector T cells express higher amounts of proapoptotic molecules Fas, P53, Bim, and P21 than Tregs and undergo apoptosis. Our data showed that there was 82% overlap between the CDR3 regions of TCR-α chain of Tregs expanded by this procedure and initial seed of Treg put in the culture (3), demonstrating that procedure expands all the Tregs irrespective of their antigen specificity. This chapter is divided into two sections: expansion of Tregs and T cell fusion.

2. Materials

2.1. Immobilization of Anti-CD3 and Anti-CD28 to Plates

1. Borate buffer (0.1 M pH 8.5) – Prepare 0.1 M solution of boric acid in water and adjust pH to 8.5 with sodium hydroxide.
2. Anti-CD3ε (clone 145-2C11).
3. Anti-CD28 (clone 37.51).
4. Petri dishes 60 mm #8603–0160 (USA Scientific).

2.2. Expansion of Tregs

1. Tissue culture medium: RPMI1640, with 10% fetal calf serum, 1 mM sodium pyruvate, 4 mM L-glutamine, penicillin and streptomycin, 10 mM HEPES (pH 7.4), 1× MEM essential amino acids 1× MEM non essential amino acids (Invitrogen), and 50 μM 2-mercaptoethanol.
2. Recombinant m-IL-2 (Peprotech or BD Biosciences).

2.3. T Cell Fusion

All solutions and media should be made to the standard required for long term in vitro culture. Use molecular biology-grade reagents.

1. TCRα-β-variant of BW5147 thymoma (1).
2. 2–3 ml aliquots of PEG 1540 (Sigma p7181).

3. HAT solution (hypoxanthine, aminopterin, thymidine, 50×, Sigma).
4. HT solution (hypoxanthine, thymidine, 50×, Sigma).
5. Dulbecco's Modified Eagle Medium (DMEM).
6. Fetal calf serum.
7. 96-Well flat-bottomed plates.

3. Methods

3.1. Immobilization of Anti-CD3 and Anti-CD28 to Plates

1. Prepare a fresh dilution of anti-CD3 and anti-CD28 (5 μg/ml each) in borate buffer and add 2 ml to a 60-mm plate, swirl the plate few times to let antibody solution stick to the plate. Incubate on a flat surface for 16 h at room temperature.
2. Tilt the plate and aspirate the coating solution, keep the plate tilted for 10 s and aspirate the residual solution. Add the complete medium (2–3 ml) to the plate and swirl the plate. Incubate it for ~1 min.
3. Repeat the above step three times.

3.2. Culture of Tregs (Adapted from Ref. (2))

1. Prepare single cell suspension from spleen and/or lymph nodes from donor mice (see Note 1) using standard methods.
2. Label cells using antibodies against CD4 and CD25. Sort $CD4^+CD25^+$ cells using FACSAria or Mo-Flo cell sorters.
3. Wash cells using complete medium three times. *Optional*: At this step cells may be stored overnight in complete medium containing 2 ng/ml IL-2 at 4°C.
4. Resuspend the cells in complete medium containing 10 ng/ml IL-2 and plate $\sim 0.1 \times 10^6$ cells in 6 ml to one plate coated with anti-CD3 and anti-CD28 as above (see Note 2).
5. At day 5 add additional 5 ml of medium containing IL-2 (10 ng/ml).
6. At day 7, there will be $\sim 10^7$ cells that can be recovered from one plate. Most of the cells (>90%) will be $Foxp3^+$. Harvest cells and proceed for the fusion (see Note 3).
7. *Optional:* If cultures were started with less number of Tregs, harvest cells, on around day 7, wash and replate the harvested T cells on newly coated plate as in step 4. These reexpanded T cells can be harvested on day 10 (3 days later) for fusion with BW5147.

3.3. T Cell Fusion (Adapted from Ref. (4))

3.3.1. Preparation of 50% PEG Solution

Melt 2–3 ml of PEG (MW = 1,500) (in 15 ml polypropylene tube) in boiling water, and once the PEG has melted quickly add the same volume of serum free DMEM, mix and filter through a 0.45-µm syringe filter into a new 15 ml tube. Place tube in 37°C water bath.

3.3.2. Fusion

3.3.2.1. Preparation of BW Thymoma

1. Count BW5147s and collect 1.0×10^7 viable cells, pellet in 50-ml tube, resuspend in 5–10 ml of DMEM, and leave at room temperature. Separately, pour 50 ml DMEM into 50 ml canonical tube and place the tube in a 37°C water bath.
2. Count expanded Foxp3 T cells to be fused. We successfully fused and produced hybridomas from as few as 0.5×10^6 up to 3.0×10^7 of expanded Treg cells. The ratio of thymoma to blasts should be approximately 5:1 (but BW5147 cells must be no less than ten million). Collect all Treg blasts, pellet them, remove supernatant, and resuspended cells in 5–10 ml of DMEM. Move resuspended blasts to 50 ml tube with BW5147 cells and bring volume to 50 ml with DMEM. Pellet combined cells and wash twice with DMEM (see Notes 4 and 5).
3. After the final spin aspirate off the DMEM, do not disturb the pellet, and spin the tube for 1 min at $250 \times g$. Carefully aspirate remaining medium with pipette to get the pellet as dry as possible. Place tube with combined BW5147 and expanded Tregs into a clean, small beaker filled to one third with 37°C tap water collected from water bath.
4. Hold the tube with the cells and tap firmly with finger to distribute the pellet of cells over the conical bottom of the tube. Once the pellet is distributed, rest the tube in the beaker with warm water. Draw the 1 ml of prewarmed (37°C) PEG solution into a sterile pipette, and dribble the PEG over the cells over a period of approximately 45 s while shaking and rolling the tube gently against the side of the beaker. Leave the tube with cells soaked in PEG for additional 45 s (total 90 s cells stay resuspended in 50% PEG). Continue to slowly turn the tube to ensure equal distribution of the PEG and cells. The lower part of the tube containing cells should remain immersed in water.
5. Start to dilute out the PEG by adding the 10 ml of prewarmed DMEM, dropwise and gently swirling to mix the PEG with the DMEM. First, add 1 ml MEM over the course of 30 s, then add 2 ml more of MEM over the course of 30 s. Continue by adding slowly 3 ml MEM over the course of 30 s and finally add the remaining 4 ml MEM over the next 30 s. As medium is added to dilute PEG, try to minimize the shear forces due the fragile nature of the hybridomas at this time. When all 10 ml have been added, gently fill the tube with the

rest of prewarmed 40 ml of DMEM, put the cap, slowly flip the tube upside down to gently mix its content, and place it for 5 min in 37°C water bath.

6. Pellet cells (250 × *g* for 5 min), remove supernatant by aspiration and resuspended pellet in MEM with 10% FCS. Then make appropriate dilutions for plating and distribute cells into 96-well plates at 0.1 ml of cell suspension per well, depending on the anticipated number of hybridomas. Generally, the number of plates should ensure that at least at one concentration no more than one third of wells will be growth-positive, indicating that growing hybridomas likely originate from single Treg cell. This serial dilution of plated cells can be used because it is difficult to predict how many hybridomas will appear and it is desirable to avoid plating the hybrids at a density of more than 1/well. The plating conditions may vary, depending upon the number of input Treg cell blasts and efficiency of fusion.

7. Approximately 24 h after the fusion, add the HAT supplement (blocks the synthesis of NA that tumor cells require for growth; however, hybridomas may grow because T cells are able to survive independent of this) to the plates by preparing 40–50 ml of DMEM/10% FBS with 3× the final concentration of HAT. 50 μl of HAT is added per well (diluted with culture medium) to make a 3× solution. The medium should be changed 7 days later with 1×HAT in culture medium.

8. If the fusion was successful, hybridomas growth will be apparent at days 8–10 by examination with an inverted microscope. The hybridomas will be ready for transfer to 24-well plates (0.5 ml of culture medium supplemented with HT) approximately 10–14 days after the fusion. At that time, hybridomas should also be evaluated for TCR and CD4 expression using specific MoAbs and flow cytometry. Only double positive $CD4^{+}TCR^{+}$ hybridomas should be further propagated.

   From this point, the culture medium can be supplemented only with HT, and the same medium should be used for the following two passages before normal culture medium can be used.

9. Following fusion, Treg hybridomas loose Foxp3 expression but produce IL-2 upon TCR stimulation. Thus Treg hybridomas antigen specificities can be examined using the same assay for IL-2/IL-4 production that is used to test antigen specificities of T hybridomas derived from conventional (originally $Foxp3^{-}$) $CD4^{+}$ T cells. We used HT-2 T cell line that is an IL-2 responsive (5) and the 3-(4,5-dimenthylthiazol-2-yl)-2,5-diphenyltetrazolium bromide (MTT)-based colorimetric assay to determine HT-2 proliferation (6).

4. Notes

1. Since apoptosis of effector T cells and expansion of Tregs under the conditions described above depend on Fas, P53, Bim, P21, and CD28, it is not advisable to expand Tregs using this procedure from mice lacking these molecules.
2. In conventional CD3 and CD28 stimulation of T cells, after 2–3 days of culture, T cells are transferred to a new plate devoid of anti-CD3 and anti-CD28 that terminates CD3 and CD28 signaling and results in growth of effector T cells.
3. The procedure described above has been optimized such that effective amount of antibody is attached to the plates for the time of the culture and T cells continuously receive CD3 and CD28 signaling, resulting in growth of Tregs and apoptosis of effector T cells (2).
4. Expansion of Tregs does not alter the clonal distribution of TCR repertoire, demonstrating that this method is not dependent on TCR specificity. No bias in TCR repertoire was examined by the direct analysis of TCRs expressed by individual Treg cells prior to the fusion (freshly sorted Tregs and after 1 week expansion in vitro), as well as after the fusion on individual Treg cell hybridomas (3).
5. Because in T cell hybridomas derived from Treg cells the expression of Foxp3 is terminated, thus sorting of Foxp3^{+} T cells is recommended to avoid contamination with non-Treg cells. The method described above disfavors the expansion andproliferation of Foxp3^{-} T cells that further ensures that pool of T cell blast used for fusion represent Foxp3^{+} T cells.

References

1. White, J., M. Blackman, J. Bill, J. Kappler, P. Marrack, D. P. Gold, and W. Born. (1989) Two better cell lines for making hybridomas expressing specific T cell receptors. *J. Immunol.* **143**:1822–1825.
2. Singh, N., M. Yamamoto, M. Takami, Y. Seki, M. Takezaki, A. L. Mellor, and M. Iwashima. CD4^{+}CD25^{+} regulatory T cells resist a novel form of CD28^{-} and Fas-dependent p53 induced T cell apoptosis *J. Immunol.* **184**:94–104.
3. Pacholczyk, R., J. Kern, N. Singh, M. Iwashima, P. Kraj, and L. Ignatowicz. (2007) Nonself-antigens are the cognate specificities of Foxp3(+) regulatory T cells. *Immunity* **27**:493–504.
4. Kappler, J. W., B. Skidmore, J. White, and P. Marrack. (1981) Antigen-inducible, H-2-restricted, interleukin-2-producing T cell hybridomas. Lack of independent antigen and H-2 recognition. *J. Exp. Med.* **153**:1198–1214.
5. Watson, J. (1979) Continuous proliferation of murine antigen-specific helper T lymphocytes in culture. *J. Exp. Med.* **150**:1510–1519.
6. Mosmann, T. (1983) Rapid colorimetric assay for cellular growth and survival: application to proliferation and cytotoxicity assays. *J. Immunol. Methods* **65**:55–63.

Chapter 4

In Vitro and In Vivo Analyses of Regulatory T Cell Suppression of CD8+ T Cells

Kim J. Hasenkrug and Lara M. Myers

Abstract

The study of regulatory T cells (Treg) requires methods for both in vivo and in vitro analyses, both of which have different limitations, but which complement each other to give a more complete picture of physiological function than either method alone. Our analyses have focused on Treg-mediated suppression of CD8+ T cells, and in particular Tregs induced by viral infection. One of the unique characteristics of virus-induced Tregs is that they can suppress CD8+ T cell function in vitro without the requirement for additional stimulation. This ability correlates with their suppressive capacity and activated status in vivo. Interestingly, while virus-induced Tregs suppress CD8+ T cell function in vitro and in vivo, they do not suppress proliferation unless they are further activated in vitro.

Key words: Regulatory T cells, CD8+ T cells

1. Introduction

The model system we use for the study of virus-induced Tregs is Friend virus (FV) infection of adult immunocompetent mice (1). FV is an oncogenic mouse retrovirus that induces acute infections leading to lethal leukemia in most strains of mice (2). However, some strains of mice recover from acute infection, but remain chronically infected for life (3). It is these chronically infected mice that have revealed a role for Tregs in suppressing CD8+ T cell responses (4).

Interestingly, depletion of CD8+ T cells during acute infection abolishes the ability of high recovery strains of mice to prevent leukemia (5), but depletion during the chronic phase has relatively little effect (3). This finding suggested that chronic FV had escaped CD8+ T cell control. Studies then showed that

George Kassiotis and Adrian Liston (eds.), *Regulatory T Cells: Methods and Protocols*, Methods in Molecular Biology, vol. 707, DOI 10.1007/978-1-61737-979-6_4, © Springer Science+Business Media, LLC 2011

chronically infected mice had defective mixed lymphocyte reactions in vitro, and also decreased $CD8^+$ T cell-mediated rejection of FV-induced tumors in vivo (6). These results suggested a change in T cell function rather than in the virus. Interestingly, further experiments showed that suppression of in vivo $CD8^+$ T cell responses could be adoptively transferred to naïve mice with $CD4^+$ T cells, but not $CD8^+$ T cells, from chronically infected mice. Analysis of the $CD4^+$ T cells revealed that the $CD25^+$ regulatory T cell subset was significantly more activated in chronically infected mice than in naïve mice, the same subset of cells that Shimon Sakaguchi had shown to be involved in suppressing anti-self reactivity to prevent autoimmune diseases (7). These studies led to the development of in vivo and in vitro analysis techniques to further study the suppressive activity of virus-induced Tregs (4, 8, 9).

2. Materials

2.1. In Vitro Suppression Assays

1. Complete medium: Iscove's modified Dulbecco's medium (IMDM) (Lonza) with 25 mM Hepes, 10% heat-inactivated (56°C for 30 min) FBS, 100 U/ml penicillin and streptomycin, 2 mM L-glutamine.
2. Coating buffer: 0.05 M $NaCO_3$ pH 9.6.
3. 5 mM stock solution carboxyfluorescein succinimidyl ester (CFSE) (Molecular Probes).
4. Brefeldin A (Sigma): 10 μg/ml [final].
5. Buffer A for bead purification: 1× PBS, 0.5% BSA, 2 mM EDTA.
6. Fixative for target cells: 1× PBS, 0.5% paraformaldehyde (PFA) if fixing overnight or 1× PBS, 2% PFA if fixing for 30 min.
7. Permeabilization buffer for intracellular staining: 1× PBS, 0.1% saponin, 0.1% NaN_3, 1% FBS.
8. 96-Well flat or round bottom tissue culture plates.

   Option 1: anti-CD3 coated, each well incubated overnight with 100 μl/well coating buffer containing 1 μg anti-CD3.

   Option 2: Peptide-loaded APCs (concentration is determined empirically depending on peptide and TCR, but we have used 4.5 μM peptide to load APCs). A gamma irradiator is required for this option.

   Option 3: Use tetramers to stimulate target cells during assay (concentration must be determined empirically for individual tetramers, but we have used 2 μl of stock tetramer solution from Beckman Coulter for 3–4 × 10^6 $CD8^+$ T cells).

2.2. Cell Harvest from the Spleen

1. Phosphate buffered balanced salt solution (PBBS) (such as Dulbecco's).
2. RBC lysis buffer (ACK): 0.16 M NH_4Cl, brought to pH 7.2 with drops of 1 M K_2CO_3.
3. Nylon 100 μm cell strainer (Fisher Scientific). We use the plunger from a 3 or 5 gauge syringe to grind the tissue through the strainer. You need one strainer and plunger per mouse tissue. The strainers can be washed, sterilized, and reused multiple times.

2.3. Treg Cell Harvest from the Liver

1. Perfusion solution: 1× PBS and 75 U/ml heparin (Fisher Scientific).
2. Necessary perfusion equipment: 10-ml syringe, 23 gauge needle, tissue scissors and tweezers, anesthesia.
3. PBBS.
4. ACK RBC lysis buffer.
5. Percoll stock for making solutions of various concentrations: To Percoll (Amersham) add 8% 10× PBS (keep sterile).
6. 35% Percoll solution: dilute Percoll stock to 35% in PBBS with 6.5 μM Hepes and 100 U/ml heparin.
7. Nylon 100 μm cell strainers (Fisher Scientific), plungers from a 3 or 5 ml syringes.

2.4. Treg Cell Harvest from the Lung

1. Perfusion solution: 1× PBS and 75 U/ml heparin.
2. Necessary perfusion equipment: 10-ml syringe, 23 gauge needle, tissue scissors and tweezers, anesthesia.
3. PBBS.
4. 1.3 mM EDTA solution: PBBS with EDTA disodium salt, pH adjusted to 7.2.
5. PBBS containing 5% heat-inactivated FBS.
6. Collagenase solution (make fresh): MEM (Invitrogen) containing 5% heat-inactivated FBS, 1 μM $CaCl_2$, 1 μM $MgCl_2$, and 150 U/ml collagenase (Gibco).
7. Percoll stock for making solutions of various concentrations: To Percoll (Amersham) add 8% 10× PBS (keep sterile).
8. For 44% Percoll solution, dilute stock solution to 44% in MEM (Invitrogen).
9. For 67% Percoll solution, dilute stock solution to 67% in MEM (Invitrogen).
10. Nylon 100 μm cell strainers (Fisher Scientific) and plungers from 3 or 5 ml syringes.

2.5. Adoptive Transfers and In Vivo Suppression Assays

1. PBBS containing 15 U/ml of heparin sodium (SoloPak Laboratories).
2. 3-ml Syringes with 23 gauge needles for i.v. injections.
3. Nylon 100 μm cell strainer (BD Bioscience).

3. Methods

3.1. Harvesting Tregs from the Spleen

Perform at room temperature.

1. To harvest Tregs from the spleen, remove the spleen and crush through a nylon 100 μm cell strainer into a 50-ml conical tube using 30 ml of PBBS.
2. Centrifuge for 5 min at 200 × *g* and decant supernatant.
3. Add 2 ml ammonium chloride and incubate 5 min to lyse RBCs. Add 30 ml PBBS solution to wash.
4. Centrifuge for 5 min at 200 × *g* and decant supernatant.
5. Wash cell pellet with 30 ml balanced salts solution.
6. Centrifuge for 5 min at 200 × *g* and decant supernatant. Resuspend in appropriate buffer for the next step.

3.2. Harvesting Tregs from the Liver

Perform at room temperature.

1. To harvest Tregs from the liver tissue, first perfuse the anesthetized mouse with PBS/heparin perfusion solution to displace blood from the tissue. Use surgical scissors to make a small incision in the right atrium for the blood to flush out and then insert a 23 gauge needle on a 10-ml syringe into the left ventricle. Slowly push 10 ml of perfusion solution through the heart. To further displace blood from the liver, push an additional 5 ml of perfusion solution through the liver via the portal vein at the base of the liver.
2. After removing the gall bladder from the liver, crush the liver through a nylon 100 μm cell strainer into a 50-ml conical tube using 30 ml of balanced salts solution.
3. Centrifuge for 10 min at 850 × *g* with no brake.
4. Aspirate the supernatant and resuspend the pellet in 15 ml of 35% Percoll by vortexing. Centrifuge for 10 min at 850 × *g* with no brake.
5. Without disrupting the cell pellet, carefully aspirate the top layer of hepatocytes and the supernatant liquid.
6. To ensure no residual hepatocytes contaminate the lymphocyte cell pellet, transfer the pellet into a fresh 15-ml tube.

7. Add 2 ml ammonium chloride and incubate 5 min to lyse residual RBCs. Wash with 13 ml balanced salts solution and continue on with the purified lymphocytes.

3.3. Harvesting Tregs from the Lung

1. To harvest Tregs from lung tissue, first perfuse the anesthetized mouse with PBS/heparin perfusion solution as described in Subheading 3.2.1.
2. Cut the lungs into small pieces with scissors and with a magnetized bar, stir at 450 rpm for 30 min at 37°C in 40 ml of 1.3 mM EDTA in a 50-ml flask.
3. Transfer to a 50-ml conical tube, vortex, then centrifuge at 500 × *g* for 5 min at room temperature.
4. Carefully aspirate the supernatant from the lung pieces.
5. Wash twice with 40 ml PBBS with 5% FCS, carefully aspirating the supernatant each time while avoiding lung pieces.
6. Transfer to a clean 50-ml flask with magnetized bar and stir for 1 h at 550 rpm at 37°C in 30 ml collagenase solution.
7. Pour and crush through a nylon 100 μm cell strainer into a 50-ml conical tube. Rinse cell strainer using an additional 15 ml collagenase solution.
8. Centrifuge for 5 min at 500 × *g* at room temperature.
9. Wash with PBBS/5% FCS and if large lung pieces remain, repeat crushing through a nylon 100 μm cell strainer into a clean 50-ml conical tube.
10. Centrifuge for 5 min at 500 × *g* at room temperature.
11. Suspend the cell pellet in 8 ml of 44% Percoll and then carefully pipet 5 ml of 67% Percoll solution under the cell suspension.
12. Centrifuge for 20 min at 500 × *g* at room temperature with the brake off.
13. Carefully aspirate the top layer of Percoll above the visible lymphocyte layer (buffy coat). Next carefully collect the lymphocyte layer.
14. Transfer the lymphocytes into a clean 15 ml conical tube and wash once with 13 ml balanced salts solution and continue on with the purified lymphocytes.

3.4. In Vitro Suppression Assays

Alternative materials are given in item 8 in Subheading 2.1 that will be used depending on the type of assay to be performed. Typically both the target cells and the Tregs are stimulated with anti-CD3-coated plates (10, 11). In such cocultures, cell division of target cells is expected to be suppressed. Figure 1 shows an example of CD8⁺ T cells stimulated with anti-CD3 to induce proliferation and expression of granzyme B. Coculturing the CD8⁺

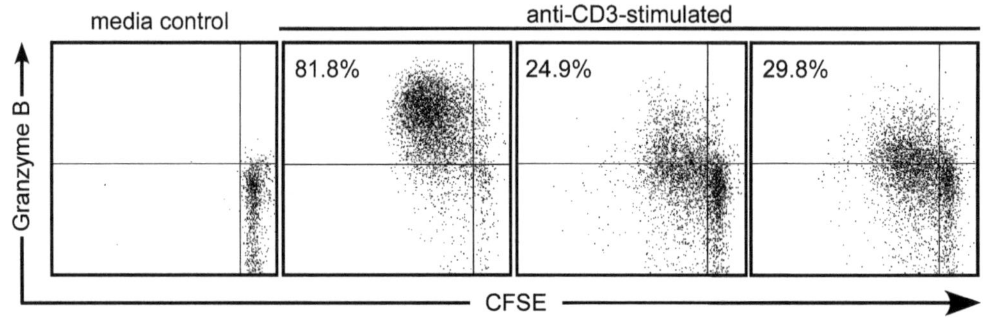

CD8⁺ Targets	+	+	+	+
Spleen Tregs	–	–	+	–
Liver Tregs	–	–	–	+

Fig. 1. In vitro suppression of CD8+ T cell proliferation and function by Tregs from the spleen and liver. The *left panel* shows the lack of proliferation and granzyme B production by unstimulated CD8+ T cells while the *next panel* shows that greater than 80% of the cells proliferate and produce granzyme B following anti-CD3 stimulation. Coculture with Tregs from either the spleen or liver significantly reduced both proliferation and granzyme B production.

T cells with Tregs from either the spleen or liver significantly reduced both proliferation and expression of granzyme B. In some situations, such as the study of virus-induced Tregs, it may be of interest to determine the suppressive capacity of the Tregs directly ex vivo, without further stimulation. In such cases rather than stimulating with anti-CD3, which would also activate the Tregs, the target cells may be stimulated with specific peptides, especially if TCR transgenic cells are used as targets. Control cells from naïve mice should be used for comparison with infected mice.

1. To assay suppression of CD8+ T cells, purify CD8+ splenocyte targets from naive mice using MACS beads (Miltenyi MACS system) according to the manufacturer's recommendations. Alternatively, TCR transgenic CD8+ cells may be used as targets if available.
2. Label the target cells with CFSE in culture media without FCS at a concentration of 5×10^7 cells/ml and a concentration of 5 μM CFSE for 10 min at 37°C with gentle agitation every 2 min (see Note 1). Block CFSE binding by adding a saturating volume of ice-cold media containing 10% FCS and wash twice to dilute out unabsorbed CFSE.
3. Purification of the Treg population is more difficult since the most definitive marker, Foxp3, is intracellular. Tregs can be enriched using biotinylated anti-CD25 and then using streptavidin MACS beads following the manufacturers' recommendations. This method yields high percentages of Foxp3+ Tregs, typically over 90%. Alternatively, Tregs can be

obtained by FACS sorting on CD4 and CD25. First, stain lymphocytes with anti-CD4 and anti-CD25 and sort for double positive live cells on a FACS Aria by gating on the CD25^{hi} cells falling within the appropriate forward scatter/side scatter CD4^{+} population. Since most but not all CD4^{+} CD25^{hi} cells are Foxp3^{+} Tregs, stain with intracellular anti-Foxp3 (eBioscience) following the manufacturers recommendations to determine purity. If the mice are available, Tregs can also be sorted from Foxp3^{GFP} reporter mice (12) by sorting on CD4^{+} GFP^{+} double positive lymphocytes (see Notes 2 and 3).

4. *Option 1*: Set up cultures in 200 μl fresh complete IMDM in a 96-well flat bottom, anti-CD3-coated tissue culture plate. Use $1–4 \times 10^5$ cells of each type per well. Using fewer (10^4) cells generally gives greater variability in the assay. Set up cultures at a 1:1:1 ratio with target cells:Tregs:helper CD4^{+} T cells. CD4^{+} T helper cells are purified from a naïve mouse by anti-CD4 MACS beads following manufacturers recommendations (see Notes 4 and 5).

   Option 2: If peptide to activate the target cell is available, target cells can be activated using peptide-pulsed APCs rather than anti-CD3-coated plates. This allows the target cells to be activated without activating the Tregs. Use the negative fraction of anti-CD4 and anti-CD8 bead purified splenocytes from a naïve mouse as the APCs. Resuspend the APCs in complete IMDM media with 10% Normal Mouse Serum. Add the peptide of interest at a concentration predetermined to activate the cells of interest and mix well by gentle agitation. Incubate at 37°C for 30–60 min and then irradiate with 3,000 rad. Wash APCs twice using media. Use the APCs at a 1:1 ratio with the target cells. Otherwise, cultures are set up as in option 1.

   Option 3: In cases where it is desirable to use target cells activated in vivo, such as from infected mice, harvest cells as described above. We have used activated CD8^{+} T cells harvested 4 days postadoptive transfer into acutely infected mice, but the activation status of target cells should be determined empirically for each system. Target cells activated in vivo by infection can be kept stimulated in vitro by addition of tetramers to cocultures with Tregs in plain plates rather than using CD3-coated plates (9). Otherwise, cultures are set up as in option 1 (see Note 6).

5. Analyze the cultures by flow cytometry and collect the supernatants for ELISA (e.g., assay for IFNγ) after 48–60 h in vitro. The target cells can be surface stained for anti-CD8 or anti-CD4 and analyzed for CFSE dilution (proliferation) and intracellular IFNγ and granzyme B following a 30 min fix at 4°C in PBS 2% PFA or an overnight fix in PBS 0.5% PFA. Permeabilize the cells in a 0.1% saponin-PBS containing

0.1% sodium azide and 0.5% BSA. If intracellular IFNγ will be tested, add Brefeldin A at 10 μg/ml for the last 5 h of culture.

3.5. In Vivo Suppression Assays Using Adoptive Transfers

Suppression of target cells in vivo may be followed using adoptive transfer of labeled cells (see Note 7). For example, in Friend virus infections there is a burst of activated Tregs at 2 weeks postinfection (13). By adoptively transferring labeled $CD8^+$ T cells into infected mice around the time of this burst, the effects of suppression on the transferred cells may be observed (4). In addition, adoptive transfer of Tregs from infected mice into naïve mice can be done to monitor their effects in, for example, a naïve mouse (6). We also use this technique to activate $CD8^+$ T cells during acute infections before Treg activity begins. These physiologically activated cells can then be recovered for use in in vitro suppression assays (9).

1. Obtain the desired transfer subpopulations as described in Subheading 3.1 and suspend them at a concentration of no greater than 10^8/ml (see Note 8) in PBBS containing15 U/ml of heparin.
2. Label the cells with CFSE as in Subheading 3.1.2 if they are to be followed for cell division (see Note 1).
2. Bring the cells to room temperature and filter through a 100 μm cell strainer or nylon mesh to remove clumps. At this point, it is usually desirable to check the purity of the cells by flow cytometry. Inject the cells slowly via the intravenous route in a volume of 0.5 ml (see Note 9).

4. Notes

1. CFSE concentrations between 2 and 10 μM can be used to adjust the brightness of the labeled cells. Cells used for in vivo transfers will often lose a significant amount of label in vivo, so concentrations at the higher end should be used.
2. The use of CD25 expression to purify Tregs has disadvantages because even in the spleen there are $CD25^{lo}$ $Foxp3^+$ Tregs that will not be acquired using MACS beads or cell sorting using CD25 as the marker. This is especially problematic when purifying Tregs from nonlymphoid tissues, like the liver and gut, where the majority are $CD25^{lo}$ and cannot be purified by these processes. The use of Foxp3-GFP reporter mice is necessary when obtaining $CD25^{lo}$ Tregs from a nonlymphoid tissue or to get the total Treg population from the spleen. In this way, you can stain with anti-CD4 and by FACS cell sorting obtain >95% pure $CD4^+GFP^+$ Tregs.

3. Remember not to stain cells with FITC or other fluorochromes that are detected in the CFSE channel when using GFP reporter mice. Cell Trace Violet (Invitrogen) works well to track proliferation of GFP^+ cells.
4. For lower cell numbers, use a 96-well round bottom plate to maximize cell-to-cell contact. Also centrifuge for 2 min at $50 \times g$ after the cultures are set up. Lower cell numbers in the well required a longer in vitro culture (72 h) because the targets were slower to proliferate and upregulate effector molecules.
5. Helper cells greatly enhance the proliferation and function of the $CD8^+$ T cells providing a better signal-to-noise ratio. It should be noted that Tregs from naïve mice that are not stimulated in vitro can actually provide help in some circumstances (9).
6. $CD8^+$ T cell function typically stops following harvest and in vitro culture unless the cells are kept stimulated, so some type of stimulation is usually required.
7. Donor cells can be followed by genetic markers such as Thy1, CD45, or expression of GFP. However, donor cells expressing GFP may be rejected as foreign in experiments lasting more than a week. CFSE-labeled cells can also be used but will lose signal following cells division. They can be followed for at least 1 month if they do not divide (see Note 1).
8. The cell concentration will vary depending on numerous variables such as whether the cells will divide, where they will home, how long they will be left in the animal, etc. We have had success with adoptive transfers of as few as 50 cells to as many as 5×10^7. It should be noted that using high numbers of cells may give results not reflective of the true in vivo situation.
9. Using a volume of 0.5 ml will help assure that the needle is in a vein and not in tissue. If the needle is in a vein, the suspension should flow with very little pressure and should not distend the surrounding tissue. Better results may be obtained using the forefinger rather than the thumb on the plunger of the syringe. The retro-orbital sinus is a convenient site to do intravenous inoculations. Inclusion of heparin sodium in the injection solution is key for the prevention of clotting and pulmonary embolisms that will rapidly kill the recipient mice.

Acknowledgments

This research was supported by the Division of Intramural Research of the National Institutes of Health, National Institute of Allergy and Infectious Diseases.

References

1. Hasenkrug, K. J. and Dittmer, U. (2007) Immune control and prevention of chronic Friend retrovirus infection. *Front. Biosci.* **12,** 1544–1551.
2. Hasenkrug, K. J. and Chesebro, B. (1997) Immunity to retroviral infection: the Friend virus model. *Proc. Natl. Acad. Sci. USA* **94,** 7811–7816.
3. Hasenkrug, K. J., Brooks, D. M. and Dittmer, U. (1998) Critical role for CD4+ T cells in controlling retrovirus replication and spread in persistently infected mice. *J. Virol.* **72,** 6559–6564.
4. Dittmer, U., He, H., Messer, R. J., et al. (2004) Functional impairment of CD8(+) T cells by regulatory T cells during persistent retroviral infection. *Immunity* **20,** 293–303.
5. Hasenkrug, K. J. (1999) Lymphocyte deficiencies increase susceptibility to Friend virus-induced erythroleukemia in Fv-2 genetically resistant mice. *J. Virol.* **73,** 6468–6473.
6. Iwashiro, M., Messer, R. J., Peterson, K. E., Stromnes, I. M., Sugie, T. and Hasenkrug, K. J. (2001) Immunosuppression by CD4+ regulatory T cells induced by chronic retroviral infection. *Proc. Natl. Acad. Sci. USA* **98,** 9226–9230.
7. Sakaguchi, S., Sakaguchi, N., Asano, M., Itoh, M. and Toda, M. (1995) Immunologic self-tolerance maintained by activated T cells expressing IL-2 receptor alpha-chains (CD25). Breakdown of a single mechanism of self-tolerance causes various autoimmune diseases. *J. Immunol.* **155,** 1151–1164.
8. Myers, L., Messer, R. J., Carmody, A. B. and Hasenkrug, K. J. (2009) Tissue-specific abundance of regulatory T cells correlates with CD8+ T cell dysfunction and chronic retrovirus loads. *J. Immunol.* **183,** 1636–1643.
9. Robertson, S. J., Messer, R. J., Carmody, A. B. and Hasenkrug, K. J. (2006) In vitro suppression of CD8+ T cell function by Friend virus-induced regulatory T cells. *J. Immunol.* **176,** 3342–3349.
10. Shevach, E. M. (2002) CD4+ CD25+ suppressor T cells: more questions than answers. *Nat. Rev. Immunol.* **2,** 389–400.
11. Von Boehmer, H. (2005) Mechanisms of suppression by suppressor T cells. *Nat. Immunol.* **6,** 338–344.
12. Bettelli, E., Carrier, Y., Gao, W., et al. (2006) Reciprocal developmental pathways for the generation of pathogenic effector TH17 and regulatory T cells. *Nature* **441,** 235–238.
13. Zelinskyy, G., Kraft, A. R., Schimmer, S., Arndt, T. and Dittmer, U. (2006) Kinetics of CD8+ effector T cell responses and induced CD4+ regulatory T cell responses during Friend retrovirus infection. *Eur. J. Immunol.* **36,** 2658–2670.

Chapter 5

Flow Cytometric Profiling of Mature and Developing Regulatory T Cells in the Thymus

Donald M. Simons and Andrew J. Caton

Abstract

Natural Regulatory T (Treg) cells are a subset of CD4+ T cells characterized by expression of the transcription factor Foxp3 and the ability to suppress immune responses. Treg cells develop in the thymus in response to highly specific interactions between the T cell receptor (TCR) and self-antigens. These processes can be recapitulated in antigen-specific systems using transgenic mice that coexpress a TCR with its cognate peptide as a neoself-antigen. Here, we describe a method for using such a system to establish a flow cytometric profile of phenotype markers expressed by developing and mature Treg cells in the thymus. Our approach is to compare antigen-specific thymocytes developing in the presence or absence of Treg cell-selecting ligands to identify phenotypic changes that characterize thymocytes undergoing selection into the Treg cell lineage.

Key words: Thymocyte, Foxp3, Immune regulation, Treg progenitor cell, Immunophenotyping

1. Introduction

T cell development in the thymus can be broadly categorized into four stages based on expression of the coreceptors CD4 and CD8 (1). The most immature thymocytes express neither of the coreceptors and are termed double negative (DN). Double positive (DP) cells have passed the β-selection checkpoint and express both CD4 and CD8. Thymocytes that have been selected on class II major histocompatibility complex (MHC) downregulate CD8 and are termed CD4 single positive (CD4SP); their class I MHC-selected counterparts become CD8SP. Mature SP thymocytes exit the thymus and join the pool of naïve CD4+ and CD8+ T cells that circulate between the blood and peripheral lymphoid organs.

Natural regulatory T (Treg) cells are a distinct subset of CD4+ T cells that develop in the thymus and are required for the maintenance

George Kassiotis and Adrian Liston (eds.), *Regulatory T Cells: Methods and Protocols*, Methods in Molecular Biology, vol. 707,
DOI 10.1007/978-1-61737-979-6_5, © Springer Science+Business Media, LLC 2011

of immune tolerance in the periphery (2). Briefly, Treg cells express a number of cell-surface markers associated with activated T cells including CD25, GITR, and CTLA-4, and require the γ_c-chain cytokine IL-2 for their development in the thymus and survival in the periphery. Maintenance of Treg cell phenotype and function requires expression of the lineage-specific transcription factor Foxp3, and mutation of the Foxp3 gene leads to Treg cell deficiency and autoimmunity in both mouse and man.

Thymic selection of Treg cells is thought to occur by a two step process requiring both T cell receptor (TCR)-dependent and -independent signals (2). Evidence from our lab and others indicates that Treg cell selection occurs by a TCR-instructive process requiring highly specific interactions between the TCR and self-antigens (3,4). Maturation of committed Treg cell precursors into Foxp3$^+$ cells, however, may be TCR independent and instead require γ_c-chain-dependent signals downstream of IL-2 (5).

While there is now convincing evidence that Foxp3$^+$ CD4SP thymocytes represent mature cells that have acquired regulatory function, the identification of progenitors that will give rise to Foxp3$^+$ CD4SP cells is less well advanced. Here, we describe a method for using flow cytometry to analyze subsets of thymocytes that are undergoing Treg cell selection in order to identify phenotypic markers that characterize mature and developing Treg cells. Our approach is to use an antigen-specific transgenic mouse system to identify phenotypic changes that characterize thymocytes developing in the presence or absence of Treg cell-selecting ligands. These characteristics are then used as a basis for the identification of mature and developing Treg cells in a nontransgenic system.

2. Materials

2.1. Isolation of Thymocytes

1. Mice. The experiments outlined in this protocol make use of single transgenic (ST) TS1 mice and double transgenic (DT) TS1×HA28 mice; however, any suitable transgenic mouse model (see Note 1) can be used. For simplicity, we will refer to TS1 as "ST mice" and TS1×HA28 as "DT mice." A non-transgenic BALB/c mouse (Charles River, Wilmington, MA) will be used for analysis in the last section of this protocol.
2. Dissection bed with restraining pins. The lid to a Styrofoam shipping container and 27 G needles can be used for this purpose.
3. Dissection scissors and fine-point forceps (Fisher Scientific, Pittsburg, PA).
4. Phosphate buffered saline (PBS): 1.9 mM NaH_2PO_4, 8.1 mM Na_2HPO_4, 154 mM NaCl; prepared in dd-H_2O.

5. Stainless steel mesh; 316 stainless steel, 200 × 200 mesh (W.S. Tyler, Mentor, OH). The stainless steel mesh should be cut into 1 in. squares, washed by submersion in 100% ethanol, rinsed with double distilled water, and autoclaved prior to use.
6. 1 cc Syringes (BD, San Jose, CA).

2.2. Immunostaining Antigen-Specific Thymocytes for Flow Cytometry

1. 96-Well microtiter plate with v-shaped wells (Costar, Corning, NY).
2. Antibodies for the Treg cell profiling panel shown in Table 1. With the exception of 6.5, all of the antibodies used in this procedure can be obtained from eBioscience (San Diego, CA), Biolegend (San Diego, CA) or BD (San Jose, CA). The antibodies should be titrated prior to use to determine the optimal dilution for staining. The 6.5 antibody is produced and biotinylated in-house following standard procedures.
3. A fluorescent conjugate of streptavidin for detection of biotin-6.5 in the secondary detection step.

Table 1
Antibody panels for profiling of TCR-transgenic and nontransgenic Treg cells

Antibody staining panels

Treg cell profiling panel				BALB/c analysis panel	
Developmental markers		Profiling markers[a]			
Antigen	Clone	Antigen	Clone	Antigen	Recommended fluorochrome
Foxp3[b]	FJK-16s	CD25	PC61.5	Foxp3[b]	efluor450
CD4	GK1.5	CD62L	MEL-14	CD4	APC-efluor780
CD8	53–6.7	CD69	H1.2F3	CD8	efluor650
TS1-TCR	6.5[c]	CTLA-4[b]	UC10-4B9	TS1-TCR	Sav-APC
		TNFRII	TR75-89	TNFRII	PE
		GITR	DTA-1	GITR	PE-Cy7
		Isotype[d]	N/A	CD25	PerCP-Cy5.5
				CD69	FITC

[a]All of these antibodies including the isotype controls should be on the same fluorochrome. Seven staining panels should be prepared for this experiment. Each staining panel consists of all of the phenotype markers plus one of the comparison markers

[b]CTLA-4 and/or Foxp3 should be stained during the intracellular staining step

[c]The 6.5 antibody used in this procedure is biotinylated and is detected with a fluorescent conjugate of Streptavidin in the secondary detection step

[d]The isotype control for these stains will be a single sample that is stained with the phenotype marker panel plus rat IgG_1, IgG_{2a}, IgG_{2b}, and Armenian hamster IgG

4. PBS supplemented with 2% heat inactivated FBS (Tissue Culture Biologicals, Tulare, CA) and 5 mM EDTA (FACSwash).
5. 1% Paraformaldehyde (PFA, USB, Cleveland, OH) in PBS.
6. FACSwash supplemented with 0.1% Triton X-100 (Intracellular staining wash, ICSwash).
7. A flow cytometer capable of at least 5-channel fluorescence.

2.3. Data Analysis: Gating Strategies and Comparisons for Phenotypic Profiling Transgenic Thymocytes

1. FlowJo (Tree Star, Ashland, OR) or similar software (see Note 2) for analysis of flow cytometric data.

2.4. Analysis of Nontransgenic Thymocytes by Flow Cytometry

1. 96-Well microtiter plates, FACSwash, PFA, and ICSwash, as in Subheading 2.2.
2. Antibodies for the BALB/c analysis panel shown in Table 1. See note for suppliers in the previous section.
3. A flow cytometer capable of at least 8-channel fluorescence.

3. Methods

In transgenic mice that coexpress a defined TCR with its cognate peptide as a neoself-antigen, thymocytes expressing the transgenic TCR can undergo enhanced selection to become Treg cells (3, 4, 6, 8, 9). The TS1 transgene encodes a MHCII-restricted TCR recognizing the site 1 (S1) determinant of PR8 influenza hemagglutinin, and can be identified by the clonotypic antibody 6.5 (10). HA28-transgenic mice constitutively express low levels of the S1 peptide, and in DT TS1×HA28 mice a significant fraction of thymocytes expressing the TS1-TCR are selected to become Treg cells (7). Using this system, we can track populations of antigen-specific thymocytes from Treg cell-selecting (DT) or nonselecting environments (ST). In the first section of this protocol we make direct comparisons between these two populations of cells in order to determine the phenotypic profile of thymocytes undergoing selection into the Treg cell lineage. In the final section of this procedure, we apply this profile to a BALB/c mouse to show that an equivalent population can be identified in a nontransgenic system.

3.1. Isolation of Thymocytes

1. The thymus (see Note 3) is a bilobed organ located in the thoracic cavity resting on top of the heart. Due to its proximity to the cardiac vasculature, it is essential to make a clean

dissection to avoid contamination of the thymocytes with peripheral blood leukocytes.

2. All dissection tools should be washed and autoclaved prior to use. One stainless steel mesh and one 1 cc syringe will be needed per thymus. For each thymus prepare a Petri dish containing 5 ml PBS.
3. Euthanize the mouse according to animal welfare regulations
4. Immobilize the mouse for dissection by pinning each paw to the dissection bed.
5. Using surgical scissors make subdermal cuts from groin to jaw and from the groin to each hind paw as illustrated in Fig. 1a. Using forceps to grasp the skin on either side of the abdominal incision pull the skin away from the body of the mouse and pin it to the dissection bed to expose the ribcage and peritoneal membrane.
6. Expose the thymus by cutting through the ribcage as shown in Fig. 1b. First, make a centerline cut through the peritoneal membrane and into the sternum. This cut will puncture diaphragm and the heart and lungs should be just visible through the incision. Second, make two lateral cuts running between the ribs and diaphragm. Third, using forceps to push the lungs aside, cut through the ribs as near to the base of the thoracic cavity as possible. Fourth, trim the pectoral muscles away from the ribcage so that the ribcage is cantilevered from the

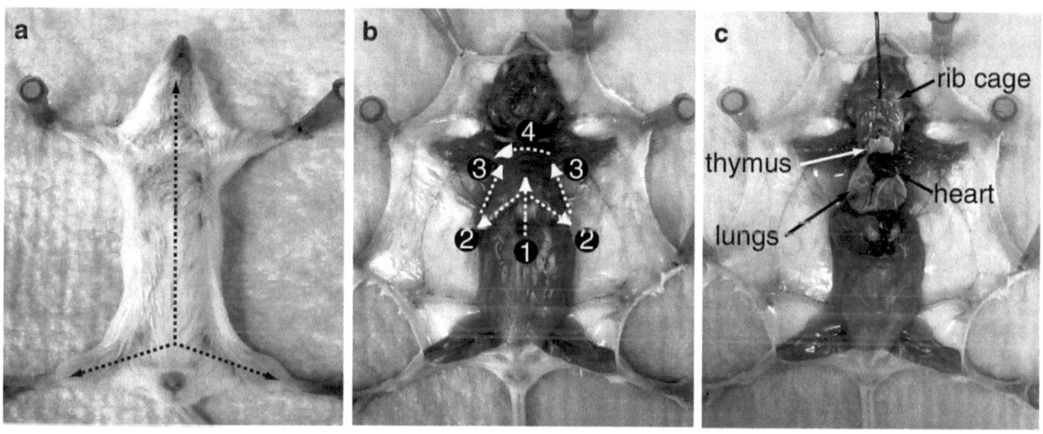

Fig. 1. Isolation of the mouse thymus. The procedure illustrated here allows removal of the thymus with minimal exposure to peripheral blood. (**a**) Euthanize and immobilize the mouse. Make three subdermal incisions as illustrated by the *dashed lines* and peel the skin away from the abdominal cavity and ribcage. (**b**) Cut through the ribs and pectoral muscles as shown and using forceps pull the ribcage up and away from the thoracic cavity to reveal the heart and thymus. (**c**) The thymus will be pulled away from the heart by the ribcage. Using scissors, cut the ribcage and thymus away from the thoracic cavity, and subsequently remove the thymus from the ribcage.

sternum and can be lifted upwards to expose the thoracic cavity as shown in Fig. 1c.

7. The thymus will be pulled up and away from the heart along with the ribcage (see Note 4). Using scissors cut the ribcage and thymus away from the thoracic cavity. This cut should be made in one stroke to minimize contact between the thymus and cardiac blood.
8. Using forceps tease the thymus away from the ribcage and place in a Petri dish containing PBS until further processing.
9. Place a stainless steel mesh over the mouth of a 15 ml centrifuge tube and seat in place using the butt of a 1 cc syringe to indent the mesh into the mouth of the tube.
10. Transfer the thymus onto the mesh and use the plunger from a 1 cc syringe to gently mash the organ through the mesh. Wash the mesh and plunger with 2 ml of PBS and repeat the process until the thymus is completely disrupted and only white connective tissue remains on the mesh.
11. Pellet the cells by centrifugation at $400 \times g$ for 4 min.
12. Decant the supernatant and resuspend the pellet in 10 ml of FACSwash. Repeat this step once more for a total of two washes. During the second wash count the cells using a hemocytometer.
13. Following the last wash resuspend the cells at 20×10^6/ml in FACSwash.

3.2. Immunostaining of Antigen-Specific Thymocytes for Analysis by Flow Cytometry

1. Experimental setup. Purify thymocytes as described above from a single- and a double-transgenic mouse. Transfer 200 μl/well (4×10^6 cells) of each cell suspension into a 96-well plate for staining. Plate seven replicate wells from each cell suspension. In this experiment, cells will be stained with seven different antibody panels for analysis by 5-color flow cytometry (Table 1). Each panel will contain the same four antibodies for determining developmental stage (developmental markers), but will vary in the final antibody (profiling marker) that will be used in comparisons. One of these panels will contain a mixture of isotype control antibodies and will be used as the negative control for staining. To simplify analysis and cytometer setup, the profiling antibodies should all be conjugated to the same fluorochrome. This protocol assumes that readers are familiar with the requirements for setting-up and compensating a flow cytometer so, the preparation of single-color compensation control samples will not be explicitly addressed here.
2. This is a three-step staining protocol. Cells are first stained for surface markers. The 6.5 antibody used here is biotinylated, and the second stain is with streptavidin-conjugated APC.

In the final step, the cells are fixed, permeabilized, and stained for the intracellular markers Foxp3 and CTLA-4.

3. All reagents used in this procedure should be ice-cold, and all incubations are carried out on ice and protected from light. Centrifugation steps should be at 400 × *g* for 4 min (surface staining) or at 800 × *g* for 5 min (intracellular staining) in a 4°C centrifuge.
4. Prepare the antibody panels for surface staining. Make sufficient cocktail for 200 μl/sample plus 5% excess. The antibody panels should be prepared in FACSwash.
5. Pellet the cells in the 96-well plate by centrifugation and discard the supernatant.
6. Resuspend the cells in 200 μl of the appropriate antibody panel and incubate for 30 min on ice. Note that the cells plated for CTLA-4 staining should only be stained with the developmental panel during this step.
7. Pellet the cells by centrifugation, discard the supernatant, and resuspend the pellet in 200 μl of FACSwash. Repeat three times.
8. Perform steps 8 and 9 only if using biotinylated or unconjugated antibodies that require secondary detection, otherwise skip to step 10. Following the last wash, resuspend the cells in 200 μl of FACSwash + a florescent-conjugate of streptavidin. Incubate for 30 min on ice.
9. Pellet the cells by centrifugation and wash thrice as in step 7.
10. Resuspend the cells in 200 μl of 1% PFA and incubate for at least 30 min on ice (see Note 5).
11. Pellet the cells by centrifugation, discard the supernatant, and wash twice with ICSwash. Remember that all postfixation centrifugation steps should be performed for 5 min at 800 × *g*.
12. Incubate the cells 10 min in 200 μl of ICSwash.
13. Prepare the intracellular staining antibody panels. One panel should contain both anti-CTLA-4 and anti-Foxp3, and will only be applied to the two samples plated for profiling CTLA-4 expression. The second panel should contain both anti-Foxp3 and hamster IgG, and will only be applied to the isotype control sample. The final panel will contain only anti-Foxp3 and will be applied to all the remaining samples. Make sufficient volume for 200 μl/sample plus 5% excess. The antibody panels used in this step should be prepared in ICSwash.
14. Pellet the cells by centrifugation, resuspend in 200 μl of the appropriate ICS antibody panel and incubate for 30 min on ice.

15. Pellet the cells by centrifugation and wash twice with ICSwash and once with FACSwash.
16. Following the last wash resuspend the cells in 200 μl of FACSwash. The cells are now ready for analysis by flow cytometry.

3.3. Data Analysis: Gating Strategies and Comparisons for Phenotypic Profiling of Transgenic Thymocytes

1. Acquisition. Some of the populations that will be analyzed here are present at very low frequencies in the thymus. It is essential that enough events are collected to allow statistically valid comparisons. Be sure to collect a sufficient number of events so that there are at least 100 cells (and preferably more) of the lowest frequency population to be analyzed (see Note 6). Also, set the flow cytometer to collect forward scatter height (FSC-H) as well as area (FSC-A) to allow exclusion of doublets.
2. Analysis. Import the data into an analysis program such as FlowJo. The following gating strategy should be used to make comparisons between samples. Be sure to apply these gates uniformly to all of the samples being analyzed.
3. Stringency gates (Fig. 2a). For accurate results it is important to exclude false positives that arise from clumps of cells being acquired by the cytometer as a single event (doublets). This can be accomplished using a plot of FSC-A vs. FSC-H to exclude doublets from further analysis. Single cells exhibit a 1:1 relationship between these two parameters and fall along a 45° angle from the origin. When two or more cells are acquired simultaneously the FSC-A is increased disproportionately to the FSC-H, and these cells will stray significantly from 45°. Establish a gate that includes only single cells ("singlets") and plot the FSC-A of the included events against their side scatter area (SSC-A). When plotted in this manner, singlet thymocytes will form a distinct population of cells that can be distinguished from cellular debris and many types of accessory cells. Use this plot to set a "thymocyte" gate. Only cells that fall within this gate should be included in the analyses described below.
4. Developmental gates (Fig. 2b). Establish gates to segregate the cells into developmental stages by plotting the singlet thymocyte population from the ST mouse on a graph of CD4 vs. CD8. Define four regions on this plot (see Note 7): $CD4^-CD8^-$ (DN), $CD4^+CD8^+$ (DP), $CD4^-CD8^+$ (CD8SP), $CD4^+CD8^-$ (CD4SP). At this point, the cells that fall within these gates will not be analyzed. Instead, the gates established here will be used to identify specific populations for analysis in the steps that follow.
5. Identification of a non-Treg cell forming control population. $Clonotype^+$ thymocytes from a ST mouse do not form Treg

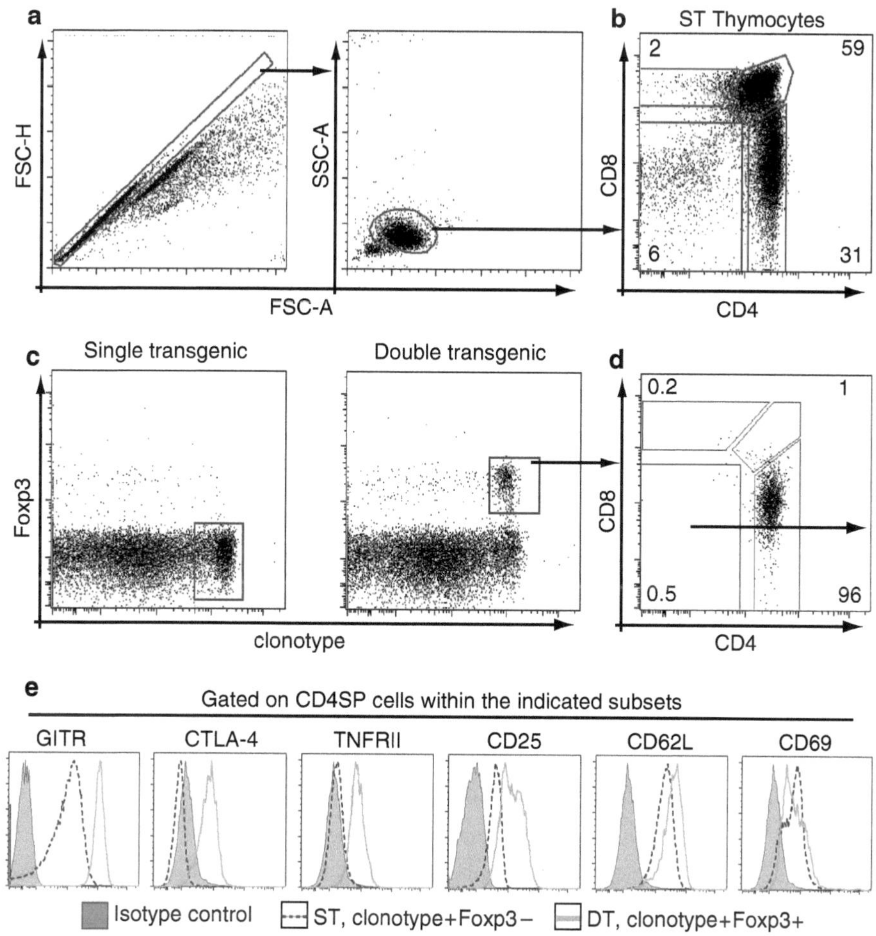

Fig. 2. Flow cytometric profiling of mature Treg cells in the thymus. Clonotype+Foxp3+ thymocytes from a DT mouse were used to develop a profile of mature Treg cell-associated phenotype markers. (**a**) Stringency gates. *Left panel* shows the gate used to exclude doublets from analysis. *Right panel* shows the thymocyte gate. (**b**) Developmental gates based upon CD4 and CD8 expression by singlet thymocytes from a ST mouse. (**c**) *Left panel* shows the gate used to identify clonotype+Foxp3− control cells from a ST mouse. *Right panel* shows the gate used to identify clonotype+Foxp3+ mature Treg cells from a DT mouse. (**d**) Developmental gates applied to clonotype+Foxp3+ Treg cells. (**e**) Histograms show the fluorescence intensity of staining by the indicated subsets for each of the profiling markers.

cells in vivo and are therefore used as a control population that is devoid of Treg cells or their progenitors. Plot singlet thymocytes from the ST mouse with clonotype and Foxp3 on the axes and gate clonotype+Foxp3− cells as shown in the left panel of Fig. 2c.

6. Identification of mature Treg cells. Here, we define clonotype+Foxp3+ thymocytes from a DT mouse as mature Treg cells. Identify these cells by plotting singlet thymocytes as Foxp3 vs. clonotype and gating the Foxp3+clonotype+ cells as indicated in the right panel of Fig. 2c.

7. Analysis of mature Treg cell development. The first step in the analysis of mature Treg cells is to determine the ontogeny of Foxp3 expression during thymic development (Fig. 2d). Plot the DT clonotype$^+$Foxp3$^+$ cells gated in the previous step with CD4 and CD8 on the axes and apply the developmental gates established in step 4. The results of this analysis clearly show that mature Foxp3$^+$ cells predominantly fall within the CD4SP subset of thymocytes, and we will therefore limit our investigation of the profiling markers to cells at this stage of development.
8. Analysis of mature Treg cell phenotype. Next, use the profiling markers to determine the phenotypic profile of mature Treg cells (Fig. 2e). The appropriate comparison to be made here is the expression of each of these markers by the CD4SP, mature Treg cells identified in steps 6 and 7, to CD4SP cells from the ST control population identified in step 5. Plot these two populations of cells as histograms of the fluorescence intensity of staining for each profiling marker. Based on this analysis we conclude that mature Treg cells in the thymus express high levels of GITR, CTLA-4, TNFRII, and CD25 relative to Foxp3$^-$ CD4SP thymocytes from a ST mouse. They also express marginally higher levels of CD62L and equal to marginally lower levels of CD69 than Foxp3$^-$ CD4SP thymocytes from ST mice.
9. Analysis of developing Treg cells. Here, we will define developing Treg cells as being enriched within the clonotype$^+$Foxp3$^-$ subset of thymocytes in a DT mouse, but absent within the same subset of thymocytes from a ST mouse. The latter was gated in step 5 of this procedure. Gate the former population by plotting singlet thymocytes from the DT mouse as clonotype vs. Foxp3 and gate on clontoype$^+$Foxp3$^-$ cells as shown in Fig. 3a. Plot the developing Treg cells with CD4 and CD8 on the axes and apply the developmental gates that were established in step 4. This analysis shows that Foxp3$^-$ cells expressing the clonotypic TCR can be found at all four stages of thymic development (Fig. 3b) in thymocytes from both ST and DT mice. The analysis of the profiling markers will be limited to DN, DP, and CD4SP cells, however, since 6.5 is a MHCII-restricted TCR, and the significance of CD8$^+$Foxp3$^+$ cells remains to be established.
10. Analysis of developing Treg cell phenotype. To establish a phenotypic profile for developing Treg cells, compare the expression of the profiling markers by each developmental subset within the DT clonotype$^+$Foxp3$^-$ cells with the corresponding population of ST control cells. Make this comparison by plotting a histogram of the fluorescence intensity of staining for each marker (Fig. 3c). Based on these plots we

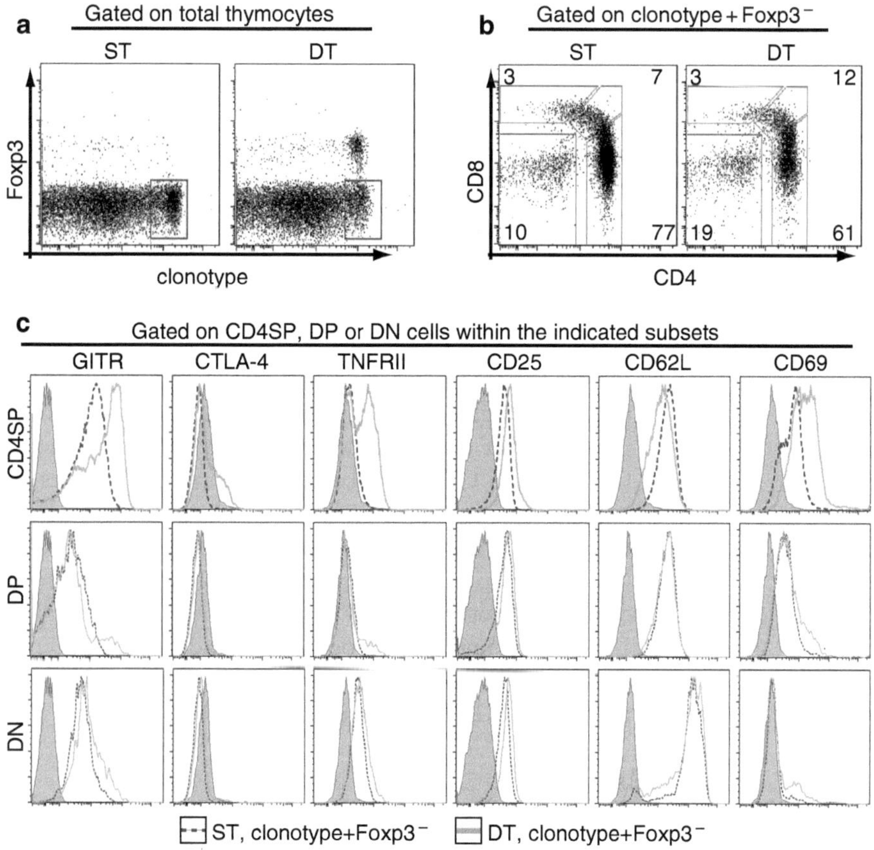

Fig. 3. Flow cytometric profiling of developing Treg cells in the thymus. Clonotype$^+$Foxp3$^-$ thymocytes from a DT mouse were used to establish a profile of phenotype markers enriched on thymocytes developing in the presence of Treg cell-selecting ligands. (**a**) Plots show the gates used to limit analysis to clonotype$^+$Foxp3$^-$ cells from ST and DT mice. (**b**) Gates used to segregate clonotype$^+$Foxp3$^-$ thymocytes from ST and DT mice by developmental stage. (**c**) Histograms show the fluorescence intensity of staining by the indicated subsets for each of the profiling markers. *Shaded histograms* represent staining by an isotype control antibody except for the CD25 data, which shows an unstained control sample.

conclude that upregulation of GITR, TNFRII, and CD69 are most strongly associated with a population of cells containing putative Treg cell progenitors that can be found at the CD4SP stage, and to a lesser degree at the DP stage, in Foxp3$^-$ thymocytes from DT but not ST mice.

3.4. Analysis of Nontransgenic Thymocytes by Flow Cytometry

1. The phenotypic profiles generated in Subheading 3.3 indicate that GITR, TNFRII, CD69, and CD25 may be useful surface markers for both mature and developing Treg cells. In this section of the protocol we will determine whether or not equivalent populations of cells can be identified in a nontransgenic BALB/c mouse.
2. Experimental setup. Purify thymocytes as described in Subheading 3.1 from a BALB/c mouse, and also from ST

and DT mice for comparison. Plate 4×10^6 cells from each mouse into a 96-well plate for immunostaining.

3. Stain the cells as described in Subheading 3.2, but replace the phenotype and profiling markers with the BALB/c analysis panel listed in Table 1. Stain thymocytes from all three mice with this panel.
4. Collect the data on a flow cytometer, taking care to record enough events for analysis.
5. Using FlowJo or equivalent analysis software apply stringency gates as described in Subheading 3.3.
6. Define the thymocyte populations for analysis. Plot the singlet thymocytes gated in the previous step with CD4 and CD8 on the axes. Set developmental gates to segregate the DN, DP, CD8SP, and CD4SP subsets. Based on the profiles generated in the previous section, analysis will be limited to CD4SP cells. Plot the CD4SP subset from each mouse on a graph of clonotype vs. Foxp3. Gate the clonotype^{+}Foxp3^{-} cells from the ST mouse, clonotype^{+}Foxp3^{-} and clonotype^{+}Foxp3^{+} cells from the DT mouse, and clonotype^{-}Foxp3^{-} and clonotype^{-}Foxp3^{+} cells from the BALB/c mouse for analysis (Fig. 4a). Plot these populations as GITR vs. TNFRII (Fig. 4b, c).

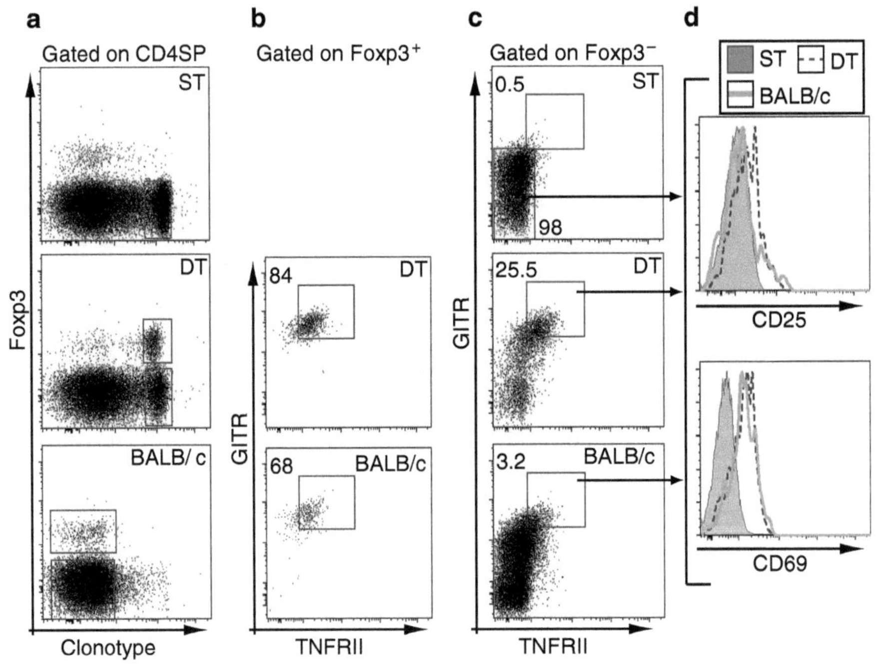

Fig. 4. GITR and TNFRII expression by developing and mature Treg cells in the thymus of a BALB/c mouse. The flow cytometric profiles of mature and developing Treg cells from transgenic mice were validated by assessing their expression on BALB/c thymocytes. (**a**) Expression of clonotypic TCR and Foxp3 by the indicated mice. The gated populations were used for analysis. (**b**) GITR and TNFRII expression by Foxp3^{+} thymocytes from DT and BALB/c mice. (**c**) GITR and TNFRII expression by Foxp3^{-} cells from ST, DT and BALB/c mice. (**d**) Histograms of the fluorescence intensity of CD25 and CD69 staining by the indicated populations.

7. Establishment of the GITR$^+$TNFRII$^+$ gate. Using the clonotype$^+$Foxp3$^-$ cells from the ST mouse as a negative control and the clonotype$^+$Foxp3$^+$ cells from the DT mouse as a positive control establish a gate for GITR$^+$TNFRII$^+$ cells.
8. Identification of mature Treg cells. Apply the GITR$^+$TNFRII$^+$ gate to the Foxp3$^+$ BALB/c thymocytes gated in step 6. Gating the Foxp3$^+$ cells in this manner clearly shows that a majority of mature Treg cells from BALB/c mice express high levels of both TNFRII and GITR (Fig. 4b).
9. Identification of developing Treg cells. Apply the GITR$^+$TNFRII$^+$ gate to the Foxp3$^-$ cells identified in step 6. When the Foxp3$^-$ cells are viewed in this manner, the GITR$^+$TNFRII$^+$ population of cells that is enriched in the DT but absent in ST mice can also be found in a nontransgenic BALB/c mouse (Fig. 4c). Plot the fluorescence intensity of staining by these cells for CD25 and CD69 to verify that these two markers are upregulated along with GITR and TNFRII (Fig. 4d). We can conclude based on these histograms that GITR and TNFRII mark a population of cells in the BALB/c thymus that also express CD25 and CD69 at similar levels to the presumptive Foxp3$^-$ Treg cell precursors identified in DT mice (see Subheading 3.3.10).

4. Notes

1. Treg cell formation using the TS1×HA system has been reported with HA expression driven by the β-globin locus control region, the β-myoglobin heavy chain promoter, and by SV40, AIRE and Igκ promoters (4,6). The DO11×OVA system can also be used to track the development of antigen-specific Treg cells using the KJ-126 clonotypic antibody. Thymic Treg cell formation has been reported in this system using both the insulin promoter to drive OVA expression and also when OVA is targeted to the nucleus (7–9).
2. All data displayed in this protocol was generated using FlowJo. Equivalent analyses can be performed using a number of alternative software suites including FCS Express by De Novo Software, Venturi One by Applied Cytometry, Cyflogic, and Weasel (developed by the Walter and Eliza Hall Institute of Medical Research).
3. The size and cellularity of the thymus can vary greatly depending on the age of the mouse being dissected. Thymic involution occurs between 8 and 10 weeks of age in mice resulting in a 50–75% reduction in thymic cellularity. DT mice will also have reduced thymic cellularity due to the presence of deleting

antigen in the thymus. The reduction in thymus size will be dependent upon the amount of antigen in the thymus and must be determined empirically for each transgenic strain.

4. If the thymus is not pulled away from the heart with the rib-cage, use fine-point forceps to gently tease away the connective tissue connecting the two organs. Be careful not to puncture the heart.
5. The sensitivity of an antibody to fixation time must be determined empirically for each clone. We have left cells in fixative for as long as overnight without significant loss of Foxp3 staining using the FJK-16s clone.
6. We typically only collect events that fall within the stringency gates described in Subheading 3.3. Using these gating criteria, the number of events required for valid analyses is typically between 300,000 and 500,000.
7. The relative proportions of the developmental subsets can be substantially skewed by the transgenic expression of TCRs and/or antigen. For example, expression of an MCH class II-restricted TCR by the TS1 mouse results in an enrichment of CD4SP cells compared to a BALB/c thymus. Although it is still relatively straight-forward to distinguish between the developmental subsets in a transgenic mouse, it may be useful to include a single well of BALB/c thymocytes in your surface stains to establish these gates.

Acknowledgments

The authors would like to thank Malinda Aitken, Christina Mergenthaler, Abigail Liebow, Alissa Basehoar, and Lori Mroz for their invaluable help in maintaining the transgenic mouse lineages described here. This work was supported by R01-AI59166 and by the Commonwealth Universal Research Enhancement Program, Pennsylvania department of Health. DMS is supported by T32 CA09171.

References

1. Starr TK, Jameson SC, Hogquist KA. (2003) Positive and negative selection of T cells. *Annu. Rev. Immunol.* **21**, 139–176.
2. Josefowicz SZ, Rudensky A. (2009) Control of regulatory T cell lineage commitment and maintenance. *Immunity* **30**, 616–625.
3. Jordan MS, Boesteanu A, Reed AJ et al. (2001) Thymic selection of CD4+CD25+ regulatory T cells induced by an agonist self-peptide. *Nat. Immunol.* **2**, 301–306.
4. Apostolou I, Sarukhan A, Klein L, von Boehmer H. (2002). Origin of regulatory T cells with known specificity for antigen. *Nat. Immunol.* **3**, 756–763.
5. Lio CW, Hsieh CS. (2008) A two-step process for thymic regulatory T cell development. *Immunity* **28**, 100–111.
6. Aschenbrenner K, D'Cruz LM, Vollmann EH et al. (2007) Selection of Foxp3(+) regulatory T cells specific for self antigen expressed and

presented by Aire(+) medullary thymic epithelial cells. *Nat. Immunol.* **8**, 351–358.

7. Picca CC, Oh S, Panarey L, Aitken M, Basehoar A, Caton AJ. (2009) Thymocyte deletion can bias Treg formation toward low-abundance self-peptide. *Eur. J. Immunol.* **39**, 3301–3306.
8. Walker LS, Chodos A, Eggena M, Dooms H, Abbas AK. (2003) Antigen-dependent proliferation of CD4+ CD25+ regulatory T cells in vivo. *J. Exp. Med.* **198**, 249–258.
9. Kawahata K, Misaki Y, Yamauchi M et al. (2002) Generation of CD4(+)CD25(+) regulatory T cells from autoreactive T cells simultaneously with their negative selection in the thymus and from nonautoreactive T cells by endogenous TCR expression. *J. Immunol.* **168**, 4399–4405.
10. Kirberg J, Baron A, Jakob S, Rolink A, Karjalainen K, von Boehmer H. (1994) Thymic selection of CD8+ single positive cells with a class II major histocompatibility complex-restricted receptor. *J. Exp. Med.* **180**, 25–34.

Chapter 6

ChIP-on-Chip for FoxP3

Ye Zheng

Abstract

Regulatory T (Treg) cells play a key role in dominant suppression of immune response and maintenance of immune homeostasis. Foxp3, a member of the forkhead transcription factor family, is indispensable for Treg cell development and function. Mice and human with Foxp3 mutations are severely impaired in Treg cell generation and develop lethal autoimmune diseases. We combined chromatin immuno-precipitation and mouse whole genome tiling array profiling (ChIP-on-Chip) to identify the direct downstream targets of Foxp3 in regulatory T cells. Our result showed that Foxp3 not only directly determines expression of a number of Treg signature molecules, but also regulates a group of transcription factors, which potentially control the expression of other Treg-specific genes.

Key words: Regulatory T cell, Foxp3, ChIP-on-Chip, Genome tiling array, Model-based Analysis of Tiling Arrays

1. Introduction

Immune system has a variety of ways to prevent harmful autoimmune responses. Recent studies established an unequivocal role of regulatory T cells in the maintenance of immune homeostasis. Foxp3, a member of the forkhead transcription factor family, is a pivotal factor involved in Treg development and function (1, 2). Mutations of Foxp3 gene in mice and human lead to paucity of Treg cells and severe autoimmune diseases (3–5). Transduction of Foxp3 into non-Treg naïve T cells endows them with in vivo suppressor capacity (6, 7). It is still not fully understood the detail of Foxp3-dependent gene expression program and its impact on Treg development and function. To this end, we combined Foxp3 antibody chromatin immuno-precipitation with mouse whole

George Kassiotis and Adrian Liston (eds.), *Regulatory T Cells: Methods and Protocols*, Methods in Molecular Biology, vol. 707, DOI 10.1007/978-1-61737-979-6_6, © Springer Science+Business Media, LLC 2011

genome tiling array profiling and identified ~700 direct Foxp3 target genes (8, 9). Our results provided a framework for future studies on molecular pathways downstream of Foxp3 in regulatory T cells.

2. Materials

2.1. Foxp3 Chromatin Immuno-Precipitation (ChIP)

1. $CD4^+CD25^+$ regulatory T cells Isolation Kit (Miltenyi).
2. RPMI medium (Invitrogen) supplemented with 10% fetal bovine serum (FBS, Hyclone).
3. Formaldehyde 36.5% solution.
4. 2.5 M glycine.
5. Phosphate buffered saline (PBS, Invitrogen).
6. Cell lysis buffer: 25 mM HEPES pH 8.0, 1.5 mM $MgCl_2$, 10 mM KCl, 0.3% NP-40 (IGEPAL CA-630, Sigma), 1 mM Dithiothreitol (DTT, Sigma), 1× protease inhibitors cocktail (Roche). Both DTT and protease inhibitors are added right before use.
7. Nuclei lysis buffer: 50 mM HEPES pH 8.0, 140 mM NaCl, 1 mM EDTA, 1% Triton X-100, 0.1% sodium deoxycholate, 0.2% SDS, 1× protease inhibitors cocktail. Protease inhibitors are added right before use.
8. TE buffer: 10 mM Tris–HCl pH 8.0, 1 mM EDTA.
9. Protein A agarose (Millipore).
10. Rabbit anti-Foxp3 IgG. We generated Foxp3 antibody by immunizing rabbit with full-length Foxp3 protein expressed in *E. coli*. Anti-Foxp3 IgG was affinity-purified from antisera using Foxp3 protein-conjugated agarose column.
11. ChIP wash buffer: 20 mM Tris–HCl pH8.0, 1 mM EDTA, 250 mM LiCl, 0.5% NP-40, 0.5% sodium deoxycholate.
12. ChIP elution buffer #1: 10 mM Tris–HCl pH 8.0, 1 mM EDTA, 1% SDS.
13. ChIP elution buffer #2: 10 mM Tris–HCl pH 8.0, 1 mM EDTA, 0.67% SDS.
14. Proteinase K 10 mg/ml (Roche).
15. 4 M LiCl.
16. Phenol/chloroform/isoamyl alcohol 25:24:1 mix (Sigma).
17. Chloroform.
18. Phase Lock Gel Light (Fisher).
19. Glycogen 20 mg/ml (Fermentas).

2.2. Analysis of Precipitated DNA by Quantitative PCR

1. Applied Biosystems 7300 Real-Time PCR System.
2. Power SYBR Green PCR Master Mix (Applied Biosystems).
3. MicroAmp™ Optical 96-Well Reaction Plate (Applied Biosystems).

2.3. PCR Amplification and Hybridization of ChIP DNA

1. GeneChip Sample Cleanup Module (Affymetrix).
2. Sequenase Version 2.0 DNA Polymerase (USB).
3. Primer A sequence: GTTTCCCAGTCACGGTC$(N)_9$ (IDT, HPLC purified).
4. Primer B sequence: GTTTCCCAGTCACGGTC (IDT, HPLC purified).
5. Tetrad DNA Engine Thermal Cycler (Bio-Rad).
6. MicroSpin G-50 Columns (GE Life Sciences).
7. dATP, dGTP, dCTP, dTTP set (Roche).
8. dUTP (Roche).
9. TITANIUM Taq DNA Polymerase (Clontech).
10. GeneChip Mouse Tiling 2.0R Array Set (Affymetrix).

3. Methods

Foxp3 ChIP-on-Chip experiment can be largely divided into four stages: Foxp3 antibody chromatin immuno-precipitation of Treg cells; quantitative PCR to test quality of precipitated DNA; PCR amplification of ChIP DNA and hybridization of genome tiling arrays; and array data analysis and visualization.

3.1. Foxp3 Chromatin Immuno-Precipitation

1. Isolate regulatory T cells from mouse spleen and lymph node by magnetic beads selection or FACS sorting (see Note 1). Approximately 2×10^7 cells are required for one Foxp3 ChIP experiment. To obtain sufficient materials for hybridization to the whole genome tiling array set (7 arrays), starting with 8×10^7 regulatory T cells is recommended.
2. In a tissue culture flask, resuspend Treg cells at 1×10^6 cells/ml in complete RPMI medium at room temperature. Add 36.5% formaldehyde to cell suspension until final concentration reaches 1.0%. Put flask on a shaker with gentle agitation for exactly 5 min to fix the cells (see Note 2).
3. Immediately stop cross-linking by adding 2.5 M glycine to the reaction to a final concentration of 0.125 M (1:20 dilution), and mix well until medium color turns yellow.
4. Transfer fixed cells into a conical tube and centrifuge at $600 \times g$ for 5 min. Discard supernatant. Resuspend cells in

10 ml ice-cold PBS. All following steps are carried out at 4°C or on ice unless mentioned otherwise.

5. Centrifuge at 600 × *g* for 5 min. Discard supernatant. At this step, cell pellet can be snap-frozen in liquid nitrogen and stored at −80°C for several weeks without affecting final result.
6. Resuspend cells in 1 ml ice-cold PBS and transfer to an Eppendorf tube. Microfuge at 5,000 rpm for 5 min. Discard supernatant (see Note 3).
7. Resuspend cells in 1 ml ice-cold cell lysis buffer with protease inhibitors. Incubate on ice for 10 min. Microfuge at 5,000 rpm for 5 min. Discard supernatant.
8. Resuspend nuclei pellet in 1 ml ice-cold nuclei lysis buffer with protease inhibitors. Incubate on ice for 10 min.
9. Sonicate chromatin until average size of DNA fragments reach ~1 kb. Sonication condition depends on sonicator model and has to be established prior to ChIP experiment (see Note 4). For Branson Sonifier 250, set power level at 10%, sonicate sample for 15 s, and chill on ice for 1 min. Repeat 6–9 sonication cycles. Avoid overheating samples during sonication.
10. Microfuge sonicated chromatin at 14,000 rpm for 10 min. Transfer supernatant into a new tube, discard pellet. At this point, sample can be snap-frozen and stored at −80°C for several weeks.
11. Preclear chromatin by adding Protein A agarose beads to sonicated chromatin. 10 μl beads are added for every 1×10^7 cells (see Note 5).
12. Rotate tubes at 4°C for 1 h. Microfuge at 10,000 rpm for 1 min.
13. Transfer supernatant to a clean tube. Freeze down 10% of the chromatin as "input" DNA control. Equally divide the rest into two tubes (250–500 μl in each tube). Foxp3 antibody (2 μg) is added into one tube, and preimmune rabbit IgG (2 μg) is added to the other tube as negative control.
14. Keep tubes in constant rotation at 4°C for 6 h or overnight.
15. Add 50 μl Protein A agarose beads into each tube. Rotate tubes at 4°C for 2 h.
16. Microfuge at 10,000 rpm for 1 min. Transfer supernatant to a new tube and freeze down for possible sequential ChIP.
17. Resuspend Protein A beads in 1 ml nuclei lysis buffer and transfer to a new tube. Microfuge at 10,000 rpm for 1 min. Discard supernatant (see Note 6).
18. Repeat step 17.

19. Resuspend Protein A beads in 1 ml nuclei lysis buffer containing 500 mM NaCl. Microfuge at 10,000 rpm for 1 min. Discard supernatant.
20. Repeat step 19.
21. Resuspend Protein A beads in 1 ml ChIP Wash Buffer. Microfuge at 10,000 rpm for 1 min. Discard supernatant.
22. Repeat step 21.
23. Resuspend Protein A beads in 1 ml TE buffer. Microfuge at 10,000 rpm for 1 min. Discard supernatant.
24. Repeat step 23.
25. Resuspend Protein A beads in 100 μl ChIP elution buffer #1. Incubate at 65°C for 15 min with occasional agitations. Microfuge at 10,000 rpm for 2 min. Transfer supernatant to a clean tube.
26. Resuspend Protein A beads in 150 μl ChIP elution buffer #2. Incubate at 65°C for 15 min with occasional agitations. Microfuge at 10,000 rpm for 2 min. Transfer supernatant to the same tube to combine both elutions.
27. For 10% input DNA control sample, add 20 μl 10% SDS and add TE buffer to a final total volume of 250 μl.
28. Incubate samples in 65°C for 5 h or overnight to reverse formaldehyde crosslink.
29. Microfuge at 10,000 rpm for 2 min. Transfer supernatant to clean tubes. Add 250 μl TE buffer and 10 μl proteinase K (10 mg/ml) to each sample. Incubate at 37°C for 1 h.
30. Add 55 μl LiCl (4 M) to each sample. Extract samples once with phenol/chloroform/isoamyl alcohol (25:24:1) and once with Chloroform (see Note 7).
31. Add 1 μl glycogen (20 mg/ml) to each sample and mix well. Add 900 μl 100% ethanol to each sample and mix well. Precipitate DNA at –20°C for 1–2 h.
32. Microfuge at 14,000 rpm at 4°C for 30 min. Discard supernatant. Microfuge briefly, take out residue buffer. Pay attention not to dislodge pellet.
33. Dissolve DNA pellet in 50 μl TE buffer at 37°C for 30 min. Make a further 1:100 dilution for 10% input DNA control (0.1% input control). Now ChIP DNA samples are ready for analysis.

3.2. Analysis of Precipitated DNA by Quantitative PCR

At this stage, the quality of ChIP DNA samples is tested by quantitative PCR (qPCR) before performing PCR amplification.

1. Make 1:20 dilution of ChIP DNA sample and 0.1% input DNA control sample in TE buffer.

Table 1
Quantitative PCR primers for Foxp3 ChIP

Gene	Forward	Reverse
Il2ra	GGGTCAGGCCAACTTAGATGAG	CTCAACAAAGACTGAGAAGCAAGGT
Ikzf2	CCGTAAATAGAGGCTGCAGAAAG	TGCTGCAGTGTTTTCCGAGTT
Ctla4	TAATAATAACCAAGATAGGTGAGGAGCTT	TCTGATACAGCTGCAACGTCAA
Nt5e	CAGGAACAGCTCAGAGGTCAGA	TGTTAGAGCCGTTCTTGCATTG
Prdm1	TTGTTTACTCTGACGCGCAAA	GATCGGCACACCCTCTGCTA
Crem	CCTATCCCGTGCACCTCGTA	CTGCAACCTGTTGGAAATTCAG
Pde3b	TTTGGGCCGCATAGAGAAAA	CAGTGAATCATCAGCAGCACAA
Gmpr	CAGCTGGAACAGCCTTGGAA	AAATGTCAAGGCCCCTGTGA

All primer pairs listed here are designed to flank verified Foxp3 binding regions except for *Gmpr*, which is used routinely as a negative control

2. Set up qPCR reactions in triplicates for each sample/primers combination: 12.5 μl 2× SYBR Green PCR Mix, 2.5 μl H_2O, 5 μl primer mix (1 μM each), and 5 μl diluted DNA sample. Total reaction volume: 25 μl. Commonly used positive control and negative control qPCR primers are listed in Table 1.
3. Perform qPCR with the following protocol:
   (a) 50°C 2 min, 1 cycle.
   (b) 95°C 10 min, 1 cycle.
   (c) 95°C 15 s ->60°C 30 s ->72°C 30 s, 40 cycles.
   (d) 72°C 10 min, 1 cycle.
4. Calculate percentage of input (%input) of each sample/primers combination by comparing signal from precipitated DNA with 0.1% input DNA control (Fig. 1) (see Note 8).

3.3. PCR Amplification of ChIP DNA

1. Clean up ChIP DNA with Affymetrix GeneChip Sample Cleanup Module according to kit instruction.
2. Set up four identical reactions (see Note 9) for each ChIP DNA sample: sample DNA 10 μl, 5× Sequenase Buffer 4 μl, Primer A (200 μM) 4 μl. Total volume 18 μl.
3. Incubate reaction mix at 95°C for 4 min, quickly transfer tube on ice.
4. Prepare master mix cocktail for random priming reaction. For each reaction: 20 mg/ml BSA 0.1 μl, 0.1 M DTT 1.0 μl,

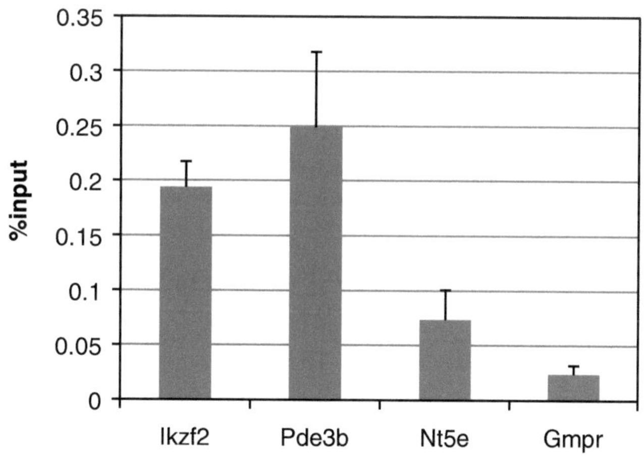

Fig. 1. Quantitative PCR analysis of Foxp3 ChIP DNA. DNA sample isolated from Foxp3 antibody chromatin immuno-precipitation of Treg cells is analyzed by quantitative PCR. *Ikzf2*, *Pde3b*, and *Nt5e* are positive controls for Foxp3 binding regions, whereas *Gmpr* serves as negative control.

25 mM dNTPs 0.5 μl, 1:10 diluted Sequenase 1.0 μl. Total volume: 2.6 μl.

5. Add 2.6 μl Sequenase cocktail to each sample, mix well, perform four round of priming as described below:
   (a) 10°C 5 min.
   (b) Ramp up temperature from 10 to 37°C over 9 min.
   (c) 37°C for 8 min
   (d) 95°C for 4 min
   (e) Put tube on ice.
   (f) Add 1.0 μl Sequenase to each sample.
   (g) Repeat (a) to (f) for 2 more cycles.
   (h) 10°C 5 min.
   (i) Ramp up temperature from 10 to 37°C over 9 min.
   (j) 37°C for 8 min.
   (k) Put samples on ice.
6. Purify primed ChIP DNA with MicroSpin G-50 columns as described below:
   (a) Spin MicroSpin column at 10,000 rpm for 1 min, discard flow-through.
   (b) Change collection tube, transfer reaction mix (~20 μl) to column.
   (c) Spin MicroSpin column at 10,000 rpm for 2 min, collect flow-through.

7. Prepare dNTP/dUTP mix: dATP 10 mM, dGTP 10 mM, dCTP 10 mM, dTTP 8 mM, and dUTP 2 mM.
8. Set up four identical PCR amplification reactions for each ChIP DNA sample: 10× PCR buffer 10 μl, 25 mM $MgCl_2$ 3 μl, dNTP/dUTP mix 3.75 μl, 100 μM Primer B 4 μl, primed ChIP DNA 20 μl, Taq Polymerase 2 μl, distilled H_2O 57.25 μl. Total volume 100 μl.
9. Run PCR program as described below:
   (a) 95°C 30 s.
   (b) 45°C 30 s.
   (c) 55°C 30 s.
   (d) 72°C 1 min.
   (e) Repeat (a) to (d) for 14 additional cycles.
   (f) 95°C 30 s.
   (g) 45°C 30 s.
   (h) 55°C 30 s.
   (i) 72°C 1 min.
   (j) Repeat (f) to (i) for 14 additional cycles. For each additional cycle, add 5 s to extension time (60, 65, 70 s, etc.) (see Note 10).
   (k) Put samples on ice.
10. Check the size and quantity of amplified DNA on an agarose gel (Fig. 2) (see Note 11).
11. Perform qPCR using primers for positive and negative controls to verify the quality of amplified DNA.
12. Amplified DNA samples are submitted to microarray facility to conduct routine fragmentation, labeling, and hybridization procedures. For Foxp3 ChIP-on-Chip, we used Affymetrix GeneChip Mouse Tiling 2.0R Array Set (see Note 12).

3.4. Bioinformatics Analysis of Mouse Genome Tiling Array Results

The full detail of bioinformatics analysis of data generated from tiling arrays is beyond the scope of this chapter. In brief, analysis can be divided into three steps.

1. Use Affymetrix GeneChip Operating Software (GCOS) to convert original tiling array data file (DAT file) to CEL file format.
2. Use Model-based Analysis of Tiling-arrays (MAT) program (10) to process tiling array CEL file, including calculating the adjusted signal for each oligo probe and mapping the Foxp3 binding regions in the mouse genome (see Note 13). MAT generates a list of binding regions with their coordinates and a *P*-value score associated with each region. Binding regions

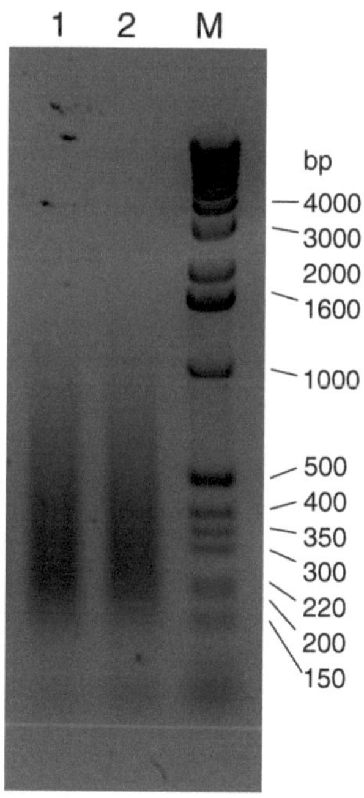

Fig. 2. Foxp3 ChIP DNA after PCR amplification. PCR-amplified Foxp3 ChIP DNA samples were analyzed on an agarose gel (2%). *1, 2*: two independent ChIP DNA samples after PCR amplification. *M*: 1 kb DNA ladder.

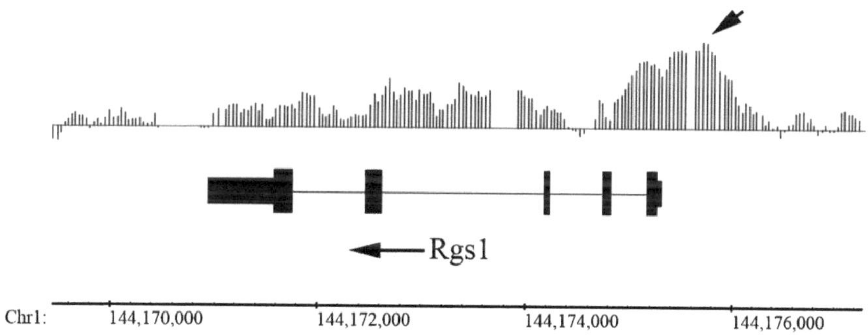

Fig. 3. Visualization of Foxp3 binding regions. Foxp3 binding region around *Rgs1* promoter is visualized using the Affymetrix Integrated Genome Browser. Each *bar* represents the signal intensity of an individual oligonucleotide probe. The *arrow* points to the peak of the binding region.

can be verified by additional qPCR with primers targeted to these regions (see Note 14).

3. To visualize binding regions, MAT program generates two files: .bar file for Affymetrix Integrated Genome Browser (IGB, Fig. 3), and .bed file for online Genome Brower developed by University of California at Santa Cruz (UCSC).

4. Notes

1. To isolate mouse regulatory T cells, we routinely use $CD4^{+}CD25^{+}$ Regulatory T Cell Isolation Kit from Miltenyi. FACS sorting can also be used to purify regulatory T cells for ChIP experiment. Because Foxp3 is specifically expressed in Treg cells, we routinely perform Foxp3 ChIP experiment with isolated Treg cells that are 80–90% positive for both CD4 and CD25 cell surface markers.
2. Instead of 10–20 min of cross-linking time for most other cells, we found 5 min is sufficient for regulatory T cells. Longer fixation time results in formation of cell clumps and poor sonication of chromatin in following steps.
3. The ChIP protocol is modified from Zhang et al. (11).
4. There are several factors affecting the outcome of sonication: power level, pulse time, and number of pulse cycles. We found setting power level at 20–25 W is usually optimal for sonication of regulatory T cells. Higher power will increase the chance of foam formation significantly. Lower power is not sufficient to break DNA into the right size. We choose to use pulse time between 10 and 15 s. Longer pulse time can generate excessive heat in sample. Pulse cycle number has been determined with a pilot experiment. Because of the scarcity of Treg cells, we used chromatin isolated from total mouse T cells for pilot experiment. A small aliquot of chromatin was taken out from the tube after each pulse and replaced with an equal volume of nuclei lysis buffer. After 15 pulses, all aliquots are reverse-cross-linked and precipitated as described in steps 27–33. The size of DNA in each aliquot is determined by running in an agarose gel. The final number of pulses is the minimum number that can break down DNA to the desired size.
5. Protein A agarose beads are stored in ethanol from supplier. Wash Protein A agarose beads three times with TE buffer and resuspend in TE buffer at 1:1 ratio before use.
6. Washing steps are crucial to reduce background signal in later quantitative PCR experiment. Make sure all washing buffers are free of mouse genomic DNA contamination and resuspend agarose beads thoroughly at each wash step.
7. The use of Phase Lock Gel during phenol/chloroform extractions can greatly improve separation of organic and aqueous phases and improve final yield.
8. DNA samples generated from a good Foxp3 ChIP experiment should be at least fivefold more enriched in regions

amplified by positive control primers (i.e., *Pde3b*, *Nt5e*) than negative control primers (i.e., *Gmpr*).

9. The PCR amplification protocol is modified from Affymetrix ChIP protocol.
10. PCR amplification cycles are determined by pilot experiment. It should be the minimum cycle number required to produce sufficient amount of DNA for hybridization. Overamplification can disproportionally increase background signal.
11. After PCR amplification, ChIP DNA size is usually reduced to 300–500 bp. Quantity of DNA can be determined by UV absorption. Typically, one PCR amplification reaction can generate 10 μg DNA.
12. There are a total of seven arrays in Affymetrix GeneChip Mouse Tiling 2.0R Array Set. We use 9 μg amplified DNA to hybridize to each individual array and reuse the labeled DNA once for a second array.
13. Several programs were developed for analysis of ChIP-on-Chip data. From our experience, MAT performed quite well in terms of generating the most relevant binding regions that can be verified by qPCR.
14. From our experience, a *P*-value cut-off threshold at 6.0 gives a list of verifiable Foxp3 binding regions. There are still a substantial number of binding regions with *P*-values between 5.0 and 6.0 that are verifiable by qPCR, so the choice of cut-off threshold has been determined according to specific downstream application.

Acknowledgments

The author would like to thank Professor Alexander Rudensky for his advice and support for this project, Steven Josefowicz for help and discussion, Arnold Kas for bioinformatics analysis, and Wei Li and Shirley Liu for assistance on MAT program. This work was supported by Cancer Research Institute and National Institute of Health (NIH).

References

1. Sakaguchi S, Yamaguchi T, Nomura T, Ono M. 2008. Regulatory T cells and immune tolerance. *Cell* 133: 775–87
2. Zheng Y, Rudensky AY. 2007. Foxp3 in control of the regulatory T cell lineage. *Nat Immunol* 8: 457–62
3. Brunkow ME, Jeffery EW, Hjerrild KA, Paeper B, Clark LB, Yasayko SA, Wilkinson JE, Galas D, Ziegler SF, Ramsdell F. 2001. Disruption of a new forkhead/winged-helix protein, scurfin, results in the fatal lymphoproliferative disorder of the scurfy mouse. *Nat Genet* 27: 68–73

4. Bennett CL, Christie J, Ramsdell F, Brunkow ME, Ferguson PJ, Whitesell L, Kelly TE, Saulsbury FT, Chance PF, Ochs HD. 2001. The immune dysregulation, polyendocrinopathy, enteropathy, X-linked syndrome (IPEX) is caused by mutations of FOXP3. *Nat Genet* 27: 20–1
5. Wildin RS, Ramsdell F, Peake J, Faravelli F, Casanova JL, Buist N, Levy-Lahad E, Mazzella M, Goulet O, Perroni L, Bricarelli FD, Byrne G, McEuen M, Proll S, Appleby M, Brunkow ME. 2001. X-linked neonatal diabetes mellitus, enteropathy and endocrinopathy syndrome is the human equivalent of mouse scurfy. *Nat Genet* 27: 18–20
6. Hori S, Nomura T, Sakaguchi S. 2003. Control of regulatory T cell development by the transcription factor Foxp3. *Science* 299: 1057–61
7. Fontenot JD, Gavin MA, Rudensky AY. 2003. Foxp3 programs the development and function of CD4+CD25+ regulatory T cells. *Nat Immunol* 4: 330–6
8. Zheng Y, Josefowicz SZ, Kas A, Chu TT, Gavin MA, Rudensky AY. 2007. Genome-wide analysis of Foxp3 target genes in developing and mature regulatory T cells. *Nature* 445: 936–40
9. Marson A, Kretschmer K, Frampton GM, Jacobsen ES, Polansky JK, MacIsaac KD, Levine SS, Fraenkel E, von Boehmer H, Young RA. 2007. Foxp3 occupancy and regulation of key target genes during T-cell stimulation. *Nature* 445: 931–5
10. Johnson WE, Li W, Meyer CA, Gottardo R, Carroll JS, Brown M, Liu XS. 2006. Model-based analysis of tiling-arrays for ChIP-chip. *Proc Natl Acad Sci U S A* 103: 12457–62
11. Zhang X, Odom DT, Koo SH, Conkright MD, Canettieri G, Best J, Chen H, Jenner R, Herbolsheimer E, Jacobsen E, Kadam S, Ecker JR, Emerson B, Hogenesch JB, Unterman T, Young RA, Montminy M. 2005. Genome-wide analysis of cAMP-response element binding protein occupancy, phosphorylation, and target gene activation in human tissues. *Proc Natl Acad Sci U S A* 102: 4459–64

Chapter 7

Live Imaging of Dendritic Cell–Treg Cell Interactions

Milka Sarris and Alexander G. Betz

Abstract

The decision to launch an immune response is made during the interaction of helper T cells and regulatory T cells with dendritic cells. Recognition of antigen leads to formation of immunological synapses at the interface between the cells and to activation of the T cells. The length of interaction between the T cells and dendritic cells influences the functional outcome. We have shown that in the absence of proinflammatory stimuli, regulatory T cells and naive helper T cells interact differently with dendritic cells. Neuropilin-1, which is expressed by most regulatory T cells but not naive helper T cells, promotes prolonged interactions with immature dendritic cells, resulting in higher sensitivity to limiting amounts of antigen. We tracked T cell dendritic cell interactions in real-time using time-lapse microscopy, assessed synapse formation by immunofluorescence, and measured regulatory T cell activation by dendritic cells using suppression assays.

Key words: Regulatory T cell, Helper T cell, Dendritic cell, Immunological synapse, Live cell microscopy, Immunofluorescence

1. Introduction

The regulation of immune responses relies on interactions between helper T cells, regulatory T cells, and dendritic cells. Dendritic cells capture antigens and present them to both helper and regulatory T cells (1, 2). Prolonged contact between a T cell and a dendritic cell leads to the formation of an "immunological synapse," during which cell surface and signaling molecules are recruited to the contact zone to form supramolecular activation complexes (SMACS) (3, 4, 5). The central area of the SMAC (cSMAC) is enriched in T-cell receptor molecules (which bind to peptide/MHC class II complexes on dendritic cells), while the peripheral area (pSMAC) is enriched in the adhesion molecule

George Kassiotis and Adrian Liston (eds.), *Regulatory T Cells: Methods and Protocols*, Methods in Molecular Biology, vol. 707,
DOI 10.1007/978-1-61737-979-6_7,

LFA-1 (which binds to ICAM-1 on dendritic cells) (4, 5), although variations of this type of structure have also been described (6, 7). Activation of helper T cells promotes the onset of immune responses (1), whereas activation of regulatory T cells leads to their suppression (2). In a situation in which both helper and regulatory T cells recognize the antigen presented, the decision to launch an immune response appears to be dependent on the presence of “danger signals” that can alter the interaction behavior of dendritic cells (8, 9).

In this chapter, we describe how the dynamics of T cell–dendritic cell interactions can be studied by time-lapse microscopy ex vivo. We provide a protocol for confocal immunofluorescence that allows assessment of the distribution of key molecules that form the immunological synapse and describe a suppression assay that can be used to evaluate regulatory T cell activation by dendritic cells.

2. Materials

2.1. Preparation of Cell Populations

1. Phosphate-buffered saline (PBS) is prepared using PBS tablets (Sigma-Aldrich) according to manufacturer’s instruction. The pH of every new batch of PBS is checked with a pH strip (pH should be between 7 and 7.5). Store at 4°C and use cold.
2. PBS/FCS: PBS supplemented with 2% fetal calf serum (Hyclone). Store at 4°C and use cold.
3. MACS buffer: PBS supplemented with 2 mM EDTA and 0.5% Bovine Serum Albumin (Sigma-Aldrich). Store at 4°C and use cold.
4. Complete RPMI culture medium with glutamax (Invitrogen), 10% FCS, 50 μM β-mercaptoethanol, penicillin (1 μg/ml), streptomycin (1 μg/ml). Store at 4°C. Warm up at 37°C before using.
5. Lympholyte M (Cedarlane). Store at 4°C.
6. Cell strainers (BD Biosciences). 1 or 5 ml syringe plungers.

2.2. Preparation of T-Cell Populations

1. Antibodies: FITC anti-CD8, FITC anti-CD19, FITC anti-CD11c, FITC anti-CD11b, FITC anti-Grl, PE-Cy5 anti-CD4, APC anti-Foxp3 (all from BD Biosciences), PE anti-CD25 (Miltenyi Biotech). Cytofix/Cytoperm Kit (BD Biosciences). Store at 4°C.
2. Cell sorting: Anti-FITC microbeads, anti-PE microbeads (Miltenyi Biotech). AutoMACS (Miltenyi Biotech). Store at 4°C.

2.3. Preparation of Bone Marrow-Derived Dendritic Cells

1. Antibodies: rat anti-CD16/32, rat anti-CD11b, rat anti-CD4, rat anti-GR1, rat anti-CD19, and rat anti-CD8 (BD Biosciences). Store at 4°C.
2. Cell sorting: anti-rat Dynabeads (Invitrogen), Dynal MPC™-L (Invitrogen). Store at 4°C.
3. IL-4, GM-CSF (Peprotech). Upon reconstitution in PBS, store in aliquots at −20°C.

2.4. Time-Lapse Microscopy

1. Imaging RPMI/HEPES medium: Phenol-red free RPMI supplemented with 20 mM HEPES (both from Gibco). Store at 4°C. Warm up at 37°C before using.
2. Lab-Tek chambered slides (Lab-Tek).
3. Cell tracker dye of choice (optional): CMFDA (green), CMTMR (orange), CMTPX (red)(Invitrogen).

2.5. Confocal Immunofluorescence

1. 0.01% poly-l-lysine solution (Sigma-Aldrich). Store at 4°C.
2. Multispot microscope slides (C.A. Hendley Essex Ltd.).
3. Paraformaldehyde (Electron Microscopy Sciences). Store stock solution (16%) at room temperature. Prepare fresh working dilution in PBS every time just before use. Use under a fume hood and discard in a hazardous container.
4. Saponin (Sigma-Adrich). Store solution at room temperature.
5. Donkey serum (Jackson Immunoresearch). Store in aliquots at −20°C.
6. Antibodies: FITC anti-CD3 (BD Biosciences), goat anti-ICAM-1 (R&D systems), donkey anti-goat IgG (Jackson Immunoresearch). Store at 4°C.
7. Vectashield H-1000 mounting medium (Vector Laboratories).
8. Borosilicate glass coverslips (BDH).
9. Nail varnish.

2.6. Image Analysis

Volocity software (Visualization and Quantitation package) (PerkinElmer).

2.7. Suppression Assays

1. Ovalbumin (Sigma-Aldrich). After reconstitution, store in aliquots at −20°C.
2. ^{3}H-Thymidine (Amersham). Store at 4°C, in an appropriate radioactivity containment unit. Take the necessary safety measures for use and disposal of radioactive material.
3. Unifilter 96 GF/C with sealing stickers (PerkinElmer).
4. Microscint-20 cocktail (PerkinElmer).
5. Equipment: 137Cs irradiator. Filtermate Harvester (PerkinElmer). TopCount microplate scintillation counter (PerkinElmer).

3. Methods

3.1. Preparation of T-Cell Populations

Irrespective of the downstream application, high purity of the various cell populations used is paramount. We routinely prepare T-cell subpopulations from mouse spleens and/or lymph nodes on the basis of cell-surface markers with a purity >95% by magnetic cell sorting. Whenever possible, we deplete unwanted cell populations, as positive selection often alters the activation status of the cells. For example, positive selection with CD4 antibodies can lead to signaling by CD4, which influences the antigen recognition process (5, 10, 11).

In a non-immunized mouse kept under pathogen-free conditions, naïve helper T cells are $CD4^{+}CD25^{-}$ while regulatory T cells are $CD4^{+}CD25^{+}$ and express the transcription factor Foxp3 (12). To obtain $CD4^{+}$ cells, we deplete total splenocytes from unwanted cells using a cocktail of CD11c, CD11b, GR1, CD8, and CD19 antibodies to remove dendritic cells, macrophages, granulocytes, $CD8^{+}$ cells, and B cells respectively. All antibodies are conjugated with FITC, allowing their depletion using anti-FITC secondary antibodies coupled to magnetic beads. Subsequently, the $CD4^{+}$-enriched cells are selected based on their expression of CD25. To obtain reliable results, it is essential to validate the purity of the cell populations after every sort by flow cytometry.

1. Under sterile conditions, dissect mice (see Note 1) that have been just sacrificed in compliance with the relevant laws and institutional guidelines. Remove the spleens carefully and place them in a small tube in PBS on ice. The number of spleens to dissect depends on the desired number of cells (see Note 2). Prepare cell suspensions in PBS by gently forcing the spleen through a cell strainer with 70 μm pores (up to five spleens per strainer) with the help of a syringe plunger using circular movements. Pass this cell suspension through a second strainer to ensure full removal of cell clumps and connective tissue (see Notes 3 and 4).
2. Split the cell suspension in 5 ml aliquots per spleen in 15 ml falcon tubes. Carefully layer 2.5 ml of Lympholyte M under the cell suspension by slowly pipetting it to the bottom of the tube using a glass Pasteur pipette. Spin the tubes at $1,200 \times g$ for 20 min, at room temperature without breaking (see Note 5).
3. Extract the lymphocyte layer that will form on top of the lympholyte M layer with a Pasteur pipette and transfer onto a new 15 ml falcon tube. Wash this suspension twice in PBS, by centrifugation at 800 and $400 \times g$, respectively, for 10 min, and resuspend in PBS/FCS at 10^8 cells/ml (see Note 6).

4. Add the following antibodies: FITC anti-CD11c, FITC anti-CD11b, FITC anti-GR1, FITC anti-CD8, FITC anti-CD19 to the cells, gently mix, and incubate for 45 min on ice (or 20 min at room temperature and 20 min on ice). All antibodies are used at a concentration range of 0.5–1 μg/ml (see Note 7).
5. Wash the cells twice in PBS (400 × *g* for 10 min at 4°C). Check the staining by FACS. 70–80% of the cells should be stained. If not, repeat the staining.
6. Resuspend the cells in MACS buffer at 10^8 cells/ml and add anti-FITC microbeads at a 1:10 dilution for 30 min at 4°C. Mix the cell/bead mixture a couple of times during the incubation by inverting the tube.
7. Wash the cells in MACS buffer (400 × *g* for 10 min at 4°C) and resuspend the cells in MACS buffer at 10^8 cells/ml. Split in 2 ml aliquots.
8. Run each sample through a DEPLETE program in the AutoMACS. Pool the depleted samples and then spin down the cells (400 × *g* for 10 min at 4°C). Resuspend the cells in 2 ml MACS buffer. Pass this sample through a DEPLETE05 program in the AutoMACS. Check the purity of the cell suspension by flow cytometry. >99% of the FITC-labeled cells should be depleted (see Note 8). Meanwhile, take a small sample of the cells and stain in an eppendorf tube with PE-Cy5 anti-CD4 for 30 min on ice. Wash the cells and check them by flow cytometry. The percentage of $CD4^+$ cells should be >90%. Alternatively, the CD4 stain can be done after the second selection, for both $CD4^+CD25^+$ and $CD4^+CD25^-$.
9. Once the purity is confirmed, proceed to the CD25 selection. Resuspend the cells in PBS/FCS in 10^8 cells/ml (You should have about a tenth of the cells you started with at this stage). Incubate with a PE anti-CD25 antibody for 30 min on ice (see Note 9).
10. Wash twice in PBS. Resuspend in MACS buffer as before and incubate with anti-PE microbeads at a 1:5 dilution for 30 min at 4°C. Wash once in MACS buffer and resuspend in 2 ml.
11. Pass the sample through a POSSELD2 program on the AutoMACS. Keep the positive fraction on ice and pass the negative sample through a DEPLETE05 program. Check the cell purity by FACS. The positive fraction should be 90–95% PE-positive. The negative fraction should be >99% PE-negative (see Note 10).
12. We recommend performing an intracellular Foxp3 stain to validate the purity of the $CD4^+CD25^+$ cell preparation. While it might not be possible to do this every time, we strongly

recommend it when changing batches of antibodies or microbeads. For the staining, use the BD Cytofix/Cytoperm kit or similar. Take an aliquot of around 10^5 cells from the $CD4^+CD25^+$ and the $CD4^+CD25^-$ fraction. Centrifuge at 400×*g* for 10 min. Resuspend in 25–50 μl of PBS/FCS, supplemented with 2 μg/ml anti-CD16/32. Wash the cells once in PBS/FCS (400×*g* for 10 min at 4°C). Resuspend in 100 μl of Fix/Perm buffer (one part Fix/Perm concentrate/three parts Fix/Perm diluent). Incubate for 2 h or overnight at 4°C. Wash once in PBS/FCS and twice in Perm buffer (one part Perm buffer/nine parts distilled water). Resuspend in PBS/FCS supplemented with APC anti-Foxp3 at a 1:100 dilution. Wash 3 times with Perm buffer (it is important to wash well after staining). Resuspend in PBS/FCS and analyze it on a flow cytometer. Typically, the percentage of $Foxp3^+$ cells in a 95% pure $CD4^+CD25^+$ cell fraction is approximately 90%.

3.2. Preparation of Bone Marrow-Derived Dendritic Cells

In this experimental set-up, dendritic cells are prepared in vitro from bone marrow hematopoietic progenitors. The main reason we chose this approach is that dendritic cells isolated directly from lymphoid tissues are phenotypically heterogeneous and their activation state is difficult to determine (13). In contrast, bone marrow-derived dendritic cells are considered to be phenotypically homogeneous immature dendritic cells (13). Although it is also possible to perform similar experiments with purified dendritic cell populations, this procedure is not covered in this chapter. The bone marrow cell suspensions are enriched in hematopoietic progenitor cells by negative selection of lymphocytes stained with CD16/32, CD11b, CD4, GR1, CD19, and CD8 antibodies. This cell preparation should be done 7 days before the T-cell preparation (see Note 11).

1. Under sterile conditions, cut the femurs and tibia from the sacrificed mice (see Note 12). Remove the muscle tissue around the bone with a pair of scissors and cut both ends of the bone close to the joints. Using a syringe with a needle (0.6×30 mm), flush out the bone marrow with PBS into a 50 ml falcon tube. Generate a cell suspension by gently pipetting up and down and pass it through a cell strainer with the help of a 1 ml or 2 ml syringe plunger. Spin down the cell suspension at 400×*g* for 10 min at 4°C and resuspend it in 5 ml PBS.
2. Purify the progenitor cells using Lympholyte M centrifugation and wash the cells twice as described in Subheading 3.1.
3. Resuspend the cells in PBS/FCS at 10^8 cells/ml. Incubate with a cocktail of rat anti-mouse CD16/32, CD11b, CD4, GR1, CD19, and CD8. All antibodies are used at a concentration range of 0.5–1 μg/ml (see Note 7).

4. Wash the cells twice in PBS and then resuspend in PBS/FCS. Incubate with anti-rat Dynabeads (approx. 2 beads/target cell and 2×10^7 beads/ml) for 45 min at 4°C, under continuous rotation.
5. Place tubes containing the cell/bead mixture into the Dynal MPC™-L magnet. Allow the beads and labeled cells to stick to the side of the tube that is in contact with the magnet. This takes about 2 min. Remove the depleted cell fraction and place in a new tube. Resuspend the cell/bead mixture in 5 ml of PBS and repeat the procedure. Pool the two depleted fractions. Place the tube with the depleted fraction in the magnet again. Allow the remaining labeled cells to stick to the side of the tube. Remove the depleted cell suspension and spin down the cells.
6. Resuspend the depleted cells in complete RPMI, supplemented with 50 ng/ml GM-CSF and 20 ng/ml IL-4 at a the of 1.5×10^6 cells/ml. Plate 0.5 ml/well in a 24-well plate.
7. Replace the supernatants with fresh medium containing IL-4 and GM-CSF every 2 days. When doing so, make sure to flick the plates very well so that the cells in suspension are removed (see Note 13). On day 7 of culture, typically more than 90% of the cultured cells are adherent, CD11c+ dendritic cells (see Note 14).

3.3. Time-Lapse Microscopy

This protocol is set up for the monitoring of T cell–dendritic cell interactions during the initial 20 min of coculture.

1. Harvest the dendritic cells by gently pipetting up and down (see Note 15).
2. Wash the dendritic cells once in PBS. Resuspend the dendritic cells in warm RPMI/HEPES imaging medium. We recommend a concentration of 10^6 cells/ml.
3. Before imaging, resuspend the purified T cells in warm, RPMI/HEPES imaging medium at a cell concentration of 2×10^6 T cells/ml.
4. Optional: The cells can be labeled with a fluorescent cell tracker dye such as CMFDA (green), CMTMR (orange), or CMTPX (red) in order to distinguish the cells by fluorescence as well as morphology. In this case, after purifying the T cells and harvesting the dendritic cells, they can be resuspended in the labeling solution, which is a PBS dilution of the dye stock in the range of 1–5 μM (see Note 16). The labeling solution should be prewarmed at 37°C. Incubate for 15 min in a water bath of 37°C. Wash with warm, complete RPMI medium and resuspend in the same medium. Leave the cells in an incubator for 30 min (see Note 17) and then spin down ($400 \times g$ at room temperature) and resuspend in the RPMI/HEPES imaging medium.

5. Gently mix 50 μl of each cell suspension in an 1.5 ml eppendorf tube and then transfer 100 μl of the mixed cell suspension into a well of an 8-well Lab-Tek chambered slide (see Note 18). With the help of the pipette tip, distribute the cell suspension evenly in the well. Incubate the remaining individual cell suspensions at 37°C in a tissue culture incubator until use.
6. Place the chambered slide immediately on an already heated stage (37°C) of a confocal microscope (see Note 19). A 40× oil-immersion objective gives good enough resolution to observe changes in the morphology of the cells. Depending on the experimental design, a 20× or a 60× objective might be more suitable. Allow the cells to settle in the bottom of the well. This takes 3–5 min (see Note 20). Meanwhile, observe the cells under transmission light and identify a region in which the cells are distributed evenly and the cell density is optimal.
7. Use the DIC (differential interference contrast) option of the confocal microscope to maximize the image contrast.
8. If the cells have been labeled in some way, use the respective laser to excite the fluorescent dye. Adjust the laser power according to the signal intensity, taking care not to saturate or photobleach the fluorescent signal. Adjust the rest of the acquisition parameters of the software in order to get an optimal signal-to-noise ratio (gain and offset of the photomultiplier tubes, pinhole diameter, resolution). If needed, use the digital zoom to further magnify an area of interest. Averaging is not needed for this kind of acquisition and is not recommended for videos of high temporal resolution.
9. Acquire a time-lapse movie. A frame interval of 10 s is in most cases of sufficient temporal resolution to follow the dynamics of T cell–dendritic cell interactions (Fig. 1a, b). The duration of the movie for this protocol should not exceed 20–30 min. For longer imaging experiments, FCS should be included in the imaging medium (see Note 21).

3.4. Confocal Immunofluorescence

1. Coat multispot slides (use a slide with 15 spots of a 6-mm diameter) with 0.01% poly-l-lysine by applying 2 μl of the poly-l-lysine solution on each spot. Place the slides in a humid box and incubate at room temperature for 30 min.
2. Rinse the slides in sterile distilled water. Air-dry thoroughly for 10–15 min.
3. Apply 10 μl of cell suspension. The cells should be suspended in RPMI/HEPES imaging medium at such a concentration so that the cell density per well is 2×10^4 for the dendritic cells and 4×10^4 for the T cells. Incubate the cells in a humid box for 25 min at 37°C.

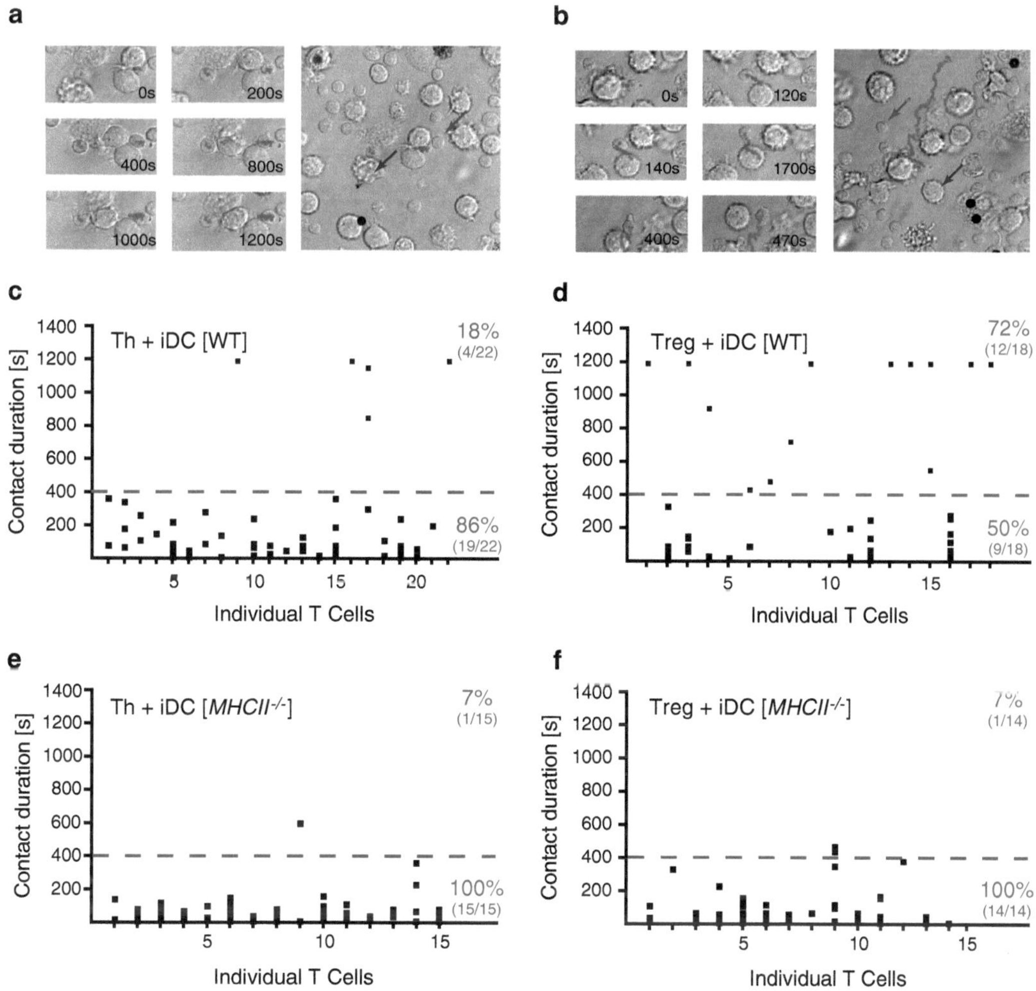

Fig. 1. Treg cells form more MHC class II-dependent long interactions with immature dendritic cells (iDC) than naive Th cells. $CD4^+CD25^+$ (Treg) or $CD4^+CD25^-$ (Th) cells were cocultured with iDCs and imaged as described in the experimental procedures. (**a**, **b**) Representative examples of T cells forming either (**a**) long interactions or (**b**) multiple short interactions with iDCs. Snapshots of the area surrounding the traced T cell at the indicated time points are shown (*left*). The complete path (black trace) traversed by the T cell in 20 min is shown (*right*). Representative T cells (*light arrows*) and iDCs (*dark arrows*) have been marked. (**c**–**f**) Interactions observed between (**c**, **e**) Th or (**d**, **f**) Treg cells and (**c**, **d**) WT or (**e**, **f**) MHC class II-deficient iDCs ($MHCII^{-/-}$) in individual experiments. Columns represent T cells with each of the dots denoting the length of an interaction made. All T cells that have made at least one contact with an iDC are included. The frequency of T cells interacting with an iDC for longer or shorter than 400 s (*dashed line*) is given as percentage and as ratio (reproduced from (9) with permission from CellPress).

4. For fixation, immerse the slide into a freshly prepared 4% paraformaldehyde solution in PBS at room temperature (all subsequent steps are done at room temperature). Incubate the slide for 15 min. Discard paraformaldehyde into a hazardous waste container (see Note 22).
5. Rinse the slides in PBS for 5 min twice, by immersing the slides in PBS (see Note 23). At this stage, the slides can be stored at 4°C for a couple of days.

6. For permeabilization, immerse the slides into PBS-0.02% saponin (see Note 22). Incubate for 10 min.
7. For blocking, immerse the slides into PBS-3% donkey serum (the serum should be from the same host as the secondary antibody) and incubate for 15 min.
8. Remove the slides from blocking solution. Wipe around the slide spots with a rolled-up tissue or Whattman paper (take care not to dry the wells) to prevent cross-contamination of the samples. Apply 10 μl of primary antibody solution per spot. The primary antibodies, FITC anti-CD3 and goat anti-ICAM-1, are used at a 1:50 dilution (see Note 24). Keep the slides in a humid box wrapped in aluminum foil to protect from light. Incubate the slides for 45 min.
9. Wash 5 times with PBS for 5 min (see Note 25).
10. Dry the slides as before. Apply 10 μl of secondary antibody solution per spot. The Alexa 647-anti-goat secondary antibody (far red fluorescence emission) is used at a 1:500 dilution (see Note 24). Incubate the slides for 45 min.
11. Wash 5 times in PBS for 10 min (see Note 25).
12. Carefully dry the slide around the wells as before and apply 1–2 drops of mounting medium (VectaShield) in the middle of the slide (not on a well). Spread the drop across the slide, in between the wells, with the help of a tip. Quickly, mount a coverslip by placing it on one side of the slide at a 45° angle and slowly lowering it on the other side. Avoid introducing bubbles. Remove any excess medium seeping out between the slide and the coverslip with Whattman paper. Seal the coverslip with nail varnish and allow it to dry in the dark.
13. Store the slides refrigerated and in the dark. Analyze as soon as possible, but no later than 48 h after preparation.
14. For analysis, place the slides on a confocal microscope, with the coverslip on the side of the objective. Use a 60× (or higher magnification) oil-immersion objective.
15. Identify areas of interest (containing T cell–dendritic cell contacts) under transmission light. Use a 488 nm Argon laser, or similar, to excite FITC fluorescence and for the transmission light acquisition. Use a 635 nm red diode laser, or similar, to excite Alexa-647 fluorescence. Adjust laser power from low to high taking care not to saturate the fluorescent signal. Adjust the gain and offset of the photomultiplier tubes and the pinhole size to get an optimal signal-to-noise ratio with the lowest possible laser power. It is recommended to use the averaging option to enhance the signal-to-noise ratio.

    Use identical parameters for all samples within an experiment. If needed, use the digital zoom to further magnify an area of interest. Set the start and end point of the volume that is to be scanned. Set a step of 0.25–0.5 μm. Acquire stacks of images.

3.5. Image Analysis

The choice of image analysis depends very much on the experimental design. Here we give a quick view of how one tracks cells, measures interaction times between cells, and processes 3D objects using the Volocity Software from PerkinElmer. Volocity is a high performance, 3D imaging software that is designed specifically for the needs of microscopic image analysis. To learn how to operate the software in detail, it is recommended to consult Volocity's comprehensive user guide and/or ask for a demonstration. There are different Volocity products that can be purchased separately or combined. For the cell tracking, it is necessary to purchase "Volocity Quantitation" package and, for the 3D image processing, it is necessary to purchase the "Volocity Visualization" package. There are additional softwares that can do this type of analysis, such as Imaris and Metamorph. More basic analysis tools can be found in ImageJ, which is freely available online and is accompanied by a large number of plug-ins, which can be used to perform specific tasks. We strongly recommend performing all image analysis in a blinded fashion so as not to bias the analysis.

1. Open Volocity and create a new library (see Note 26). If you are importing time-resolved or multichannel (for example, with a green, red, and a bright field channel) data, choose the "New Image Sequence" option in the library. Drag and drop the data into the new image sequence window (see Note 27).
2. In the pop-up window, define how the image sequence should be arranged in time points, channels, and slices.
3. The data will be displayed in multiple views in different tabs. The type of view depends on the package of Volocity used and will differ depending on the nature of the data. For example, in the "Image" view the data are represented as an XY image, which can be navigated in time. 3D data can be represented as an XY, XZ, and YZ section of the volume, or as a brightest point merge of the XY stack along the *Z*-axis.
4. From the tools menu, choose the "Change Colors" option to assign colors to channels.
5. In the navigator toolbar, select the controls for modifying the intensity in each channel and for navigating through the movie in time. You can play the movie at a fixed rate or using a slider at the bottom of the image.
6. For the purpose of tracking the cells, go to the "Measurements" tab. The first step is to mark the cells. You can either do this manually or choose the automatic option of the software.

The latter requires you to define a "protocol." For example, if the T cells are labeled green, open the measurements tab and start a new protocol by choosing a finding task, such as "Find Objects by Intensity" or "Find Objects by RGB," and selecting in which channel (in this case green) you want the objects to be found. You can try out different parameters to be used for the finding task like shape or contrast. You can also manually adjust the intensity threshold that is used to select the objects. Once you are satisfied with a protocol, save it and apply it to a selected image sequence. The selected objects will be shown as a colored overlay in the image preview and standard morphological and intensity measurements will be displayed in the "Measurements table" for each object. You should check the protocol on a number of different time points before storing any measurement. Use the time navigation controls to move between time points.

7. Add the "Track Objects" task on your protocol. This will track the centroid position of the objects. Once the "Track Objects" task is added to the protocol, click on the icon next to it to open a secondary window with options on how the tracking should be done. You can try different settings to see which tracking options are best suited to the type of movements observed.
8. Select "Make a measurement item" in the "Measurements" menu and ensure "Measure All Timepoints" is selected. This makes a "Measurements item" that contains tracks.
9. To view the results, open the "Measurement item." A raw table will appear, which shows objects found in each time point and measurements (such as speed and distance) relevant to each. You can also view a chart with all the tracks superimposed and their start point set to zero. It is possible to manually track objects, if the nature of the data is too complex to allow automated tracking.
10. Unlike the tracking task, the measuring of interactions is done entirely manually. For this, you have to analyze each T cell or each dendritic cell one by one. Focus on one cell at a time, zoom in if necessary and scroll along the movie. Once you see a cell–cell contact, count the number of frames during which the cells stay in contact and calculate the contact duration based on the number of frames. Continue the same analysis for the rest of the movie and repeat the procedure for all the cells to be analyzed. If possible, it is best to analyze all the cells in the movie. If there are too many cells to be manually analyzed, randomly choose a number of cells to be analyzed. In this case, analyze the same number of cells from each movie of an experiment.

11. The data produced by this kind of analysis can be represented in many different ways, depending on the question asked. One example is to represent how many interactions and of what length individual T cells form in a video. One possible representation is to plot the time on the *Y*-axis and individual T cells on the *X*-axis with dots displaying each interaction formed by that T cell and its duration (Fig. 1c, d). An analogous analysis can be performed for dendritic cells.
12. To process 3D volumes, import the data as an image sequence as described above. You can view the image as a rendered 3D volume, as a brightest point merge of the XY image stack, and as an XY image with an inspector that gives a view of an XZ and YZ section. A particularly useful tool for looking at synapse structures is the 3D slice tool. This tool can be used to reslice the data set in any chosen rotation of the *X*-, *Y*-, and *Z*-axis. This is very useful, as the contact zone between two cells is rarely aligned with the XY, YZ, or XZ planes and reslicing allows the viewing of the "face" of the contact zone between the two cells and the navigation through the volume in the same plane (see Note 28) (Fig. 2).

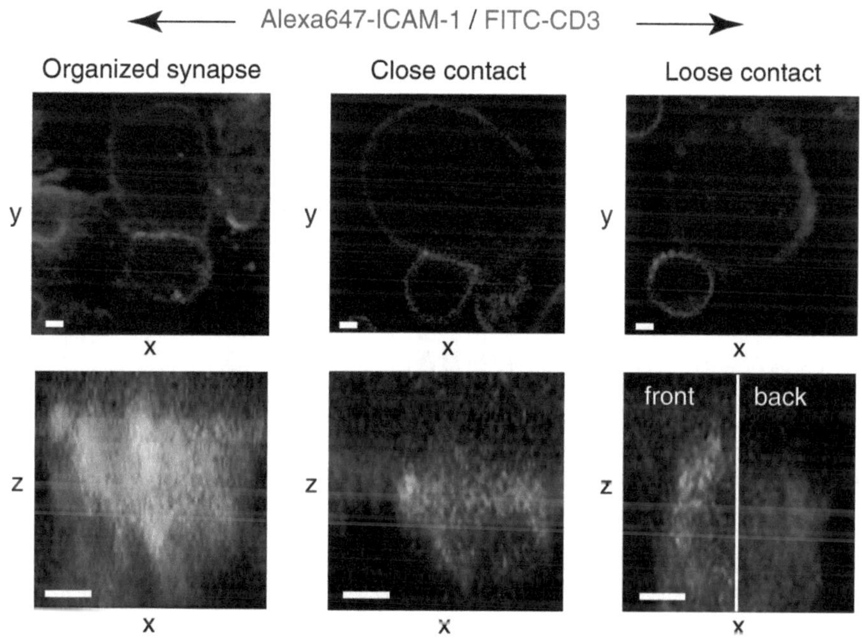

Fig. 2. Analysis of synapse formation between T cells and iDCs. Images representative of an organized synapse, close contact, and loose contact on a single confocal section on the medial xy plane (*top*) or in a projection of zx images spanning 0.5 μm in the y direction in the area of the contact zone between the T cell and the iDC (*bottom*). In the case of the representative example of a loose contact, the projection of zx images is split in two halves spanning 0.5 μm in the y direction (front/back of the contact zone) (reproduced from (9) with permission from CellPress).

3.6. Suppression Assay

The suppressive activity of regulatory T cells can be assessed in vitro by measuring the proliferation of cocultured helper T cells that recognize the antigen presented by the dendritic cells. There are many permutations of this experiment and we present a representative example (Fig. 3).

1. T cells are prepared from DO11.10 transgenic mice, which express a transgenic receptor specific for the epitope 323–339 of ovalbumin (see Note 29). Prepare $CD4^+CD25^+$, $CD4^+CD25^-$, and bone marrow-derived dendritic cells as described in Subheadings 3.1 and 3.2.
2. The evening before starting the coculture experiment, change the medium of the dendritic cells (complete RPMI medium supplemented with IL-4 and GM-CSF) with complete RPMI supplemented with ovalbumin (see Note 30). Different concentrations of ovalbumin can be tested in the range of 1–200 μg/ml.
3. The next morning, while preparing the T cells, irradiate the dendritic cells after harvesting them as described in Subheading 3.3 (see Note 31). Resuspend the cells in a 15 ml falcon. Place the falcon into a radioactive ^{137}Cesium irradiator and expose to 3,000 rad.

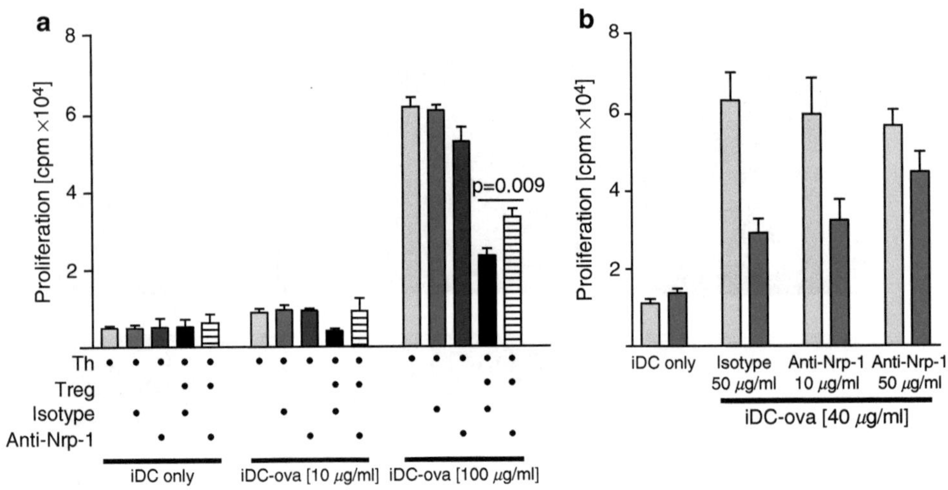

Fig. 3. Anti-Nrp-1 treatment interferes with suppressive function of Treg cells. $CD4^+CD25^-$ (Th) cells were cocultured with $CD4^+CD25^+$ (Treg) cells (both prepared from DO11.10 mice) and either untreated iDCs or ova-loaded iDCs, in the presence or absence of anti-Nrp-1 or isotype control. Proliferation was determined by ^{3}H thymidine incorporation. (**a**) Effect of anti-Nrp-1 treatment at different concentrations of antigen. $CD4^+CD25^-$ cells were cocultured with ova-loaded iDCs (indicated concentrations; 12 h), in the presence or absence of $CD4^+CD25^+$ cells, with or without anti-Nrp-1 or isotype control (10 μg/ml) (pooled data from three independent experiments performed in duplicates; $p = 0.009$, unpaired t test). Error bars represent the SEM. (**b**) Dose-dependent effect of anti-Nrp-1 treatment. $CD4^+CD25^-$ cells were cocultured with (dark bars) or without (light bars) $CD4^+CD25^+$ cells, in the presence of the indicated amounts of anti-Nrp-1 or isotype control ($n = 4$) (reproduced from (9) with permission from CellPress).

4. Wash the dendritic cells twice with warm complete RPMI medium (see Note 32).
5. Resuspend in complete RPMI medium at a density of 10^5 cells/ml and incubate them at 37°C until the preparation of the T cells is complete (see Note 33).
6. Resuspend the $CD4^+CD25^+$ cells and the $CD4^+CD25^-$ cells at 5×10^5 cells/ml (see Note 33).
7. Pipette 50 μl of each cell suspension into a U-bottom, 96-well tissue culture plate. Plate the single cell suspensions and all the combinations of cell suspensions as controls. Every well should be represented in triplicates or quadruplicates. Add complete RPMI medium up to 200 μl in each well (see Note 34).
8. Incubate in a tissue culture incubator (37°C, 5% CO_2).
9. 36–48 h later, pulse the cells with ^{3}H-Thymidine, taking the necessary precautions for working with radioactive material. Prepare a 0.04 μCi/μl dilution of ^{3}H-thymidine in pre-warmed, complete RPMI medium. Pipette 25 μl of the suspension onto each well. Avoid disturbing the cells in the bottom of the well. Return the plate in the tissue culture incubator for 16–18 h.
10. Place the plate on a Filtermate Harvester to harvest the cells and transfer the cells' DNA onto the 96 spot Unifilter membranes. Dry the membrane for 15 min at room temperature.
11. Seal the bottom of the filter membrane with a plastic sticker. Add 30 μl Microscint-20 liquid onto each well of the filter membrane. Seal with a transparent sticker and read on a TopCount microplate scintillation counter.

4. Notes

1. The mouse strain depends on the experiment. For wild type mice, we use either C57/BL6 or BALB/c mice. DO11.10 mice in a RAG-competent background can be used to purify ovalbumin-specific helper T cells and regulatory T cells. It is recommended to keep the age and gender of the mice consistent, within the 3–6 months range.
2. The number of $CD4^+CD25^+$ cells is usually the limiting factor ($1–2 \times 10^5$), as this protocol is optimized for purity rather than yield.
3. This is a passive filtration step to remove connective tissue.

4. Make sure to wash the tubes after every transfer of the cell suspension. Given the limited numbers of $CD4^{+}CD25^{+}$ cells that can be purified with this protocol, it is critical to avoid any cell losses.
5. Removal of the brake is recommended to minimize disturbance of the Lympholyte gradient and maximize cell recovery.
6. It is very important to keep the cells cold. Work fast, use cold buffers (4°C), and keep the cells on ice whenever they are not manipulated.
7. It is important to titrate the antibodies each time a new batch is used. We suggest a dilution as a general indication, but the optimal concentration has to be determined empirically for each antibody.
8. If the purity is not satisfactory, pass the sample through another round of DEPLETE05. If the problem still persists, add more microbeads and repeat the procedure.
9. It is preferable to use a CD25 antibody coupled to a different fluorophore. If FITC anti-CD25 antibody is used, impurities from the first selection are enriched during the second selection with the anti-FITC microbeads.
10. If the positive fraction does not have the desired purity, do not pass it through another POSSELD2. We have found that this does not improve the purity, but severely compromises the yield. However, if the negative sample is not pure enough, you can pass it through another DEPLETE05.
11. It is important to be consistent as to how many days after culture the dendritic cells are used. One day more or less in culture with GM-CSF and IL-4 will influence the maturation state of the dendritic cells and thus their behavior in downstream assays.
12. Usually, the femurs and tibia from two mice will give enough dendritic cells for most experiments ($1–3 \times 10^6$ cells). Unless the dendritic cells are to be used for mixed lymphocyte reactions, they should be isolated from mice syngeneic to the mice from which the T cells are isolated.
13. The removal of the nonadherent cells should be done very gently the first time and then gradually more and more thoroughly, as the cells become more and more adherent.
14. The phenotype and homogeneity of the cells can be checked by staining with anti-CD11c and anti-MHC class II antibodies, followed by flow cytometry analysis. The purity of the original bone marrow population is not critical. This sort is merely an enrichment.
15. Take care to be gentle while harvesting the dendritic cells. Mechanical stimulation can activate the dendritic cells. There will

be some cells that are more adherent than the majority and will remain attached to the bottom of the well. These are likely to be macrophages or more mature dendritic cells and should be left on the well.

16. The concentration of the labeling solution should be optimized prior to the experiment. It should be no more than 10 μM, as this compromises the viability of the cells. Each new batch of cell tracker dye should be titrated for optimum staining and cell viability. To ensure optimal performance of the dye, the stock solutions (usually in the range of 1–10 mM) should be made in DMSO and stored aliquoted at −20°C to avoid additional freeze–thaw cycles.
17. The cell tracker dyes passively diffuse through the cell membranes, but once inside the cell, are transformed into cell-trapped reaction products. This extra incubation step is important for complete modification of the label and ensures good retention of the dye in the cell.
18. For a Lab-Tek Chambered slide of 0.8 cm^2, a total cell density of 10^5 T cells and 5×10^4 dendritic cells is recommended, but can be adjusted according to the specific design of the experiment. It is important to maintain consistency in cell densities across independent, replicate experiments. The volume of the cell suspension can be between 0.2 and 0.4 ml, but we recommend 0.1 ml to minimize flow of the cells on the slide.
19. An alternative to a heated stage is a heated chamber that can be fixed to the microscope stage. This bypasses the need for HEPES-buffered medium, as the CO_2 levels can be regulated. The heated chamber is a better option for longer-term imaging experiments.
20. Be consistent in the starting time point of the acquisition after coculture. This allows a more reliable comparison between independent, replicate videos.
21. The imaging medium contains no FCS and no phenol-red to avoid background autofluorescence.
22. The conditions of fixation and permeabilization described here are suitable for the specific antibodies. Different antibodies may require modification of those conditions or use of other fixatives and detergents.
23. It is preferable to do the washes by immersing the slides into PBS rather than pouring the PBS over the slide. This minimizes detachment of the cells from the slide.
24. For each new batch, titrate the dilution of the antibody to optimize the signal over background.
25. It is important to wash thoroughly after each antibody-staining step, to minimize background staining.

26. Library is the file format created by Volocity. It is a multifile format that handles large data sets well and gives fast data access.
27. This software can support data in the format provided by the acquisition software and will preserve all the embedded information. Other analysis software may require export of the acquired files into image sequences of TIFF files.
28. The resolution of the images that are produced by reslicing in the Z axis depends on the step size between slices. Thus, it is recommended to use as small a step as possible when acquiring the stack in the confocal microscope.
29. This model was chosen to experimentally control antigen recognition. The DO11.10 mice should be in a RAG1/2 competent background, as DO11.10RAG1/2$^{-/-}$ mice do not develop regulatory T cells.
30. At this stage, it is optional to activate the dendritic cells with lipopolysaccharide or other proinflammatory stimuli such as CD40L or TNFα.
31. Differentiated dendritic cells are not expected to proliferate. Nevertheless, it is best practice to irradiate the dendritic cells in order to exclude any proliferation (for example by contaminating progenitor cells) interfering with the T-cell proliferation readout.
32. If antibody blockade is to be performed in the experiment, Fc-receptors should be blocked at this stage. Incubate the cells for 15 min on ice with 2 μg/ml anti-CD16/32 in PBS/FCS. Wash once in PBS or complete RPMI medium.
33. The cell densities can be manipulated or titrated, according to the needs of the experiment.
34. Antibodies can be added either at this stage or the cells can be preincubated with antibodies for 30 min prior to plating. In the latter case, the antibody can be left in the coculture or can be washed off before coculture, depending on the requirements of the experiment.

References

1. Banchereau J, Briere F, Caux C, Davoust J, Lebecque S, Liu YJ et al. (2000) Immunobiology of dendritic cells. *Annu. Rev. Immunol.* **18**, 767–811.
2. Steinman RM, Hawiger D, Liu K, Bonifaz L, Bonnyay D, Mahnke K et al. (2003) Dendritic cell function in vivo during the steady state: a role in peripheral tolerance. *Ann. N. Y. Acad. Sci.* **987**, 15–25.
3. Dustin ML (2003) Coordination of T cell activation and migration through formation of the immunological synapse. *Ann. N. Y. Acad. Sci.* **987**, 51–59.
4. Monks CR, Freiberg BA, Kupfer H, Sciaky N, Kupfer A (1998) Three-dimensional segregation of supramolecular activation clusters in T cells. *Nature* **395**, 82–86.
5. Grakoui A, Bromley SK, Sumen C, Davis MM, Shaw AS, Allen PM et al. (1999) The immunological synapse: a molecular machine controlling T cell activation. *Science* **285**, 221–227.

6. Brossard C, Feuillet V, Schmitt A, Randriamampita C, Romao M, Raposo G et al. (2005) Multifocal structure of the T cell – dendritic cell synapse. *Eur. J. Immunol.* **35**, 1741–1753.
7. Dustin ML, Tseng SY, Varma R, Campi G (2006) T cell-dendritic cell immunological synapses. *Curr. Opin. Immunol.* **18**, 512–516.
8. Reis e Sousa C (2001) Dendritic cells as sensors of infection. *Immunity* **14**, 495–498.
9. Sarris M, Andersen KG, Randow F, Mayr L, Betz AG (2008) Neuropilin-1 expression on regulatory T cells enhances their interactions with dendritic cells during antigen recognition. *Immunity* **28**, 402–413.
10. Dianzani U, Shaw A, Fernandez-Cabezudo M, Janeway CAJ (1992) Extensive CD4 cross-linking inhibits T cell activation by anti-receptor antibody but not by antigen. *Int. Immunol.* **4**, 995–1001.
11. Veillette A, Bookman MA, Horak EM, Bolen JB (1988) The CD4 and CD8 T cell surface antigens are associated with the internal membrane tyrosine-protein kinase p56lck. *Cell* **55**, 301–308.
12. Sakaguchi S (2005) Naturally arising Foxp3-expressing CD25+CD4+ regulatory T cells in immunological tolerance to self and non-self. *Nat. Immunol.* **6**, 345–352.
13. Wan H, Dupasquier M (2005) Dendritic cells in vivo and in vitro. *Cell. Mol. Immunol.* **2**, 28–35.

Part III

In Vivo

Chapter 8

Genetic Tools for Analysis of FoxP3⁺ Regulatory T Cells In Vivo

Nadia M. Jeremiah and Adrian Liston

Abstract

The discovery of Foxp3 as a reliable marker for murine regulatory T cells has led to an explosion in the development of genetic tools for investigating the biology of regulatory T cells. More than 25 *Foxp3*-based mouse strains have been published with a variety of characteristics. The effects of Foxp3 expression can be analyzed using null, hypomorphic, conditional, altered control, and over-expression strains. Reporter strains are available to efficiently isolate Foxp3+ cells, with various reporter designs in terms of construct (fusion, replacement, and bicistronic positioning), and reporter system (GFP, YFP, RFP, Luciferase, Thy1.1). Multifunction strain fusion, replacement, and bicistronic positionings add functional proteins under the control of the Foxp3 promoter allowing induced apoptosis or lineage-specific Cre recombinase activity. In this chapter, we discuss the uses of the cornucopia of genetic tools, in isolation and in combination, for research on $Foxp3^+$ regulatory T cells.

Key words: Treg, In vivo, Foxp3, Transgenic, Knock-in, Knock-out, Cre-Lox

1. Introduction

The immunological research coming out of the second wave of investigation into suppressor T cells is due, in large part, to the identification of Foxp3 as a reliable marker for suppressor activity. The first mouse strain useful as a genetic tool for dissecting the function of Foxp3, the *Scurfy* mutant strain, has been available since 1959 (1); however, it was only with the identification of *Foxp3/FOXP3* mutations as the causative basis for *Scurfy* (2) and IPEX (3, 4) in 2001 that genetic tools were able to be made. The level of interest in $Foxp3^+$ regulatory T cells is such that 27 different mouse strains have been developed utilizing the Foxp3 promoter, providing a diverse set of investigatory tools. The chapters on methods throughout this book demonstrate the array of

George Kassiotis and Adrian Liston (eds.), *Regulatory T Cells: Methods and Protocols*, Methods in Molecular Biology, vol. 707, DOI 10.1007/978-1-61737-979-6_8, © Springer Science+Business Media, LLC 2011

experimental techniques feasible when these mouse strains are available. Here we provide an overview of published mouse strains and their uses.

2. Materials

2.1. Foxp3 Strains Derived from Mutation

Two Foxp3 strains have been derived from mutations, the *Scurfy* and *Crusty* mouse strains (Table 1). The *Scurfy* mutation is a spontaneous mutation caused by the insertion of two adenosine base-pairs in exon 8. The insertion leads to a frameshift in the Foxp3 mRNA transcript, resulting in a truncated Foxp3 protein without a c-terminal forkhead domain (2). The *Crusty* mutation is an ENU-induced mutation, caused by a T to A transversion in exon 12, resulting in the missense mutation I350N (5).

2.2. Foxp3 Strains Developed as Designer Alleles

Multiple mouse strains have been developed with designed manipulation of the *Foxp3* locus. These broadly fall into the following categories: loss-of-function alleles, reporter alleles, knock-in-knock-out alleles, altered control alleles, and added-function alleles (Table 2).

Three loss-of-function alleles have been developed. The *Foxp3*^{flox} allele is a conditional loss-of-function allele, with loxP sites flanking exons 1–5 (6). This allele has been developed into a knockout strain (*Foxp3KO*$^{tm1.1Ayr}$) with a deletion in exons 1–5 through germline Cre activity (6). The *Foxp3KO*^{tm1Tch} allele has been generated with a premature stop codon inserted into exon 8 (K276STOP) (7).

Five reporter alleles of Foxp3 have been developed. The *Foxp3eGFP*^{tm2Ayr} allele results in an eGFP-Foxp3 fusion protein, with normal Foxp3 function (8). By contrast, the *Foxp3eGFP*^{tm1Kuch} (9), *Foxp3eGFP*^{tm2Tch} (10), and *Foxp3eGFP*^{tm1Mal} (11) alleles include bicistronic expression of Foxp3 and eGFP, while the *Foxp3eRFP*^{tm1Flv} allele (12) has bicistronic expression of Foxp3 and eRFP.

Table 1
Mutation-derived genetic tools for analysis of Foxp3⁺ regulatory T cells

Mutation-derived	Allele construction	Background	MGI number
Scurfy	Spontaneous insertion (exon 8 frameshift)	129, C57BL/6, NOD, Balb/c	1857034
Crusty	*N*-ethyl-*N*-nitrosourea (ENU)-induced point mutation (I350N)	C57BL/6	3817855

Table 2
Designer Foxp3 alleles for analysis of Foxp3+ regulatory T cells

Designer alleles	Allele construction	Background	MGI number
Loss of function alleles			
Foxp3^{flox}	Insertion of single loxP sites flanking exons 1–5	C57BL/6	2654935
Foxp3KO$^{tm1.1Ayr}$	Deletion of exons 1–5	C57BL/6	2654936
Foxp3KO^{tm1Tch}	Nonsense mutation in exon 8 (K276STOP)	C57BL/6, Balb/c	3696705
Reporter-only alleles			
Foxp3eGFP^{tm2Ayr}	Insertion of eGFP in frame into the first coding exon, resulting in N-terminal GFP-Foxp3 fusion protein	C57BL/6	3574964
Foxp3eGFP^{tm1Mal}	Insertion of IRES-eGFP following the translational stop codon of *Foxp3*, resulting in bicistronic expression of Foxp3 and eGFP	C57BL/6	3773675
Foxp3eGFP^{tm1Kuch}	Insertion of IRES-eGFP following the translational stop codon of *Foxp3*, resulting in bicistronic expression of Foxp3 and eGFP	C57BL/6	3718527
Foxp3eGFP^{tm2Tch}	Insertion of IRES-eGFP following the translational stop codon of *Foxp3*, resulting in bicistronic expression of Foxp3 and eGFP	C57BL/6, Balb/c	3699400
Foxp3eRFP^{tm1Flv}	Insertion of IRES-eRFP following the translational stop codon of *Foxp3*, resulting in bicistronic expression of Foxp3 and eRFP	C57BL/6	3576270
Knock-in knock-out alleles			
Foxp3KIKO^{Ayr}	Insertion of eGFP in frame into the first coding exon, combined with stop codon/frameshift mutations before after eGFP and in the fifth exon of Foxp3. Results in the expression of GFP only	C57BL/6	Unregistered
Foxp3KIKO^{tm2Flv}	Insertion of IRES-luciferase-IRES-eGFP following the translational stop codon of *Foxp3*. *Foxp3* mRNA expression is destabilized in this construct. See Note 1	C57BL/6	3700150

(continued)

Table 2 (continued)

Designer alleles	Allele construction	Background	MGI number
Foxp3KIKO^{tm3Tch}	Insertion of eGFP in frame into codon 396 in exon 11 of Foxp3. Results in the expression of a truncated (non-functional) Foxp3-GFP fusion protein	Balb/c	3707723
Altered control alleles			
Foxp3$^{\Delta CNS1\text{-}GFP}$	Deletion of intronic region +2003-2707 and insertion of eGFP in frame into the first coding exon, resulting in a GFP-Foxp3 fusion protein	C57BL/6	Unregistered
Foxp3$^{\Delta CNS2\text{-}GFP}$	Deletion of intronic region +4262-4787 and insertion of eGFP in frame into the first coding exon, resulting in a GFP-Foxp3 fusion protein	C57BL/6	Unregistered
Foxp3$^{\Delta CNS3\text{-}GFP}$	Deletion of intronic region +6909-7103 and insertion of eGFP in frame into the first coding exon, resulting in a GFP-Foxp3 fusion protein	C57BL/6	Unregistered
Added function alleles			
Foxp3DTR^{tm3Ayr}	Insertion of IRES-DTR/GFP following the translational stop codon of *Foxp3*, resulting in the bicistronic expression of Foxp3 and DTR-GFP fusion protein. See Note 2	C57BL/6	3698131
Foxp3Cre$^{tm4(YFP/cre)Ayr}$	Insertion of IRES-YFP/Cre following the translational stop codon of *Foxp3*, resulting in the bicistronic expression of Foxp3 and YFP-Cre fusion protein	C57BL/6	3790499
Foxp3Cre$^{tm1(Cre)Saka)}$	Insertion of IRES-Cre following the translational stop codon of *Foxp3*. A minor reduction in Foxp3 protein results from the construct. See Note 3	Balb/c	3812203
Foxp3Thy1.1^{Ayr}	Insertion of IRES-iCaspase9-T2A-Thy1.1 following the translational stop codon of *Foxp3*. Results in the bicistronic expression of Foxp3 and a self-cleaving iCaspase9-T2A-Thy1.1 fusion protein. See Note 4	C57BL/6	Unregistered

DTR diphtheria toxin receptor; *eGFP* enhanced green fluorescent protein; *eRFP* enhanced red fluorescent protein; *IRES* internal ribosome entry site

Three knock-in-knock-out alleles are available. These alleles combine loss-of-function with a reporter construct, allowing the detection of cells with an active *Foxp3* locus but without Foxp3 function. The *Foxp3KIKO^{Ayr}* allele is a pure knockout of Foxp3 expression, with only eGFP expression resulting from the allele (13). The *Foxp3KIKO*^{tm3Tch} allele, by contrast, results in an eGFP-Foxp3 fusion protein, where the Foxp3 protein is rendered non-functional due to the deletion of a 33 amino acid C-terminal peptide that includes DNA- and transcription factor-binding and the nuclear localization sequence (7)). Functionally, this is equivalent to a pure knock-out, as the protein localizes to the cytosol and the pathology of *Foxp3KIKO*^{tm3Tch} males is indistinguishable from *Foxp3*^{KO} males (7)). The *Foxp3KIKO*^{tm2Flv} allele was not designed as a knock-in-knock-out allele, with tricistronic expression of functional Foxp3, luciferase, and eGFP (14). However, an unexpected consequence of allele design is instability in the *Foxp3* mRNA transcript, resulting in loss of functional Foxp3 over time and the development of immune pathology. This allele can be considered hypomorphic rather than a pure loss-of-function, as the delayed disease progress compared with *Foxp3*^{KO} males demonstrates partial function (14).

Three designer Foxp3 alleles are essentially altered versions of the *Foxp3eGFP*^{tm2Ayr} allele. These three alleles result in an eGFP-Foxp3 fusion protein, with normal Foxp3 function, but each allele has a deletion in a conserved noncoding region of the *Foxp3* gene, region +2003-2707 in *Foxp3*$^{\Delta CNS1\text{-}GFP}$, region +4262-4787 in *Foxp3*$^{\Delta CNS2\text{-}GFP}$, and region +6909-7103 in *Foxp3*$^{\Delta CNS3\text{-}GFP}$ (15).

Four designer *Foxp3* alleles can be classified as "added function" alleles. The *Foxp3Cre*$^{tm1(Cre)Saka}$ allele results in bicistronic Foxp3 and Cre expression allowing Cre-mediated activity in Foxp3 lineage cells. Allelic construction results in a minor drop in Foxp3 levels, but unlike the *Foxp3KIKO*^{tm2Flv} allele, this reduction does not have obvious functional consequences (16). The other three "added function" alleles combine functional proteins with a reporter. The *Foxp3Cre*$^{tm4(YFP/cre)Ayr}$ allele results in the bicistronic expression of functional Foxp3 and the YFP/Cre fusion protein, allowing both Cre-mediated activity and reporter activity (17). The *Foxp3DTR*^{tm3Ayr} allele results in the bicistronic expression of functional Foxp3 and the DTR/GFP fusion protein, making Foxp3⁺ cells detectable by eGFP reporter activity and also sensitive to diphtheria toxin-mediated apoptosis (18). Finally, the *Foxp3Thy1.1*^{Ayr} allele results in the bicistronic expression of functional Foxp3 and a self-cleaving iCaspase9-T2A-Thy1.1 fusion protein. This protein is cleaved into an inducible Caspase-9 protein and the cell surface reporter Thy1.1 (19).

2.3. Foxp3 Strains Developed as Transgenics

In addition to mutant and designer alleles of *Foxp3*, seven transgenic strains have been developed, which utilize *Foxp3* sequence (Table 3). Five of these strains utilize the Foxp3 promoter to drive functional products, while two strains are designed to drive the expression of the Foxp3 coding sequence.

Table 3
Transgenic tools for analysis of Foxp3⁺ regulatory T cells

Transgenic alleles	Allele construction	Background	MGI number
Tg(Foxp3-GFP)	BAC transgenic insertion of eGFP following the translational start codon of *Foxp3*, resulting in the expression of eGFP without the production of functional Foxp3	C57BL/6	Unregistered
Tg(Foxp3-DTR-GFP) Spa	BAC transgenic insertion of DTR-eGFP following the translational start codon of *Foxp3*, resulting in the expression of DTR-eGFP fusion protein without the production of functional Foxp3	C57BL/6, Balb/c	Unregistered
Tg(Foxp3-DTR-GFP) Doi	BAC transgenic insertion of DTR-eGFP following the translational start codon of *Foxp3*, resulting in the expression of DTR-eGFP fusion protein without the production of functional Foxp3	NOD	Unregistered
Tg(Foxp3-LuciDTR)	BAC transgenic insertion of eGFP-T2A-DTR-T2A-CBGr99 luciferase precursor protein following the translational start codon of *Foxp3*, resulting in a self-cleaving product producing eGFP, DTR, and luciferase as distinct protein products without the production of functional Foxp3	C57BL/6	Unregistered
Tg(Foxp3-EGFP/cre)1cJbs	BAC transgenic insertion of eGFP-IRES-hCre following the translational start codon of *Foxp3*, resulting in bicistronic expression of eGFP and hCre without the production of functional Foxp3	NOD	3809724
Tg(Foxp3-Foxp3)	BAC transgenic insertion of genomic *Foxp3* locus, resulting in functional Foxp3 driven from *Foxp3* promoter	C57BL/6	Unregistered
Tg(Lck-Foxp3)	Transgenic insertion of *Foxp3* cDNA driven by the distal *Lck* promoter	C57BL/6	Unregistered

BAC bacterial artificial chromosome; *DTR* diphtheria toxin receptor; *eGFP* enhanced green fluorescent protein; *hCre* humanized Cre recombinase; *IRES* internal ribosome entry site

The Tg(Foxp3-GFP) transgene is a BAC transgenic insertion of eGFP under the control of the Foxp3 promoter, resulting in transgenic eGFP (20). The Tg(Foxp3-DTR-GFP)Spa transgene is a BAC transgenic insertion of the DTR-eGFP fusion protein under the control of the Foxp3 promoter, resulting in transgenic diphtheria toxin-sensitivity to apoptosis and eGFP reporter activity (21). The Tg(Foxp3-DTR-GFP)Doi transgene is the same construct, independently generated (22). The Tg(Foxp3-EGFP/cre)1Jbs transgene is a BAC transgenic insertion of eGFP-IRES-hCre under the control of the Foxp3 promoter, resulting in bicistronic transgenic expression of both the eGFP reporter and Cre recombinase activity (23). The Tg(Foxp3-LuciDTR) transgene is a BAC transgenic insertion of eGFP-DTR-CBGr99 luciferase under the control of the Foxp3 promoter, resulting in the expression of a self-cleaving fusion protein with 2A peptide sequences resulting in three discrete protein products encoding eGFP, DTR, and luciferase (24). None of these transgenic constructs produce functional Foxp3, but equally none appear to have any effect on Foxp3 expression from the endogenous loci. Reported transcription from most of the transgenes is faithful to the endogenous loci (20, 21, 23). Transcription from the Tg(Foxp3-LuciDTR) transgene has reduced fidelity (see Note 5).

By contrast, two transgenes do drive the expression of functional Foxp3. The Tg(Foxp3-Foxp3) transgene drives the transgenic expression of Foxp3 under the native promoter, restoring functionality in Foxp3-deficient hosts (2). The Tg(Lck-Foxp3) transgene drives the transgenic expression of Foxp3 under the Lck promoter, resulting in super-physiological expression of Foxp3 in all T cells, capable of inhibiting disease in Foxp3-deficient hosts (25).

3. Methods

3.1. Use of Foxp3 Loss-of-Function Strains

There are eight characterized Foxp3 loss-of-function strains. Four strains demonstrate a simple knockout, the *Scurfy* and *Crusty* mutant strains and the *Foxp3KO*$^{tm1.1Ayr}$ and *Foxp3KO*^{tm1Tch} designer alleles. The *Foxp3KIKO*^{Ayr}, *Foxp3KIKO*^{tm3Tch}, and *Foxp3KIKO*^{tm2Flv} designer alleles combine *Foxp3* deficiency with a reporter construct, and in the case of *Foxp3KIKO*^{Ayr} and *Foxp3KIKO*^{tm3Tch} alleles can be considered equivalent to a knockout. By contrast, the final strain, *Foxp3*^{flox}, is a conditional loss-of-function, acting as a wildtype allele in the absence of Cre-recombinase activity.

The most obvious uses of the Foxp3-deficient strains are as a model for IPEX (2) and to determine the effect of the absence of regulatory T cells on the immune system (6). The *Foxp3*^{flox} line also allows the dissection of the function of Foxp3 in different cellular lineages, such as lineage-specific deletion in T cells and thymic epithelium (26) or induced deletion by exposure to soluble Cre (27). *Foxp3*-deficient strains are also highly useful as a tool for in vivo regulatory T cell assays, as functional regulatory T cells have the capacity to rescue immune pathology in *Foxp3*-deficient pups or in T cell-deficient hosts after transfer of *Foxp3*-deficient T cells (6).

In addition to the obvious uses, innovative use of *Foxp3*-deficient strains allows the replication of characteristic uses of more sophisticated strains. For example, $\textit{Foxp3}^{Cre}$ alleles can be used to determine the impact of lineage-specific deletion of particular genes (see Subheading 3.3). However, a similar experiment can be performed using Foxp3-deficient mice in a mixed bone-marrow chimera situation. In a 50%:50% mixed bone-marrow chimera between *Foxp3*-deficient bone-marrow and *gene A* knockout bone-marrow, 100% of $Foxp3^+$ regulatory T cells will be derived from the *gene A* knock-out bone-marrow, while all nonregulatory lineages will show a 50:50% distribution. This allows a pseudo-"Foxp3-specific" knockout of *gene A*, as other bone-marrow-derived populations are able to exhibit *trans*-compensation for the knockout. A mixture of CTLA-4^{KO} and $Foxp3^{KO}$ bone marrow has been used to test the significance of CTLA-4 expression in $Foxp3^+$ cells in preventing autoimmune lymphoproliferation (28). In a similar way, *Rag*-deficient and *Igμ*-deficient bone-marrow has been used in mixed bone-marrow chimeras to test B cell-intrinsic effects of gene knockouts (29), and *TCRα*-deficient bone-marrow has been used in mixed bone-marrow chimeras to test T cell-intrinsic effects of gene knockouts (30).

Another example of the innovative use of *Foxp3*-deficient mice is to generate a system analogous to the DTR-mediated deletion of $Foxp3^+$ T cells (see Subheading 3.3). By generating a 50:50% mixed chimera with *Thy1.1* $\textit{Foxp3}^{wt}$ and *Thy1.2* $\textit{Foxp3}^{KO}$ bone-marrow, all $Foxp3^+$ T cells are forced to be derived from the Thy1.1 bone-marrow. This allows the $Foxp3^+$ T cell population to be deleted through the injection of anti-Thy1.1 antibody. By contrast, other lineages experience only a 50% reduction in population size (31).

3.2. Use of Foxp3 Reporter Strains

Despite the obvious benefit in using Foxp3 as a marker for regulatory T cells, it has one considerable disadvantage – it is an intracellular protein. Direct detection of Foxp3, therefore, requires intracellular straining, thereby preventing any functional analysis in vivo or in vitro. Therefore, functional analysis relies on the use of proxy marker expression, such as CD25, or the use of reporter constructs. The plethora of Foxp3 reporter strains generated since 2005 demonstrate the utility of this approach.

Sixteen Foxp3 reporter strains have been published, 12 of which use GFP, 1 uses YFP, 1 uses RFP, and 1 uses the nonfluorescent Thy1.1 reporter. The most common construct is a bicistronic GFP reporter or Foxp3-GPF transgenic reporter, with four strains. Three strains use the Foxp3-GFP reporter, with altered control regions, allowing the role of these conserved control regions to be dissected (15). Another five strains use the same reporter system but with additional functional constructs (Cre, DTR, or Luciferase). The fidelity of reporter expression appears high for each strain (except Tg(Foxp3-LuciDTR), see Note 5), yet few direct comparisons have been performed, and it is therefore not possible to indicate a particular reporter construct as being superior.

The remaining four Foxp3 reporter strains each have a unique feature worth noting. The *Foxp3eGFP*^{tm2Ayr} reporter allele is the only Foxp3-reporter fusion protein, and therefore the only allele to allow the detection of intracellular localization (8). The *Foxp3Cre*$^{tm4(YFP/cre)Ayr}$ and *Foxp3eRFP*^{tm1Flv} reporter alleles are notable for providing fluorescent reporters in non-GFP channels, with YFP and RFP expression, respectively (8, 12). Finally, the *Foxp3Thy1.1*^{Ayr} reporter allele is unique in having a nonfluorescent membrane-bound reporter system, the Thy1.1 antigen. The advantage of this strain is that anti-Thy1.1 antibody-mediated selection allows the purification of Foxp3 strains in multiple colors or by magnetic enrichment, which makes it highly useful for enriching from low frequency populations or in intercrossing strains with preexisting fluorescent reporters (19).

3.3. Use of Foxp3 Added-Function Alleles

Eight Foxp3 mouse strains can be classified as "added-function." These alleles and transgenes include the expression of function proteins, such as Cre recombinase, DTR, iCaspase9, and Luciferase.

There are three Foxp3 strains expressing Cre recombinase. *Foxp3Cre*$^{tm4(YFP/cre)Ayr}$ mice express both Foxp3 and YFP-Cre fusion protein from the designed allele, acting as both a reporter and Cre recombinase. *Foxp3Cre*$^{tm1(Cre)Saka}$ mice express both Foxp3 and Cre from the designed allele. Tg(Foxp3-EGFP/cre)1cJbs mice express GFP and Cre in a bicistronic fashion from a BAC transgene, with Foxp3 being provided from the endogenous loci. The main use of Foxp3Cre constructs is to drive the lineage-specific excision of floxed genes to allow functional analysis of these genes within Foxp3+ cells, such as those that has been done successfully with Dicer (23, 32, 33), IL-10 (17), and CTLA4 (16). Another use of Foxp3Cre strains is for fate-mapping. This is performed by crossing a Cre-expressing line to a recombinase activated reporter, such as the iYFP construct, which includes YFP under the control of a constitutive promoter, silenced by a floxed stop sequence. Any cells that express Cre then become permanently YFP+, regardless of the continued expression of Cre, a strategy that can be exploited to determine the stability of Foxp3+ cells under specific contexts (23).

Another four Foxp3 strains express added-function constructs with a different purpose, that of allowing depletion of Foxp3+ T cells. In the absence of genetic tools, the best in vivo depletion of Foxp3⁺ T cells is achieved via anti-CD25 antibodies. However, these antibodies have major short-comings: (1) CD25 is an incomplete marker for Foxp3⁺ T cells and CD25⁺ cells can survive anti-CD25 treatment by epitope-shedding, reducing the efficacy of treatment (34); and (2) CD25 is also expressed on activated T cells, resulting in the depletion of nonregulatory subsets (35). The *Foxp3DTR*^{tm3Ayr} strain, two Tg(Foxp3-DTR-GFP) strains and *Foxp3Thy1.1*^{Ayr} strain all present an alternative method of deletion of Foxp3⁺ T cells. *Foxp3DTR*^{tm3Ayr} and Tg(Foxp3-DTR-GFP) mice express DTR on all Foxp3⁺ T cells, allowing the lineage-specific killing of Foxp3⁺ T cells through injection of diphtheria toxin (18, 21). Diphtheria toxin is highly toxic, with a single molecule capable of killing a cell, and thus injection of rate-limiting amounts of diphtheria toxin allows the partial deletion of Foxp3⁺ T cells (18). The second system of induced apoptosis, present in the *Foxp3Thy1.1*^{Ayr} construct is using a modified version of caspase-9, which is inducible by a small molecule agonist (see Note 4). It is worth noting that the inducible-apoptotic capacity can be generated by using the Foxp3Cre strains, if they are crossed to a Cre-activated inducible-apoptosis allele (e.g., iDTR (36)).

A fifth Foxp3 strain warrants separate attention, for the unique combination of functions added. The Tg(Foxp3-LuciDTR) transgenic combines not only eGFP and DTR expression, acting as a fluorescent reporter and a depletable system, but also a luciferase reporter. The luciferase reporter allows whole body imaging of Foxp3⁺ T cell localization, using luciferin (24). A caveat of this transgene is incomplete fidelity of the reporter (see Note 5).

3.4. Combinatorial Use of Foxp3 Alleles

In addition to the multiple uses of the 27 genetic tools described earlier, there are emergent uses that occur through the combinatorial use of different alleles. One important consideration is that Foxp3 is located on the X chromosome; therefore, because of X chromosome inactivation in females, two populations of Foxp3⁺ cells exist, each using one allele exclusively.

Random X chromosome inactivation provides multiple opportunities when using genetic tools based on the endogenous *Foxp3* allele (mutant alleles and designer alleles). For example, in *Foxp3*$^{wt/KIKO}$ females, half the Foxp3⁺ T cells are wildtype and half are GFP⁺ but *Foxp3*-deficient, allowing the analysis of Foxp3-deficient T cells in the healthy context (13). If combined with the *Foxp3Thy1.1*^{Ayr} allele in a *Foxp3*$^{Thy1.1/KIKO}$ heterozygous female, both populations could be sorted based on the Foxp3 allele. Likewise in *Foxp3*$^{YFPCre/wt}$ heterozygous females, half the Foxp3⁺ T cells are wildtype and half are YFP⁺ but have Cre-recombinase activity. When YFP⁺ cells from *Foxp3*$^{YFPCre/wt}$ females and *Foxp3*^{Cre} males are compared, the effect of the Cre-dependent gene deletion can be compared both in the presence and absence of normal Foxp3+ regulatory T cells (32). Again with *Foxp3*$^{DTR/wt}$ females, two populations of Foxp3+ T cells exist, one GFP+ and diphtheria-sensitive, and one GFP– and diphtheria-resistant. In contrast to *Foxp3*^{DTR} males, where diphtheria-treatment results in fatal autoimmunity, diphtheria-treatment of heterozygous females only depletes half the regulatory T cell population, allowing studies on regulatory T cell regeneration in the context of a healthy mouse (19).

The transgenic constructs available, by contrast, are not located on the X chromosome and are therefore expressed by all Foxp3+ T cells. This prevents the strategies of allele combination above, but provides different opportunities. For example, most of the transgenic constructs listed earlier do not drive expression of Foxp3 from the transgene. When the transgenes are coupled with a wildtype *Foxp3* allele, the function of the transgene is coupled to Foxp3+ T cells (e.g., reporter expression, Cre-activity, diphtheria-sensitivity). However, equally the transgenes can be combined with a *Foxp3*^{KO} allele, where the function of the transgene is now coupled to those cells that attempt to express the *Foxp3* allele but do not gain functional Foxp3 protein. Thus, studies performed with *Foxp3*^{KIKO} mice, where a Foxp3 reporter is required in Foxp3-deficient cells, can equally be performed using *Foxp3*^{KO} Tg(Foxp3-GFP) mice (20). In both cases, Foxp3-deficient cells will be labeled with GFP, one from the endogenous allele and one from the transgenic allele. Alternative combinations of the Tg(Foxp3-EGFP/cre)1cJbs or Tg(Foxp3-DTR-GFP) transgenes with a *Foxp3*^{KO} allele would make *Foxp3*-deficient cells active for Cre-recombinase or DTR, respectively (37). The reciprocal of the *Foxp3*^{KO} Tg(Foxp3-GFP) cross would be a *Foxp3*^{KIKO} Tg(Foxp3-Foxp3) cross, where a *Foxp3*-deficient reporter allele would effectively be turned into a normal *Foxp3* reporter allele by transgenic complementation of *Foxp3*.

The combination of both endogenous Foxp3 alleles and transgenic alleles allows enormous diversity in potential experiments while using a limited subset of strains. The combinatorial possibilities are too numerous to list. As a single example, *Foxp3*$^{KO/Thy1.1}$ Tg(Foxp3-EGFP/cre)1cJbs heterozygous females would have two populations of "regulatory" T cells. One population would have an active Foxp3 locus that fails to produce functional Foxp3, and would be GFP+ and have active Cre-recombinase. The other population would express functional Foxp3, would be GFP+Thy1.1+, and would have active Cre-recombinase. This combinatorial strain would allow the purification and functional testing of both regulatory (GFP+Thy1.1+) and failed regulatory T cells (GFP+Thy1.1−) after Cre-mediated excision of the gene of interest. Modified arrangements would obviously be suitable for alternative experimental questions.

4. Notes

1. Thymic expression of Foxp3 is intact with no mutations in the coding sequence. However, *Foxp3* mRNA expression level is unstable, resulting in similar immune manifestations to *Foxp3KO* strains. Unlike *Foxp3KO* or *Scurfy* strains, disease is not fatal until 3 months, indicating delayed immunopathology (14).
2. The fluorescence of GFP in *Foxp3DTR*^{tm3Ayr} is much weaker than that of the *Foxp3eGFP*^{tm2Ayr} construct.
3. This construct results in a minor decrease in Foxp3 protein within Foxp3^{+} T cells; however, the cell type is stable and no pathology results (16).
4. The iCaspase9-T2A-Thy1.1 fusion protein self-cleaves at the T2A peptide, resulting in iCaspase9 and Thy1.1 (19). The iCaspase9 protein is a fusion of caspase-9 to a mutated FKBP12 domain, to allow the induction of caspase-9 activity by the cell-permeable compound AP20187 (38). For iCaspase9 in Foxp3^{+} T cells, efficacy of deletion results have not been published.
5. Three founder lines for the Tg(Foxp3-LuciDTR) BAC transgenic have been analyzed for fidelity to the endogenous locus. Tg(Foxp3-LuciDTR)3 exhibits ~65–75% fidelity, Tg(Foxp3-LuciDTR)4 exhibits ~90–95% fidelity, and Tg(Foxp3-LuciDTR)3 exhibits >95% fidelity, as measured by the percentage of Foxp3^{+} cells surviving DT-mediated deletion (24).

References

1. Russell WL, Russell LB, Gower JS. (1959) Exceptional inheritance of a sex-linked gene in the mouse explained on the basis that the X/O sex-chromosome constitution is female. *Proc Natl Acad Sci U S A*; **45**: 554–60.
2. Brunkow ME, Jeffery EW, Hjerrild KA, Paeper B, Clark LB, Yasayko SA *et al.* (2001) Disruption of a new forkhead/winged-helix protein, scurfin, results in the fatal lymphoproliferative disorder of the scurfy mouse. *Nat Genet*; **27**: 68–73.
3. Bennett CL, Christie J, Ramsdell F, Brunkow ME, Ferguson PJ, Whitesell L *et al.* (2001) The immune dysregulation, polyendocrinopathy, enteropathy, X-linked syndrome (IPEX) is caused by mutations of FOXP3. *Nat Genet*; **27**: 20–1.
4. Wildin RS, Ramsdell F, Peake J, Faravelli F, Casanova JL, Buist N *et al.* (2001) X-linked neonatal diabetes mellitus, enteropathy and endocrinopathy syndrome is the human equivalent of mouse scurfy. *Nat Genet*; **27**: 18–20.
5. Pirie E, Beutler B, Mutagenetix. (2008) Record for "crusty", updated November 14, 2008 J:141212. *MGI Direct Data Submission*.
6. Fontenot JD, Gavin MA, Rudensky AY. (2003) Foxp3 programs the development and function of CD4+CD25+ regulatory T cells. *Nat Immunol*; **4**: 330–6.
7. Lin W, Truong N, Grossman WJ, Haribhai D, Williams CB, Wang J *et al.* (2005) Allergic dysregulation and hyperimmunoglobulinemia E in Foxp3 mutant mice. *J Allergy Clin Immunol*; **116**: 1106–15.
8. Fontenot JD, Rasmussen JP, Williams LM, Dooley JL, Farr AG, Rudensky AY. (2005)

Regulatory T cell lineage specification by the forkhead transcription factor foxp3. *Immunity*; **22**: 329–41.

9. Bettelli E, Carrier Y, Gao W, Korn T, Strom TB, Oukka M *et al.* (2006) Reciprocal developmental pathways for the generation of pathogenic effector TH17 and regulatory T cells. *Nature*; **441**: 235–8.
10. Haribhai D, Lin W, Relland LM, Truong N, Williams CB, Chatila TA. (2007) Regulatory T cells dynamically control the primary immune response to foreign antigen. *J Immunol*; **178**: 2961–72.
11. Wang Y, Kissenpfennig A, Mingueneau M, Richelme S, Perrin P, Chevrier S *et al.* (2008) Th2 lymphoproliferative disorder of LatY136F mutant mice unfolds independently of TCR-MHC engagement and is insensitive to the action of Foxp3+ regulatory T Cells. *J Immunol*; **180**: 1565–75.
12. Wan YY, Flavell RA. (2005) Identifying Foxp3-expressing suppressor T cells with a bicistronic reporter. *Proc Natl Acad Sci U S A*; **102**: 5126–31.
13. Gavin MA, Rasmussen JP, Fontenot JD, Vasta V, Manganiello VC, Beavo JA *et al.* (2007) Foxp3-dependent programme of regulatory T-cell differentiation. *Nature*; **445**: 771–5.
14. Wan YY, Flavell RA. (2007) Regulatory T-cell functions are subverted and converted owing to attenuated Foxp3 expression. *Nature*; **445**: 766–70.
15. Zheng Y, Josefowicz S, Chaudhry A, Peng XP, Forbush K, Rudensky AY. (2010) Role of conserved non-coding DNA elements in the Foxp3 gene in regulatory T-cell fate. *Nature*; **463**: 808–12.
16. Wing K, Onishi Y, Prieto-Martin P, Yamaguchi T, Miyara M, Fehervari Z *et al.* (2008) CTLA-4 control over Foxp3+ regulatory T cell function. *Science*; **322**: 271–5.
17. Rubtsov YP, Rasmussen JP, Chi EY, Fontenot J, Castelli L, Ye X *et al.* (2008) Regulatory T cell-derived interleukin-10 limits inflammation at environmental interfaces. *Immunity*; **28**: 546–58.
18. Kim JM, Rasmussen JP, Rudensky AY. (2007) Regulatory T cells prevent catastrophic autoimmunity throughout the lifespan of mice. *Nat Immunol*; **8**: 191–7.
19. Liston A, Nutsch KM, Farr AG, Lund JM, Rasmussen JP, Koni PA *et al.* (2008) Differentiation of regulatory Foxp3+ T cells in the thymic cortex. *Proc Natl Acad Sci U S A*; **105**: 11903–8.
20. Kuczma M, Podolsky R, Garge N, Daniely D, Pacholczyk R, Ignatowicz L *et al.* (2009) Foxp3-deficient regulatory T cells do not revert into conventional effector CD4+ T cells but constitute a unique cell subset. *J Immunol*; **183**: 3731–41.
21. Lahl K, Loddenkemper C, Drouin C, Freyer J, Arnason J, Eberl G *et al.* (2007) Selective depletion of Foxp3+ regulatory T cells induces a scurfy-like disease. *J Exp Med*; **204**: 57–63.
22. Feuerer M, Shen Y, Littman DR, Benoist C, Mathis D. (2009) How punctual ablation of regulatory T cells unleashes an autoimmune lesion within the pancreatic islets. *Immunity*; **31**: 654–64.
23. Zhou X, Jeker LT, Fife BT, Zhu S, Anderson MS, McManus MT *et al.* (2008) Selective miRNA disruption in T reg cells leads to uncontrolled autoimmunity. *J Exp Med*; **205**: 1983–91.
24. Suffner J, Hochweller K, Kuhnle MC, Li X, Kroczek RA, Garbi N *et al.* (2010) Dendritic cells support homeostatic expansion of Foxp3+ regulatory T cells in Foxp3 LuciDTR mice. *J Immunol*; **184**: 1810–20.
25. Guo L, Tian J, Marinova E, Zheng B, Han S. (2010) Inhibition of clonal expansion by Foxp3 expression as a mechanism of controlled T-cell responses and autoimmune disease. *Eur J Immunol*; **40**: 71–80.
26. Liston A, Farr AG, Chen Z, Benoist C, Mathis D, Manley NR *et al.* (2007) Lack of Foxp3 function and expression in the thymic epithelium. *J Exp Med*; **204**: 475–80.
27. Williams LM, Rudensky AY. (2007) Maintenance of the Foxp3-dependent developmental program in mature regulatory T cells requires continued expression of Foxp3. *Nat Immunol*; **8**: 277–84.
28. Chikuma S, Bluestone JA. (2007) Expression of CTLA-4 and FOXP3 in cis protects from lethal lymphoproliferative disease. *Eur J Immunol*; **37**: 1285–9.
29. Schmidt KN, Hsu CW, Griffin CT, Goodnow CC, Cyster JG. (1998) Spontaneous follicular exclusion of SHP1-deficient B cells is conditional on the presence of competitor wild type B cells. *J Exp Med*; **187**: 929–37.
30. Almeida AR, Legrand N, Papiernik M, Freitas AA. (2002) Homeostasis of peripheral CD4+ T cells: IL-2R alpha and IL-2 shape a population of regulatory cells that controls CD4+ T cell numbers. *J Immunol*; **169**: 4850–60.
31. Scott-Browne JP, Shafiani S, Tucker-Heard G, Ishida-Tsubota K, Fontenot JD, Rudensky AY *et al.* (2007) Expansion and function of Foxp3-expressing T regulatory cells during tuberculosis. *J Exp Med*; **204**: 2159–69.

32. Liston A, Lu LF, O'Carroll D, Tarakhovsky A, Rudensky AY. (2008) Dicer-dependent microRNA pathway safeguards regulatory T cell function. *J Exp Med*; **205**: 1993–2004.
33. Chong MM, Rasmussen JP, Rudensky AY, Littman DR. (2008) The RNAseIII enzyme Drosha is critical in T cells for preventing lethal inflammatory disease. *J Exp Med*; **205**: 2005–17.
34. Kohm AP, McMahon JS, Podojil JR, Begolka WS, DeGutes M, Kasprowicz DJ *et al.* (2006) Cutting edge: anti-CD25 monoclonal antibody injection results in the functional inactivation, not depletion, of CD4+CD25+ T regulatory cells. *J Immunol*; **176**: 3301–5.
35. Couper KN, Blount DG, de Souza JB, Suffia I, Belkaid Y, Riley EM. (2007) Incomplete depletion and rapid regeneration of Foxp3+ regulatory T cells following anti-CD25 treatment in malaria-infected mice. *J Immunol*; **178**: 4136–46.
36. Buch T, Heppner FL, Tertilt C, Heinen TJ, Kremer M, Wunderlich FT *et al.* (2005) A Cre-inducible diphtheria toxin receptor mediates cell lineage ablation after toxin administration. *Nat Methods*; **2**: 419–26.
37. Lahl K, Mayer CT, Bopp T, Huehn J, Loddenkemper C, Eberl G *et al.* (2009) Nonfunctional regulatory T cells and defective control of Th2 cytokine production in natural scurfy mutant mice. *J Immunol*; **183**: 5662–72.
38. Nor JE, Hu Y, Song W, Spencer DM, Nunez G. (2002) Ablation of microvessels in vivo upon dimerization of iCaspase-9. *Gene Ther*; **9**: 444–51.

Chapter 9

In Vivo Treg Suppression Assays

Creg J. Workman, Lauren W. Collison, Maria Bettini, Meenu R. Pillai, Jerold E. Rehg, and Dario A.A. Vignali

Abstract

To fully examine the functionality of a regulatory T cell (T_{reg}) population, one needs to assess their ability to suppress in a variety of in vivo models. We describe five in vivo models that examine the suppressive capacity of T_{regs} upon different target cell types. The advantages and disadvantages of each model including resources, time, and technical expertise required to execute each model are also described.

Key words: T_{reg}, Homeostasis, IBD, Experimental colitis, EAE, Tumor, B16 melanoma, In vivo, Foxp3

1. Introduction

The suppressive activity of regulatory T cells (T_{regs}) is most conveniently assessed using standard in vitro T_{reg} assays (see Chapter 2). Although performing these assays is an important step in deciphering the function of a regulatory population, in vitro culture conditions cannot replicate the complex in vivo microenvironment. Consequently, assessing T_{reg} function in vivo is more physiologically relevant. Indeed, in vivo assays provide a more significant regulatory challenge for T_{regs} than in vitro assays. For instance, IL10-deficient T_{regs} are fully functional in vitro but defective in a variety of in vivo models (1–3). Despite the importance of in vivo assays to assess T_{reg} function, they are clearly more technically challenging as they tend to require time to complete, more resources, and often more T_{regs} than in vitro assays. However, in vivo T_{reg} suppression assays represent an important tool in assessing the function of this critical immune population.

George Kassiotis and Adrian Liston (eds.), *Regulatory T Cells: Methods and Protocols*, Methods in Molecular Biology, vol. 707,
DOI 10.1007/978-1-61737-979-6_9, © Springer Science+Business Media, LLC 2011

Here, we describe five different in vivo models that assess T_{reg} function: (1) homeostasis model, (2) inflammatory bowel disease (IBD) recovery model, (3) experimental autoimmune encephalomyelitis (EAE) model, (4) B16 melanoma model, and (5) *Foxp3*⁻ rescue model. These models are very effective at elucidating T_{reg} function while only requiring between 0.5 and 1×10^6 T_{regs} per mouse. The requirements and pros and cons of the five models are illustrated in Table 1. We would recommend the use of at least three in vivo models to assess the regulatory activity of a test population, although additional models would clearly provide a more detailed examination. It should be noted that this is not intended to be an exhaustive list, but rather a collection of methods that have been frequently used to assess T_{reg} function in vivo. Other models have been described, but many are less well-characterized (4–6).

2. Materials

2.1. Common to all Protocols

1. All of the models require mice for donor T cell populations as well as *Rag1*$^{-/-}$ or *Foxp3*$^{-}$ mice for recipients. The number of mice required differs depending upon the model, the number of experimental groups, and the number of replicate experiments.
2. Blocking solution: 10% sterile mouse serum in PBS + 5% FBS.
3. Murine cell culture medium: RPMI [Mediatech] supplemented with 10% FBS [optimal manufacturer and lot to be determined empirically], 2 mM l-glutamine [Mediatech], 1 mM Sodium Pyruvate [Mediatech], 100 mM Non-Essential Amino Acids [Mediatech], 5 mM HEPES free acid [Mediatech], 10 ml of 5.5×10^{-2} 2-mercaptoethanol [Invitrogen], and 100 U/ml Penicillin/Streptomycin [Mediatech] (see Note 1).
4. Gey's solution for red blood cell lysis: 12 mM potassium bicarbonate ($KHCO_3$), 156 mM ammonium chloride (NH_4Cl), diluted in water. Filter sterilize the solution through a 0.2-μm filter.
5. V-bottom 96-well tissue culture plate [Nunc].
6. 70 μM nylon cell strainer [Beckton Dickinson].
7. 50 ml conical tubes.
8. 15 ml conical tubes.
9. Sterile normal mouse serum [Gibco].
10. Phosphate buffered saline (PBS) [Mediatech].

Table 1
Overview of five in vivo T_{reg} suppression models

Model	Target cells[a]	Resources required[b]	Time to results[c]	Time requirements[d]	Technical procedures[e]	Technical complexity[f]
Homeostasis	Naïve homeostatically expanding $CD4^+$ T cells	Minimal	7 days	Minimal	i.v. Injections	Simple
IBD Recovery	Th1 T cells (Th17)	Moderate/substantial (large number of recipient mice and significant access to sort facilities on demand)	56 days	Weekly monitoring and weighing Frequent monitoring upon sickness Detailed analysis and histology	i.v. Injections i.p. Injections Optional mucosal analysis Histological analysis	Moderate
EAE	Th17 and Th1 T cells	Moderate (model-specific reagents including peptides, adjuvants, and toxins)	30 days	6 injections Daily monitoring	Emulsions s.c. Injections i.p. Injections	Moderate
B16 Melanoma	$CD8^+$ T cells	Substantial (large number of mice, significant amount of sorting and many model-specific reagents including inoculation and surgical reagents)	1° tumor: 15–20 days 2° tumor: additional 15–20 days	Daily monitoring of tumors Multiple injections Multihour surgery	i.v. Injections i.d. Injections Measurement of tumors Surgical resection of tumors Isolation of tumor infiltrating lymphocytes	Difficult

(continued)

Table 1 (continued)

Model	Target cells[a]	Resources required[b]	Time to results[c]	Time requirements[d]	Technical procedures[e]	Technical complexity[f]
Foxp3⁻rescue	Primarily lymphocytes	Moderate (large number of *Foxp3*⁻ breeders, moderate number of donor mice and access to sort facilities on demand)	25–30 days	Timed/monitored pregnancies Long sorts Difficult injections Time consuming analysis	Marking/ genotyping 1-day-old pups i.p. Injections into 2-day-old pups Histological analysis	Moderate

[a] The cell populations that are primarily suppressed by T_{regs} in the model listed
[b] An indication of the amount of mice, sort time, and materials required for an average 3 group experiment as described in the methods
[c] Time required to complete one experiment starting from the initial injections of the mice. Time required for analysis is not included and will be in addition to time noted
[d] Stages in the protocols that may be time demanding
[e] Procedures that are required in the protocol that may require some level of training depending upon the investigator's level of expertise. This does not include sorting and flow cytometry, which are required techniques in all of the models
[f] The overall level of complexity for each protocol, taking into consideration time, resources, and techniques required

11. Hanks balanced salt solution (HBSS) [Mediatech].
12. Sterile 1 ml syringes, use plunger for homogenization [Beckton Dickinson].
13. Sterile 3 ml syringes [Beckton Dickinson].
14. Sterile 27G needles [Beckton Dickinson].
15. Fluorescently tagged antibodies (CD4, CD25, CD45RB, Thy1.1, Thy1.2, Foxp3).
16. 40 μM nylon cell strainer [Beckton Dickinson].
17. Fluorescent activated cell sorter (FACS) buffer: PBS + 0.05% NaN_3 + 5% FBS.
18. Trypan Blue.
19. Scissors and forceps suitable for tissue collection.
20. 24-well cell culture plate [Corning].

2.2. IBD Model

1. Sterile 23G needles [Beckton Dickinson].
2. Sterile 10 ml syringes [Beckton Dickinson].
3. Digital weighing scale.
4. Plastic container such as a pipette tip box lid (Not absolutely required but useful as a reference for accurate weight measurement and also used to place the mouse while weighing).
5. Tissue cassettes for histology [ThermoFisher Scientific].
6. 10% Neutral buffered formalin solution [ThermoFisher Scientific].

2.3. EAE Model

1. Incomplete Freund's adjuvant (IFA) [ThermoFisher Scientific].
2. *Mycobacterium tuberculosis* H37Ra (killed and desiccated) [ThermoFisher Scientific] (see Note 2).
3. Solution of MOG35-55 (MEVGWYRSPFSRVVHLYRNGK) peptide diluted to 1 mg/ml in PBS.
4. *Bordetella pertussis* toxin, diluted to 1 μg/ml in PBS [ThermoFisher Scientific] (see Note 3).
5. Two 2-ml glass Hamilton syringes with double-ended locking hub (Luer-lock) connector or 3-way stopcock [ThermoFisher Scientific].
6. Sterile 1 ml tuberculin slip tip syringes [Beckton Dickinson].
7. Sterile 25G needles [Beckton Dickinson].
8. Isofluorane anesthesia apparatus (optional).
9. Mouse ear clipper.
10. Frosted glass tissue homogenizer [ThermoFisher Scientific].
11. Percoll [Amersham Bioscience].

2.4. B16 Melanoma Model

1. Sterile blunt needles.
2. Sterile 30G needles [Beckton Dickinson].
3. B16 culture media: RPMI [Mediatech] supplemented with 7.5% FBS [optimal manufacturer and lot to be determined empirically], 2 mM l-glutamine [Mediatech], 100 mM Non-Essential Amino Acids [Mediatech], and 100 U/ml Penicillin/Streptomycin [Mediatech].
4. T175 flasks [ThermoFisher Scientific].
5. Trypsin-EDTA [Mediatech].
6. Isofluorane anesthesia apparatus.
7. Heating pad or heat lamp.
8. Dial caliper [Bel-Art Products].
9. RPMI media without any additives [Mediatech].
10. 2 ml cryo vials [Nunc].
11. Small electric razor [Oster].
12. Q-tips.
13. Surgical providone iodine solution [Applicare Inc.].
14. Single use alcohol pads [ThermoFisher Scientific].
15. Blunt forceps [ThermoFisher Scientific].
16. Surgical scissors [Roboz].
17. Neosporin triple antibiotic ointment [ThermoFisher Scientific].
18. Buprenorphine or Rimadyl [must be obtained through a pharmacy].
19. Steel wound clips and Autoclip wound clip applicator [Beckton Dickinson].
20. Autoclip wound clip remover [Beckton Dickinson].
21. Percoll [Amersham Bioscience].
22. 5% H_2O_2 in PBS.

2.5. Foxp3⁻ Rescue Model

1. Insulin syringe fitted with a 30-G needle [Beckton Dickinson].
2. Camera.
3. Ruler or other scale bar.
4. Soft tissue organ cassettes [ThermoFisher Scientific].
5. 24-well cell culture plate [Corning].
6. Tissue cassettes for histology [ThermoFisher Scientific].
7. 10% Neutral buffered formalin solution [ThermoFisher Scientific].

3. Methods

3.1. Purification of Mouse T_{conv}/T_{reg} for In Vivo T_{reg} Suppression Assays

Mouse T_{conv} and T_{reg} can be separated using fluorescently conjugated antibodies, based on their expression of cell surface proteins. Mouse T_{conv} and T_{reg} can be separated using only CD4 and CD25 markers. However, by also staining with CD45RB, naïve T_{conv} can be separated from memory T_{conv} and T_{reg}, resulting in better purity of both populations. A similar strategy can be utilized by staining cells with CD44 and CD62L, where CD44^{low}/CD62L^{high} populations represent the naïve, T_{conv} cells. To maximize purity and recovery, one would ideally utilize a Foxp3 reporter strain, such as *Foxp3*^{GFP} (7), crossed with the mutant strain of interest.

Fluorescence activated cell sorting (FACS) is the preferred method of cell purification because of the purity of cell populations obtained. Greater than 95% purity can routinely be obtained by FACS. If FACS is not possible or available, an alternative method of purification utilizes antibodies coupled with magnetic or paramagnetic particles for cell sorting. Cells should be prepared using the manufacturer's guidelines (e.g., MACS -http://www.miltenyibiotec.com/en/NN_21_MACS_Cell_Separation.aspx, Dynabeads -http://tools.invitrogen.com/content/sfs/manuals/114%2063D.Dynabeads%20FlowComp%20Mouse%20CD4±CD25±Treg%20Cells(rev001).pdf). Under optimal conditions, one can obtain purities of 85–90% by MACS. If an induced regulatory population is being assessed, methods appropriate for their generation and purification should be used. These methods are also detailed in the companion Chapter 2. Additionally, it is advisable to enrich for T cells prior to sorting to reduce the amount of sorting time required. T cell enrichment can be done by removing the B cells by a standard panning protocol, Dynabeads or by MACS (see Note 4). Regardless of the purification method used, it is imperative that the purity of all sorted populations are confirmed by flow cytometry prior to commencing in vivo assays.

1. Harvest spleen and lymph nodes from mice.
2. Tease apart tissue with the plunger from a 1 ml syringe through a 70-μm cell strainer into a 50-ml conical tube. Rinse strainer twice with HBSS to recover all cells. Alternatively, splenocytes may be teased apart between two frosted glass microscope slides.
3. Centrifuge homogenate at 300 × *g* (1200 rpm) for 10 min.
4. Resuspend homogenate in 1 ml Gey's solution per spleen. Gently swirl for 2 min and then quench reaction by adding 12 ml of HBSS.
5. Centrifuge at 300 × *g* for 10 min (see Note 4).

6. Resuspend cells in blocking solution at 0.5 ml per spleen.
7. Incubate cells for 10 min at 4°C.
8. Add fluorescently conjugated antibodies at a final concentration of 1:200 at 0.5 ml per spleen for 20–30 min at 4°C. For example, anti-CD4 Alexa 647 (or APC), anti-CD45RB (PE), and anti-CD25 FITC (see Note 5).
9. Wash cells with 5 ml PBS + 5% FBS. Centrifuge cells at 300 × *g* for 10 min.
10. Resuspend cells in PBS + 5% FBS and strain through 40 μm filter.
11. Purify cells by FACS according to the profile shown in Fig. 1.
12. Determine the purity of the sorted cells by flow cytometry.

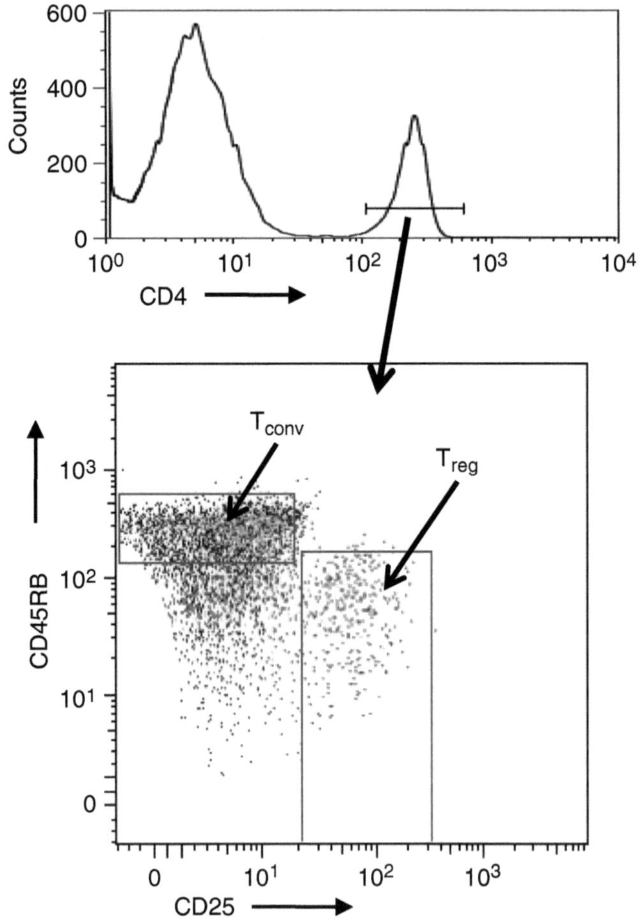

Fig. 1. Gating profile for sorting T_{regs} and T_{conv} cells. The cells are first gated on live lymphocytes (not shown) and then a second gate is placed on the CD4+ cells (histogram). The CD4+ cells are further separated into either a CD45RB^{high}/CD25$^-$ (T_{conv}) gate or CD45RB^{low}/CD25$^+$ (T_{reg}) gate.

In some cases, it is desirable to expand T_{regs} to generate greater cell numbers. Murine T_{regs} can be expanded using the following protocol: Murine T_{regs} are activated at 5×10^5cells/ml in a 96-well round bottom plate in complete RPMI medium containing 1 ng/ml PMA, 200 ng/ml Ionomycin, and 100 IU/ml murine IL-2. Following 4–5 days of activation, cells should be washed and resuspended in culture media containing 50 IU/ml IL-2 at a density of 5×10^5/ml in a 24-well culture plate. Cells can be maintained in IL-2 supplemented media and passaged to maintain a cell density of 5×10^5 cells/ml. Following 10 days in culture, T_{reg} expansion is approximately tenfold. Expanded T_{regs} maintain Foxp3 expression and suppressive capacity.

3.2. Statistical Analysis of Results

In all the models, it is important to determine the statistical significance between groups. A variety of statistical methods can be used. When comparing two independent samples of continuous data, a two-sample *t*-test is recommended when the normality assumption is reasonable. If the data are heavily skewed, contain outliers or the normality assumption is not valid for any reason, the Wilcoxon-Mann-Whitney test is the preferred nonparametric alternative. Three or more independent groups should be compared using one-way ANOVA or a nonparametric analysis such as the Kruskal-Wallis test. Two related samples (paired) should be compared using the paired t-test or the Wilcoxon signed rank test. In all parametric analyses, means should be reported with a 95% confidence interval or the standard error. Results from nonparametric analyses should include the median, minimum, and maximum. *P*-values should be reported in all cases. In the experiments that require analyses at certain points over time such as EAE disease progression, weight change over time in the IBD model, and kinetics of tumor growth in the B16 melanoma model, more advanced statistical analyses are required because of the correlation between the data points. Therefore, the type and number of statistical analyses should be determined empirically.

3.3. Homeostasis Model

This model assesses the ability of T_{regs} to suppress the homeostatic expansion of T_{conv} cells upon transfer into a lymphopenic *Rag1*$^{-/-}$ host. In this model, T_{conv} cells are sorted from B6.PL mice that express the congenic marker Thy1.1, and the T_{regs} are sorted from C57BL/6 mice that express Thy1.2. The T_{conv} cells (Thy1.1) are transferred alone or with T_{regs} (Thy1.2) into *Rag1*$^{-/-}$ mice. Seven days later, the number of T_{conv} cells is determined in the spleens of the recipient mice. Typically, there is a 50% reduction in the number T_{conv} when they are transferred with T_{regs} (8, 9). The CD4^{+} T cells that are controlled by T_{regs} have a memory-like phenotype but are otherwise naïve and are not activated (10, 11). Thus, this model assesses the capacity of the test T_{reg} population to control homeostatically expanding "naïve" T cells.

3.3.1. Injection of T_{conv} and T_{regs}

1. Sort T_{regs} and T_{conv} cells from mice with different congenic markers as described in Subheading 3.1. It is advisable to use different congenic strains to distinguish T_{reg} from the T_{conv} cells during analysis. This protocol describes the use of B6.PL mice (mice that express the congenic marker Thy1.1) for the isolation of the T_{conv}. However, B6.SJL-*Ptprca Pep3b*/BoyJ mice, which express the congenic marker, CD45.1, as opposed to CD45.2 (expressed on cells from C57BL/6 mice) can also be used.
2. Following the sort, centrifuge cells at 300 × *g* (1,200 rpm) for 10 min. Resuspend the T_{regs} in 1 ml of PBS + 0.1% FBS and the T_{conv} cells in 2 ml of PBS + 2% FBS.
3. Count the cells using a hemocytometer and trypan blue to exclude dead cells
4. Dilute the T_{regs} to 5×10^5 cells/ml and the T_{conv} to 2×10^6 cells/ml with PBS + 0.1% FBS.
5. Determine the number of *Rag1*$^{-/-}$ recipient mice that will be used per group based upon the total number of T_{regs} and T_{conv} (see Note 6).
6. Use one 15 or 50-ml conical tube per group and add the following: for T_{conv} only group add 1 ml of T_{conv} cells per mouse in the group (e.g., 5 mice = 5 ml of T_{conv}), for the T_{conv} plus T_{reg} groups add 1-ml each of T_{regs} and T_{conv} per recipient mouse in the group and vortex cells (e.g., 5 mice = 5 ml of T_{conv} + 5 ml of T_{regs}) (see Note 7).
7. Centrifuge cells for 300 × *g* for 5 min.
8. Resuspend cells in X ml of PBS + 0.1% FBS (where X = the number of mice in the group multiplied by 0.5 ml) (see Note 8).
9. Load the cells into a 3-ml syringe and inject *Rag1*$^{-/-}$ mice intravenous (i.v.) into the tail vein with 0.5 ml/mouse using a 27-G needle (see Note 9).

3.3.2. Analysis of Experimental Mice

1. Seven days later euthanize mice, dissect spleens and place into separate labeled tubes of HBSS (see Note 10).
2. Process the individual spleens as detailed in Subheading 3.1.
3. Resuspend cells in 1 ml of RPMI + 10% FBS.
4. Count the cells using a hemocytometer and trypan blue to exclude dead cells
5. Stain 200 μl of cells first with 10% normal mouse sera in FACS buffer for 5 min on ice to block the Fc receptors (see Note 11) and then CD4 and the appropriate congenic markers (e.g., Thy1.1 [distinguish T_{conv}] and Thy1.2 [distinguish T_{regs}]) in a 96-well V-bottom plate for 20 min on ice.
6. Centrifuge cells at 300 × *g* for 2 min and wash twice with FACS buffer.

7. Resuspend cells in 100 µl of FACS buffer and analyze by flow cytometry to determine the percentages of T_{regs} ($Thy1.2^+$ cells) and T_{conv} ($Thy1.1^+$ cells).
8. Calculate the number of T_{conv} and/or T_{reg} cells per recipient by multiplying the total number of live cells in the spleen by the percentage of T_{conv} and T_{regs}.

3.4. Inflammatory Bowel Disease (IBD) Recovery Model

The mucosal surface of the intestine is exposed to a variety of antigenic insults from dietary intake and the commensal flora. Regulatory T cells are important in maintaining intestinal homeostasis and preventing inflammatory bowel disease (IBD) in both humans and mice (12, 13). Experimental colitis in mice closely mimics many of the symptoms of human IBD and is a very useful model to assess the function of T_{regs} in a mucosal environment.

Experimental colitis is induced by transfer of naïve $CD4^+CD45RB^{high}$ T cells into immunodeficient mice resulting in wasting disease within 4–6 weeks (14). Injection of T_{regs} following the onset of disease symptoms leads to recovery from the disease (15). In this murine model of IBD, the disease is induced by the expansion of autoreactive T cells in combination with antigenic factors present in the intestinal flora. One example is *Helicobacter hepaticus*, which is a common pathogen found in many mouse facilities. This pathogen normally colonizes the cecum and colon and causes disease in susceptible hosts (16). Our laboratory has adopted this recovery model of colitis as it provides a robust method for assessing T_{reg} function. As an alternative, some labs use a preventative model in which T_{regs} and T_{conv} are injected at the same time. IBD is mediated by $CD4^+$ Th1 and Th17 cells (14, 17), and thus this model assesses the capacity of the test T_{reg} population to control these T cell populations.

3.4.1. Induction of Colitis

1. Determine the number of *Rag1*$^{-/-}$ mice needed for the experiment (see Note 12).
2. On the day of the injection (Day 0) weigh the *Rag1*$^{-/-}$ mice using a digital scale (see Note 13).
3. Purify T_{conv} cells ($CD4^+$ $CD45RB^{high}$ $CD25^-$) cells from C57BL/6 mice by FACS as described in Subheading 3.1.
4. Following the sort, centrifuge cells at 300 × *g* for 10 min and resuspend the T_{conv} in 2 ml of PBS + 2% FBS.
5. Count the cells using a hemocytometer and trypan blue staining to exclude the dead cells. Resuspend the T_{conv} cells in PBS + 2% FBS at 1×10^6 cells/ml.
6. Load the cells into a 3-ml syringe and inject *Rag1*$^{-/-}$ mice i.v. through tail vein with 5×10^5 T_{conv} cells (0.5 ml/mouse) using a 27-G needle (see Note 9).

3.4.2. Monitoring Body Weight

1. Weigh mice on the day of injection of T_{conv} cells and then once a week for 2–3 weeks. Once the mice start losing weight (over 2% body weight loss), monitor the mice daily for a sudden weight loss of up to 5% body weight, which is usually within a couple of days of the initial weight loss (see Note 13). In addition to weighing the mice, it is important to screen for clinical symptoms. Typical symptoms include lethargy, dehydration, hunched appearance, and diarrhea.
2. Percent weight change is calculated by comparing the current weight to the initial weight at day 0 as follows: percent weight change = ((weight at day 0 − current weight)/weight at day 0) × 100 × −1.0. For example, if the starting weight of the mouse at day 0 was 20 g and the current weight is 19 g, then percent weight change is calculated as follows: percent weight change = ((20–19)/20) × 100 × −1.0 = −5%. This indicates that the mouse has lost 5% of its body weight. Typically, the mice start losing weight around 3–4 weeks post T_{conv} transfer.
3. When the mice have lost 5% of their body weight, prepare T_{regs} for transfer (see Notes 14 and 15). Purify T_{regs} ($CD4^+$ $CD45RB^{low}$ $CD25^+$) as described in Subheading 3.1. Count the cells using a hemocytometer and trypan blue staining to exclude dead cells. Centrifuge cells at 300 × *g* for 10 min and resuspend the T_{regs} cells in PBS + 2% FBS at 1.5×10^6 cells/ml.
4. Load the cells into a 3-ml syringe and inject *Rag1*$^{-/-}$ mice intraperitoneally (i.p.) with 7.5×10^5 T_{regs} (0.5 ml/mouse) using a 27-G needle.
5. Tabulate the body weight of mice at the time of T_{reg} injection. Separate mice into experimental groups (i.e., wild type T_{reg}, experimental T_{reg} or no T_{reg} group) with similar percent weight loss among groups prior to T_{reg} injection.
6. The body weight of the mouse at the point of T_{reg} injection is taken as the starting weight for further assessment of disease progression or recovery. Thus, the percent weight change following T_{reg} injection is calculated as follows: percent weight change = ((weight at the time of injection of T_{regs} − current weight)/weight at the time of injection of T_{regs}) × 100 × −1.0. Accurate monitoring of body weight provides an indication of whether the mouse has recovered from colitis or not. Weigh mice every 7 days from the day of T_{reg} injection for 4 weeks and tabulate the weights (see Note 13).

3.4.3. Analysis of Experimental Mice

1. Four weeks following injection of T_{regs}, the mice are euthanized, and the spleen and mesenteric lymph nodes are collected into separate wells of a 24-well plate containing 1 ml HBSS for flow cytometric analysis.

2. Tease apart each spleen and mesenteric lymph nodes separately as described in Subheading 3.1 and stain 200 μl of cells first with 10% normal mouse sera in FACS buffer for 5 min on ice to block the Fc receptors and then stain for CD4, CD25, CD44, CD62L (to distinguish memory and naïve T cells), CD69 (early activation marker), and Foxp3 (to detect T_{regs}) in a 96-well V-bottom plate for 20 min on ice.
3. Centrifuge cells at 300 × *g* for 2 min and wash twice with FACS buffer.
4. Resuspend cells in 100 μl of FACS buffer and analyze by flow cytometry to determine the percentages of T_{regs} and T_{conv}.

3.4.4. Preparing the Colon for Histological Analysis

The colon can be prepared for histological analysis at the same time the spleen and mesenteric lymph nodes are collected.

1. Cut the colon from just above the rectum using scissors.
2. Using forceps gently tweeze out the colon so that it separates from the attached connective tissues.
3. Cut again just below the cecum to obtain the colon which is now untangled from the connective tissue. It is important that this procedure is carried out as consistently as possible among the individual mice as the severity of the disease shortens the length of the colon, which can be measured and tabulated.
4. Hold the colon straight at one end using a forceps. Use a 10-ml syringe filled with 10% neutral buffered formalin <!> (<!> Caution: Irritant and suspected carcinogen. Perform with caution when flushing out the fecal matter as the formalin can spray over the personnel performing this procedure. Use eye protection or perform this step in a fume hood. Dispose of the formalin waste as per your institutional guidelines. Refer to the manufacturer's MSDS for more details.) attached to a 23-G needle to flush out the fecal matter through the length of the colon into an empty waste container.
5. Once the colon is clear of fecal matter, the tissue is placed in a numbered tissue cassette and stored in formalin until all the different groups are collected 4 weeks post T_{reg} injection. Samples should be paraffin-embedded, sectioned at 5 μm, and stained with Haemotoxylin and Eosin (H&E) following standard histological protocols (see Note 16).

3.4.5. Microscopic Analysis and Scoring of the Colonic Tissue

It is important that the severity of the inflammation is assessed and scored in a blinded manner. Typically the score ranges between 0 and 5, where a score of 0 is given when there is no inflammation and a score of 5 denotes severe ulceration, diffuse transmural inflammation, and crypt loss. Details of the different scores are as follows:

Score 0: No Inflammation.

Score 1: Minimal inflammation, multifocal infiltrates in the lamina propria.

Score 2: Mild inflammation in the lamina propria and submucosa.

Score 3: Moderate inflammation in the lamina propria, sub mucosa, focally transmural, mucosal hyperplasia, minimal necrosis, focal ulcers, and mucin depletion.

Score 4: Severe focally extensive inflammation, transmural, crypt necrosis/loss, epithelial hyperplasia, erosions, some ulcers, mucin depletion.

Score 5: Ulceration, loss of crypts, severe diffuse transmural inflammation.

3.4.6. Representing Weight Loss and Histological Scores

1. Weight loss is usually graphed using the mean of the weights plus the standard error of the mean from the different groups (i.e., wild type T_{reg}, no T_{reg} or experimental T_{reg} group). For the purpose of monitoring, the recovery of mice from weight loss, the starting weight is taken as the weight at which the mice are given T_{regs}. Mice given wild type T_{regs} will start recovering with evident weight gain. In contrast, T_{regs} defective in their function will not be able to alleviate weight loss and the mucosal inflammation. The control group, which did not receive T_{reg} (no T_{reg} group), will also continue to lose weight (see Note 17).
2. Histological score (mean and standard error of mean) between 0 and 5 is plotted for each group.

3.5. Experimental Autoimmune Encephalomyelitis (EAE) Model

EAE is a useful and well developed murine model of the human autoimmune disease, multiple sclerosis. Since T_{regs} can contribute significantly to the reduction and control of the disease in mice (18, 19), EAE is a valuable system to assess the function of T_{regs} in vivo. Although the protocol requires daily disease monitoring, the data obtained can potentially reveal small differences in T_{reg} efficacy either through disease score, disease incidence, or disease kinetics. EAE can be induced with several peptide and protein antigens derived from the CNS of mice. However, this protocol is limited to the description of MOG35-55 immunization of C57BL/6 mice, as it allows for the use of multiple genetically modified mouse strains available on the C57BL/6 background. EAE is mediated by $CD4^+$ Th1 and Th17 cells (20–22), and thus this model assesses the capacity of the test T_{reg} population to control these T cell populations.

3.5.1. Injection of T_{regs}

1. T_{regs} are injected the day before EAE disease induction. Separate the mice into experimental groups (i.e., mice that will not receive T_{regs}, mice that will receive control T_{reg}, and

mice that will receive experimental T_{regs}), and mark each mouse either by ear tag or ear clipping. Normally, five mice per experimental group are used in an experiment.

2. Sort T_{regs} and T_{conv} cells as described in Subheading 3.1. It is advisable to use mice with different congenic markers such as B6.PL mice (mice that express the congenic marker Thy1.1) or B6.SJL-*Ptprca Pep3b*/BoyJ mice (mice that express the congenic marker, CD45.1), if brain infiltrating T cells will be analyzed.
3. Following the sort, centrifuge cells at 300 × *g* for 10 min and resuspend the T_{regs} in 1 ml of PBS + 2% FBS.
4. Count the cells using a hemocytometer and trypan blue to exclude dead cells.
5. Dilute the T_{regs} to 5×10^6 cells/ml with PBS + 0.1% FBS.
6. Inject 200 μl T_{regs} (1×10^6) i.v. (see Notes 9 and 18).

3.5.2. Preparing the CFA/MOG35-55 Peptide Emulsion

1. Prepare 4 mg/ml Complete Freund's Adjuvant (CFA) <!> (<!> Caution: CFA is an inflammatory reagent. Avoid skin or eye exposure. Self injection can cause a positive PPT test and lead to a granulomatus reaction and skin lesion. Use gloves and protective eyewear while handling CFA. Refer to the manufacturer's MSDS for more details.) by diluting 100 mg of heat killed *Mycobacterium tuberculosis* in 25 ml of IFA. Mix the solution using a frosted glass tissue homogenizer. The solution can be stored at 4°C for at least 1 month. Prior to each use, mix CFA thoroughly as the bacterium tends to settle to the bottom of the vial.
2. The emulsion is made the day before the injections. Make the emulsion at 1:1 ratio of CFA to peptide diluted in PBS. The final concentration of *M. tuberculosis* in the emulsion will be 2 mg/ml. To make the emulsion, load the appropriate amount of CFA (0.5–1 ml) into one 2-ml glass syringe (the volume of CFA should not exceed one half of the syringe), expunge the air and lock with connector, set aside.
3. Load an equal amount by volume of the peptide into another 2 ml glass syringe, expunge the air and connect to the other syringe from step 2 through the connector.
4. Carefully mix the two solutions. Always start by completely pushing the peptide into the CFA. Then continue pushing the mixture back and forth between the two syringes for at least 10 min. Cool the emulsion at −20°C for 5 min, mix again for 10 min, and leave at 4°C overnight.
5. The next day remix the emulsion prior to injections. The sudden increase in resistance in the syringe during the mixing indicates that an emulsion has formed (see Note 19).

6. Expunge the emulsion completely into one of the syringes, exchange the empty syringe with a 1-ml syringe, and carefully load the emulsion into the 1 ml syringe for injections.
7. Attach the 25G needle to 1 ml syringe containing the emulsion and force out any air bubbles.

3.5.3. Immunization of Mice for EAE Induction

1. Prepare 1 ml syringes loaded with peptide emulsion for subcutaneous (s.c.) injections and pertussis toxin for i.p. injections. For example, for injection of 15 mice, prepare five syringes each loaded with 600 μl of emulsion and five syringes loaded with 600 μl of pertussis toxin.
2. To inject the mice, anesthetize mice in an isofluorane chamber or have a second person hold the mouse by the nap of the neck and at the base of the tail and gently stretch the mouse over the cage bar lid, taking care not to injure or suffocate the mouse.
3. Inject 50 μl emulsion s.c. into both shoulder pads and both flanks (a total of 200 μl containing 100 μg of peptide and 400 μg of CFA).
4. Inject 200 μl of 1 μg/ml *Bordetella pertussis* toxin diluted in PBS i.p..
5. After 48 h administer another 200 μl of pertussis toxin i.p..

3.5.4. Monitoring Disease

Monitor mice daily starting at day 8 post immunization (see Note 20). Assign clinical scores based on the following criteria (see Note 21):

Score 0: No obvious physical motor differences are observed when compared with the unimmunized mouse. When the mouse is picked up, the tail has tension and the feet are separated.

Score 1: Complete flaccidity of the tail or hind limb weakness (not both). A weak tail and an unsteady gait are the initial signs of paralysis. When the mouse is placed on top of the cage bar lid, the tail will fall between the bars or hang flaccidly over the edge of the cage. To verify complete flaccidity, the tail can be flicked in the upward direction. In a healthy mouse, the tail will stay partially erect and will not immediately fall down. Additionally, when the mouse is picked up by the tail, a paralyzed mouse will hang straight, with no tail rigidity or curving of the tail base. The hind limb weakness usually presents as an unsteady walk and slipping of the mouse's hind limbs between the bars of the cage lid. Hind limb weakness can be present in the absence of the flaccid tail and should be scored as 1 or 1.5, if there is partial tail paralysis.

Score 2: Both limp tail and hind limb weakness or partial paralysis. In addition to monitoring hind limb weakness, another early sign of paralysis is the loss of the righting reflex. When a healthy mouse is put on its back, it quickly flips to the upright position. A sick mouse may have slow to complete impairment in the righting reflex. In the absence of other signs, impairment of the righting reflex of any grade is scored as 2.

Score 3: Total hind limb paralysis. The mouse can no longer use hind limbs to maintain rump posture or walk. The mouse is able to move hind legs to some degree, but if put on top of the cage bar lid, the feet will fall through and it will be unable to pick them back up.

Score 4: Hind limb paralysis and front limb weakness/paralysis. With the total loss of movement in hind limbs, the mouse drags itself only on its forelimbs. Mice appear alert and feeding, but do not move around the cage. Mice at this stage should be given food on the cage floor, water bottles with long sipper tubes, and daily subcutaneous saline injections to prevent death by dehydration.

Score 5: Moribund. Mice at this stage are not feeding, not alert, and close to death. If the mouse is scored 5, it should be immediately euthanized. After a mouse is given a score of 5, the same score is entered for the rest of the duration of the experiment (see Note 22).

Half scores can be given, if the clinical symptoms fall in between the two scores (i.e., if the symptoms appear to affect only one side of the mouse). Expect the experimental group to have scores ranging between 2 and 3 at the peak of disease. The normal or wildtype T_{reg} treated group should have scores between 1 and 2 (see Note 23).

3.5.5. Data Analysis

Data can be graphed as the average of the clinical scores of all mice in one experimental group (*y*-axis) against the day post immunization (*x*-axis). Additionally, incidence can be graphed as percent of mice presenting any clinical symptoms (*y*-axis) vs. days post immunization (*x*-axis).

3.5.6. Analysis of Brain-Infiltrating Lymphocytes

The brain and spinal cord are both targets of cellular infiltration. A significantly larger number of cells can be obtained from the brain than the spinal cord with limited technical difficulty when compared with spinal cord dissection. If one wishes to analyze the phenotype or perform functional analyses with the lymphocytes infiltrating the brain, the following protocol can be performed.

1. Sacrifice mice by CO_2 inhalation or a similar method as approved by IACUC guidelines.
2. Place the mouse on its stomach and spray with 70% ethanol.
3. Using surgical scissors, make a small incision through the skin on the back below the neck area and remove the skin revealing the scalp.
4. Gently cut the skull bone around the perimeter of the scalp starting at the back base of the skull and moving forward

toward the front of the head. Flip the top part of the skull from the back toward the front of the head and expose the brain.

5. Remove the brain and transfer into a conical tube containing PBS or HBSS.
6. Create a single cell suspension of the brain tissue by homogenizing it through a 40-μM cell strainer into a 50-ml conical tube with the plunger of a 1-ml syringe.
7. Centrifuge homogenate at 300 × *g* for 10 min at 4°C.
8. Resuspend homogenate in 7 ml of room temperature HBSS.
9. Dilute 100% Percoll to 90% and 70% by volume in PBS.
10. Add 3 ml of 90% Percoll to tubes containing 7 ml of homogenate and invert to mix and make a 27% Percoll solution.
11. Carefully underlay with 70% Percoll.
12. Centrifuge at 415 × *g* (2,500 rpm) for 25 min at 18°C without brakes.
13. Transfer the cells at the 27/70% interface to a new 15 ml tube.
14. Fill the tube with culture media and centrifuge at 300 × *g* for 10 min at 4°C.
15. At this point the brain cellular infiltrate is ready for analysis by flow cytometry or in vitro assays.

3.6. B16 Melanoma Model

B16 cells are weakly immunogenic owing to their reduced MHC I expression (23). The parent B16 line (B16-100K) is nonmetastatic and develops a well encapsulated intradermal (i.d.) tumor. Metastatic variants of the dermal parent line including lung and liver metastatic cell lines have been developed to study eradication of metastatic tumors (24, 25). The B16F10 mouse melanoma cell line was originally provided by Isaiah Fidler (MD Anderson Cancer Center, Houston, TX) and passaged intradermally in mice four times at a dose of either 100,000 cells (referred to as B16-100K) or 25,000 cells (referred to as B16-25K) (26) to ensure reproducible and aggressive i.d. tumor growth at the specified cell dose. The B16-25K cell was found to grow more reproducibly as lung metastases, so this line was chosen for future experiments involving intravenous tumor cell inoculation. Previous studies have shown that T_{regs} prevent anti-tumor immunity against the poorly immunogenic B16 melanoma (26, 27). Wild type naïve $CD4^{+}CD25^{-}$ and $CD8^{+}$ T cells alone or in combination with T_{regs} are adoptively transferred into *Rag1*$^{-/-}$ mice. The following day, mice are challenged with an i.d. inoculation of B16-100K cells. Tumor size is monitored daily to determine the effect of the T_{regs} on tumor burden. Because of variations in tumor size, we suggest having at least five mice per group. Concomitant immunity can be further assessed by surgical excision of the primary tumor, followed by secondary challenge with B16 i.d. at a remote site, or a metastatic variant of B16, B16-25K, i.v. to assess lung metastases. Tumor clearance in the B16 model is mediated by CD8⁺ T cells (26, 27), and thus this model assesses the capacity of the test T_{reg} population to suppress $CD8^+$ T cells.

3.6.1. Culture of B16 Melanoma Cells

B16 cells should be thawed 4–5 days prior to tumor challenge and maintained at a low passage to ensure good viability. B16 cells are an adherent cell line and should adhere to culture flasks within 1 day. Cells that are slow to adhere or do not adhere should not be cultured or used in assays. As cell density is critical to the proper growth and viability of B16 cells, it is important to seed cells at multiple concentrations to ensure that at least one flask will be optimal for inoculation.

1. Add 26, 27, 28, and 29 ml of B16 culture media to each of four T175 flasks (see Notes 24 and 25).
2. Remove 1 vial of B16 cells containing approximately $4–5 \times 10^6$ cells in 1 ml from liquid N_2.
3. Thaw cells by holding and shaking in a 37°C water bath for approximately 30 s.
4. As soon as the freeze media thaws, transfer the contents of 1 vial into a 50-ml conical tube containing 19 ml of B16 culture media.
5. Invert tube to mix.
6. Transfer 1, 2, 3, or 4 ml of cells into each of the four T175 flasks for a total volume of 30 ml per flask.
7. Shake flasks to mix, making sure the medium completely covers the bottom of each flask.
8. Culture cells in an incubator at 37°C, 5% CO_2 for 4–5 days. One day before tumor challenge, aspirate media from flasks and replenish flasks with fresh, B16 culture media (prewarmed in a 37°C water bath).

3.6.2. Adoptive Transfer of T cells

Rag1$^{-/-}$ mice are reconstituted with 9×10^6 $CD4^+CD25^-$ T cells, 6×10^6 $CD8^+$ T cells, and 1×10^6 T_{regs} (in desired groups).

1. Purify $CD4^+CD25^-$ T cells, $CD8^+$ T cells, and T_{regs} from desired source as described in Subheading 3.1 (see Note 26).
2. Count all cells and adjust in murine T cell culture medium (see materials Subheading 2.1) to 9×10^6/ml ($CD4^+CD25^-$ T cells), 6×10^6/ml ($CD8^+$ T cells), and 1×10^6/ml (T_{regs}).
3. For each group, combine 1 ml of cells per mouse into a 15-ml conical tube. For example, an experiment that includes five mice per group with two groups (Group A: with T_{regs} and Group B: without T_{regs}) will be divided into two tubes. Each tube will contain 5 ml of $CD4^+CD25^-$ T cells and 5 ml of $CD8^+$

T cells. In addition, add 5 ml of T_{regs} to the tube containing Group A cells.

4. Centrifuge cells at 300×*g* for 10 min and aspirate supernatant.
5. Resuspend cells in sterile PBS + 2% FBS at 0.3 ml per mouse to be injected. Include an additional 10% of volume PBS+2% FBS to account for minor losses that occur when loading the syringes for injections. For example, Groups A and B will each be resuspended in 1.65 ml PBS+2% FBS. (0.3 ml/mouse×5 mice)+0.15 ml=1.65 ml
6. Attach a blunt needle to a sterile 3 ml disposable syringe and pull cells into syringe by drawing up plunger.
7. Remove blunt needle and replace with sterile 27G needle. Maintain sterility of cells at all times.
8. Inject cells i.v. into the tail vein of the mice (see Note 9).

3.6.3. B16 Melanoma Cell Preparation

One day following adoptive transfer of T cells into *Rag1*$^{-/-}$ mice, challenge the mice with the B16 melanoma (see Note 27). Each mouse will receive 1.2×10^5 B16 cells intradermally in the rear flank.

1. Place sterile PBS and frozen Trypsin-EDTA aliquots (7 ml per T175 flask) in 37°C water bath for 15–20 min to thaw and warm.
2. Place flasks of B16 cells under microscope to determine health and confluency of cells. To ensure good viability, cells should be harvested when flasks reach no more than 75–85% confluence.
3. Determine the best dilution(s) of cells for harvest. Cells should be about 70% confluent and be well adhered to the flask. A flask that contains cells that are clumpy or have died and are floating should not be used.
4. Aspirate media from flasks and wash cell monolayer with 15 ml warm PBS. Repeat.
5. Add 7 ml of Trypsin-EDTA per flask and swirl to coat the cells.
6. After about 30 s, forcefully tap flasks to release cells from flask. To confirm that the cells have been released, visualize cells under the microscope.
7. Immediately add 12 ml cold B16 culture media to quench the Trypsin-EDTA. It is important to follow this trypsinization and quenching protocol exactly as over trypsinizing cells will decrease cell viability.
8. Transfer cells to 50 ml conical tube(s) and centrifuge at 300×*g* for 5 min at 4°C.

9. Wash cells twice in cold RPMI without any additives.
10. Before final spin, count cells by trypan blue exclusion using a hemocytometer. Viability should be >95%. If viability is <95%, the cells should not be used.
11. Resuspend cells at 2.4×10^6/ml in cold RPMI without any additives. This equates to 1.2×10^5 B16 cells per 50 μl.
12. Aliquot cells into 2 ml cryo vials or 1.5 ml Eppendorf tubes for easy loading of syringes and to avoid over-mixing of cells during inoculation.
13. Place cells on ice and maintain on ice for duration of the inoculation.

3.6.4. B16 Inoculation and Measurement of Primary Tumor Growth

Because of the technical difficulty of the tumor inoculation, mice must be anesthetized with isofluorane for the procedure (see Note 28). In theory, alternate anesthetics can be used if so desired; however, their use must be determined empirically. The presence of fur hinders proper measurement of the tumor size; therefore, fur must be removed from the site of tumor injection, using a small electric razor. The use of Nair™ or other depilatory can cause a minor inflammatory reaction on the skin and is therefore not recommended.

1. Mix cells by vortexing briefly. Draw cells into 1 ml syringe fitted with 30G needle. Remove air bubbles and set syringe aside.
2. Anesthetize mouse with isofluorane using a chamber design stationary anesthesia machine. When a mouse is sufficiently anesthetized, its respiratory rate decreases and has no reaction to a pinch of the toe.
3. Transfer anesthetized mouse from chamber to a clean work space nearby. Place the mouse on its left side. Using a small electric razor, shave the fur from the hind leg up to just below the front leg (approximately 12 × 12 mm patch). See photograph in Fig. 2a.
4. Inject 1.2×10^5 B16 cells in 50 μl of RPMI i.d. on shaved right flank Fig. 2a. It is important that cells are injected i.d. and not subcutaneously. Subcutaneous tumors cannot be surgically excised; therefore, no secondary challenge (or analysis of tumor infiltrating lymphocytes) is possible if inoculated subcutaneously. If i.d. inoculations are done properly, a small, liquid-filled bubble will appear on the skin, and the bubble will lift up with the skin when the skin is pulled away from the mouse.
5. Transfer mouse into a clean cage that was prewarmed with a heating pad and monitor recovery. Mouse should be mobile and active within a minute or so following tumor inoculation (see Note 29).

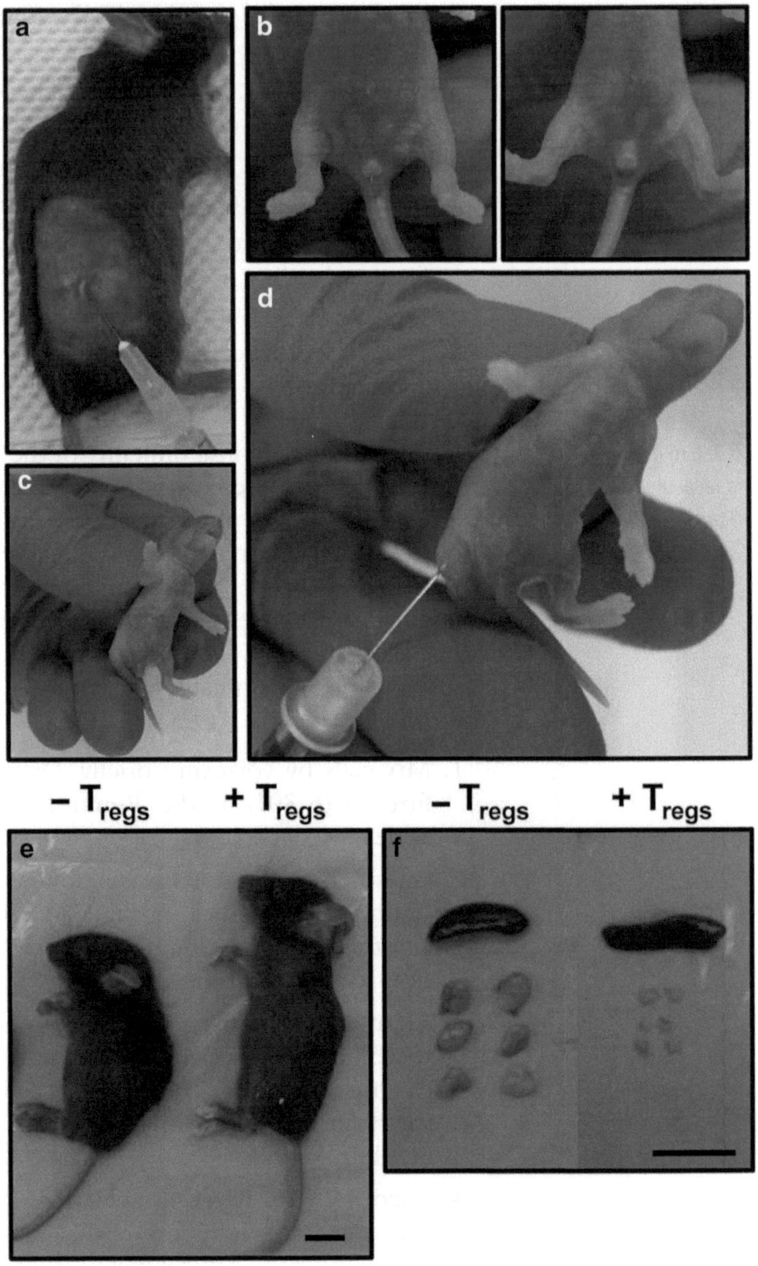

Fig. 2. Procedures and analysis pertaining to the B16 melanoma and *Foxp3⁻* rescue models. (**a**) The area on the flank of the mice that requires shaving and the location of the i.d. injection for the B16 melanoma model. (**b**) Sexing of a female (*left*) and male (*right*). (**c**) The proper technique recommended to hold a 2-day-old pup for i.p. injections. (**d**) The i.p. injection technique for a 2-day-old pup in the *Foxp3⁻* rescue model. (**e**) Exterior of *Foxp3⁻* mice that received no T_{regs} (*left*) or wild-type natural T_{regs} (*right*). (**f**) Spleens and lymph nodes of *Foxp3⁻* mice that received no T_{regs} (*left*) or wild-type natural T_{regs} (*right*). *Black bar*, 1 cm.

6. Tumors will develop 5–7 days post challenge. At the first sight of a tumor, all tumors must be measured daily using a dial caliper.
7. Tumor size can be reported in diameter (mm) or in total area (mm^3). Measure length and width of tumor in mm to report diameter or calculate total area of tumor using the following equation: ($a^2 \times b/2$, where a is the smaller caliper measurement and b the larger).
8. Tumors should be excised when a tumor ulcerates, reaches a maximal diameter of 10 mm (500 mm^3), when discomfort or impaired mobility is noted or as set by IACUC guidelines.

3.6.5. Surgical Excision of Tumor and Secondary Challenge

Intradermal primary tumors should be surgically excised at 5–10 mm diameter. Less than 5% of primary tumors should recur following surgery. Any mice with recurrent primary tumors should be omitted from the concomitant tumor study (27). As with the primary tumor challenge, anesthetize mouse with isofluorane and monitor breathing rate and reflexes throughout the procedure. To maintain sterility, surgery is performed in a laminar flow hood. Mice must remain anesthetized during surgery; therefore, a mobile isofluorane machine attached to a breathing tube must be set up within the hood.

1. Anesthetize mice with isofluorane. When mice are sufficiently anesthetized, transfer mice (1 mouse at a time) from the isofluorane chamber into a laminar flow hood.
2. Place the mouse on its back with its nose securely fitted into the breathing tube.
3. Place gauze or a paper towel under mouse to prevent movement on the slick surface of the hood during surgery.
4. Swab tumor and surrounding skin with a Q-tip coated in surgical iodine solution. Beginning in the center of the tumor and working toward the edges, swab the tumor in a circular motion.
5. Swab area with a single use alcohol pad.
6. Repeat iodine and alcohol cleaning of the area.
7. Loosely grasp the tumor with blunt forceps.
8. Using surgical scissors make a small incision through the skin near the tumor and gently cut around the tumor to remove the tumor together with a 2-mm perimeter of healthy skin. This margin will ensure that the tumor does not recur.
9. The tumor can be kept for analysis of infiltrating lymphocytes as detailed in Subheading 3.6.6.
10. Join remaining skin with forceps and close the wound with 2–3 steel wound clips.

11. Liberally apply Neosporin triple antibiotic ointment (or any triple antibiotic ointment) to wound using a fresh Q-tip.
12. To help control pain, inject 0.1 mg (0.1 ml of 0.015 mg/ml) of buprenorphine or 0.15 mg (0.05 ml of 2.5 mg/ml) of rimadyl subcutaneously in the upper dorsal skin between the shoulders in the back of the head.
13. Transfer mouse to a clean cage that has been prewarmed with a heating pad or is under a heat lamp.
14. Alternatively, if a tumor challenge is desired at a remote site in the same animal, transfer the mouse to a sterile work space, and shave the fur off of the mouse on opposite side of primary tumor.
15. Administer secondary i.d. tumor challenge as described above for primary tumor (Subheading 3.4.4). Alternatively, the B16-25K metastatic cell line can be injected i.v. (1.2×10^5 cells in 100 μl PBS) to initiate a metastasis study.
16. Monitor mice for recovery from anesthesia. Recovery from the prolonged anesthesia that is required for surgical excision is slower than that following the primary challenge. Mice should be mobile and active within 2–3 min following cessation of anesthesia (see Note 29).
17. After 5–7 days, surgical wounds will be healed and wound clips can be removed. Remove clips using a wound clip remover.

3.6.6. Analysis of Dermal Tumor-Infiltrating Lymphocytes

If one wishes to analyze the phenotype or perform functional analyses with the tumor infiltrating lymphocytes, the following protocol can be performed. Tumors can be surgically excised as detailed in Subheading 3.6.5 or can be analyzed as an endpoint following sacrifice of the animal.

1. Excise tumor as described earlier and transfer into a sterile 24-well tissue culture plate containing PBS.
2. Create single cell suspension of tumor tissue by teasing tumor between frosted microscope slides. Wash plate and slides with 1 ml PBS and transfer into a 15-ml tube.
3. Centrifuge homogenate at 300 ×*g* for 10 min at 4°C.
4. Resuspend homogenate in 2 ml of room temperature 80% Percoll [100% Percoll: 90 ml neat Percoll + 10 ml 10× HBSS (both sterile). Dilute 100% Percoll in 1× HBSS to create 80% and 40% solutions].
5. Transfer cells in 80% Percoll to a new 15-ml tube.
6. *SLOWLY* layer 2 ml of 40% Percoll on top of the 80% solution. Be careful not to disturb the interface.
7. Centrifuge at 415 ×*g* for 25 min at 18°C without brakes.

8. Transfer the cells on the 40/80% interface to a new 15-ml tube using a glass pipet or a longer pipet tip.
9. Fill the tube with B16 culture media and centrifuge at 300×*g* for 10 min at 4°C.
10. Resuspend in PBS+2% FBS containing antibodies of interest and incubate for 20 min on ice.
11. Centrifuge cells at 300×*g* for 10 min at 4°C and proceed with purification by FACS.

3.6.7. Analysis of Lung Metastases

Lung metastases will develop 22–28 days post inoculation. Since metastases are not outwardly visible, a sentinel mouse must be sacrificed to determine presence of metastases. We therefore recommend inoculating 2–3 extra mice (in addition to the experimental mice) that can be used to assess the presence of metastases.

1. Prepare 10 ml of 5% H_2O_2 in PBS in a 5-cm tissue culture dish and two, 5 cm tissue culture dishes with 10 ml PBS only.
2. Sacrifice mice by CO_2 inhalation or a similar method as established by IACUC guidelines.
3. Place the mouse on its back and spray with 70% ethanol. Cut mouse open to reveal the lungs in the thoracic cavity.
4. Open the rib cage, revealing the lungs. Remove the lungs and place directly in a dish containing PBS for 1 min, transfer to dish containing 5% H_2O_2 in PBS for 1 min, and then transfer to dish containing fresh PBS for 1 min. This process will inflate the lungs and render metastases more visible.
5. Dissect the lungs into lobes and count both sides for black pigmented spots, which are metastases. Metastases are readily visible by eye. If desired, a dissection microscope can be used to aid in counting.
6. If more than 40 metastases are visible, the lungs can also be weighed as a quantitative measurement.

3.7. *Foxp3⁻* Rescue Model

Mice that lack Foxp3 develop a rapid, multiorgan lymphoproliferative disease that results in lethality 16–25 days after birth (28). Transfer of 1×10^6 natural T_{regs} into 1–3 day old *Foxp3⁻* pups can protect them from the lethal autoimmune disease for at least 5 weeks (29). However, the use of more than 1×10^6 T_{regs} may provide longer, more significant protection. Optimal cell number must be determined by the investigator. Because of the limited window of opportunity for this transfer (T_{regs} injected beyond 3 days of life will not prevent disease), timed breeding, rapid genotyping of litters, and open access to cell sorting facilities to isolate T_{regs} are critical components of a successful experiment.

Although initiating a large experiment is challenging, we suggest having at least 5 mice per group. This model assesses the capacity of the test T_{reg} population to control $CD4^+$ and $CD8^+$ T cell lymphoproliferation and autoimmune destruction that underlies the disease manifested in *Foxp3*$^-$ mice (29).

3.7.1. Breeding and Genotyping Considerations

Male *Foxp3*$^-$ mice do not survive long enough to breed; therefore, *Foxp3*$^{+/-}$ (heterozygous) female mice crossed to male C57BL/6 mice must be used as breeders. Genotypic distribution of the mutant allele does not follow expected Mendelian genetics. Only approximately 10–20% of the offspring are male knockouts; therefore, a large number of breeder pairs may be needed to generate the number of mice needed for even a small experiment. For example, from 5 breeder pairs, one might expect 5–10 pups per month. This must be taken into consideration when planning experiments.

Careful monitoring and timing of breeding is critical for the success of this experiment. Our experience suggests that litters are born approximately 19 days after a plug is visible in *Foxp3*$^{+/-}$ mice; however, this can vary from institution to institution, so gestational length must be determined by the investigator. On the day that they are born, litters must be sexed and males must be tagged (typically 1 toe is removed as ear tagging at this age is not possible). Newborn male mice are distinguished by the presence of a small black dot at the base of the tail on the front side of the mouse (Fig. 2b). In addition, tails must be clipped, digested, DNA extracted, and genotyping performed within 24 h. These necessities must be considered and accounted for in the planning of these experiments as T_{regs} must be adoptively transferred into day 2 (ideally) or day 3 old mice.

3.7.2. Adoptive Transfer of T_{regs}

1. Isolate T_{regs} as described in Subheading 3.1.
2. Count cells by trypan blue exclusion using a hemocytometer and resuspend 1×10^6 T_{regs} in 30 μl per mouse sterile PBS + 2% FBS.
3. Because of the small volume to be injected (30 μl per mouse), it is important to transfer cells into a tube/vial into which a small needle can be inserted to load the syringe for inoculations. The lid of a 2-ml cryo vial works very well for this purpose.
4. Pull cells from the lid of a 2-ml cryo vial into an insulin syringe fitted with a 30-G needle by drawing up plunger. Maintain sterility of cells at all times.
5. Cells should be injected i.p. into mice. Cells injected i.p. directly though the abdominal wall are prone to leaking out when the needle is removed. Instead, injecting into the peritoneal cavity through the front side of the hind leg, parallel to the quadriceps, works well (Fig. 2c and d).

3.7.3. Macroscopic Analysis of Foxp3⁻ Mice

By 10–14 days of life, untreated *Foxp3⁻* mice are distinguishable from their wild-type littermate controls, based on their small stature, hunched posture, scruffy fur, and scaly erythematous skin on the ears and tail. The disease progresses such that 20–25 day old mice are terminally ill and moribund. Histological analysis of lungs, liver, and skin (ear) reveals severe inflammation and destruction of tissue architecture marked by inflammatory cell infiltrations in the liver, lungs, and skin dermis. At the cellular level, untreated *Foxp3⁻* mice have splenomegaly and severe lymphadenopathy with a two to threefold increase in T cells compared with wild type littermates. To fully dissect the pathological manifestations of the disease and the ability of T_{regs} to prevent such disease, both macroscopic and microscopic analysis is needed. Typically, we perform this analysis at 3.5–4 weeks of age.

1. Macroscopic analysis of mice can be performed by inspection of the mouse's external appearance. A 6 point scoring system is used to describe a mouse's external appearance.
   (i) Is the mouse runted (small)? No: 0 point; Yes: 1 point
   (ii) Is the mouse's tail scaly and/or with lesions? No: 0 point; Yes: 1 point
   (iii) Are the mouse's ears small in size, scaly, and/or with lesions?; No: 0 point; Yes: 1 point
   (iv) Are the mouse's eyelids scaly, not fully open?; No: 0 point; Yes: 1 point
   (v) What is the activity level of the mouse? Normal: 0 point; Moderately impaired mobility: 1 point; Immobile: 2 points
2. As the number of male *Foxp3⁻* mice born per litter is small, it is often necessary to analyze mice on multiple days. Therefore, it is important to have a means to compare mice. This is accomplished by taking the following photographs next to a ruler or other scale bar: (a) whole mice from the relevant groups laying on their sides and (b) spleen and lymph nodes (inguinal, axillary and cervical) lined up next to each other from the relevant groups (Fig. 2e and f) (see Notes 30 and 31).
3. Microscopic analysis involves both histological analysis and analysis of cellularity of the spleen and lymph nodes (see Note 31).
4. Prepare lung, liver, and ear pinna for H&E analysis. Liver and ear can be directly placed into tissue cassettes for processing. Lungs must be inflated with 10% neutral buffered formalin (see Note 32) prior to placing in cassettes. To inflate lungs, place the mouse on its back and dissect away skin on the front side of mouse from mid section to head. Make a small incision through the muscle, just below the neck, revealing the trachea. Using a 5-ml syringe fitted with a 25-G needle filled with 10%

formalin, gently insert needle into trachea. Slowly fill lungs with formalin solution until they are visibly expanded. Remove needle, excise lungs, and place in tissue cassette. All cassettes must be stored in 10% neutral buffered formalin to fix and preserve samples. Samples should be paraffin embedded, sectioned at 5 μm, and stained with H&E, according to standard histological methods.

5. Transfer spleens and lymph nodes into a 24-well cell culture plate containing 1 ml HBSS or PBS and maintain on ice at all times during processing.
6. Process and assess mice individually. Tease apart each spleen and pooled set of lymph nodes separately with a 1-ml syringe plunger through a 70-μm cell strainer into a 50-ml conical tube. Rinse cell strainer 2 times with HBSS to recover all cells. Alternatively, spleens and lymph nodes may be teased apart between two frosted glass microscope slides.
7. Centrifuge homogenate at 300 × *g* for 10 min.
8. Resuspend homogenate in 1 ml Gey's solution per spleen. Gently swirl for 2 min and then quench reaction by adding 12 ml of HBSS.
9. Centrifuge at 300 × *g* for 10 min.
10. Resuspend cells in 1 ml of FACS buffer containing 10% normal mouse serum for 5 min on ice. Add an additional 1 ml of FACS buffer containing fluorescently conjugated anti-CD4 and anti-CD25 antibodies and incubate for 20 min on ice.
11. Count cells by trypan blue exclusion using a hemocytometer.
12. Centrifuge cells at 300 × *g* for 2 min and wash twice with FACS buffer
13. Resuspend cells in 100 μl of FACS buffer and analyze by flow cytometry to determine the percentages of total $CD4^+$ and $CD4^+$ $CD25^+$ T cells.
14. Calculate the number of total $CD4^+$ and $CD4^+$ $CD25^+$ T cells by calculating the total number of live cells in the spleen or lymph nodes multiplied by the percentage of total $CD4^+$ and $CD4^+$ $CD25^+$ T cells determined by flow cytometry.

3.7.4. Microscopic Analysis and Histological Scoring

The severity of inflammation should be assessed and scored in a blinded manner by an experienced veterinary pathologist. The scoring system used for assessing inflammation in *Foxp3*$^-$ mice is based on a simple algorithm for expressing inflammatory infiltrates in the lungs, liver, and ear. The scores allotted to these three tissues were 0–9, 0–11, and 0–8, respectively, giving a maximum possible total of 28.

Lung: The lumen of the airways and alveoli do not have any inflammatory infiltrates. However, in *Foxp3*$^-$ mice lacking T_{regs}, inflammation is present surrounding the bronchioles, the pulmonary blood vessels, and with expansion of the interstitium (septa) (Fig. 3). Each of these is separately assigned a numerical number of 0, 1, 2, or 3. A score of 0 is assigned if minimal or no inflammatory infiltrates are associated with the interstitium, the tissue surrounding the bronchioles and the blood vessels. A score of 1, 2, or 3 is assigned if inflammation is associated with <10%, 10–50%, or >50% of the bronchioles, blood vessels, or interstitium, respectively.

Liver: Portal tract inflammation, periportal/periseptal interface hepatitis, and hepatic lobular inflammatory foci are the three criteria for scoring the degree of inflammation in the liver of *Foxp3*$^-$ mice lacking T_{regs} (Fig. 3). Portal inflammation is scored 0–3 while interface hepatitis and lobular inflammatory foci are allotted a score of 0–4. *Portal inflammation:* A score of 0 is assigned when portal tracts do not have any inflammatory cells. A score of 1, 2, or 3 is assigned if inflammation is associated with <25, 25–75, or >75% of the liver portal tracts, respectively. *Periportal/periseptal interface hepatitis:* A focus of interface hepatitis associated with either a few or most of the portal tracts are scored 1 and 2, respectively. Two or more foci of interface hepatitis surrounding <50 or >50% of the portal tracts or periseptae is scored 3 and 4, respectively. *Lobular inflammation:* Foci of granulocytes and/or lymphocytes with or without necrotic heptocytes that expand the sinusoid are considered foci of inflammation while foci of granulocytes without necrotic hepatocytes that do not expand the sinusoid are excluded as foci of extramedullary hematopoiesis. The number of inflammatory foci in 10 contiguous 10× objective fields are counted and recorded as the average number of foci per 10× field and given a score of 0–4. A score of 0 is assigned when sinusoidal foci of inflammatory cells are absent. One focus or less per 10× field, 2–4 foci per 10× field, 5–10 foci per 10× field, and more than ten foci per 10× field are scored 1, 2, 3, and 4, respectively.

Ear pinna: The percent of the ear dermis with inflammatory infiltrates and the intensity of the dermal inflammation are the criteria for determining the degree and severity of the ear involved with inflammation in *Foxp3*$^-$ mice lacking T_{regs} (Fig. 3). *Percent of ear with inflammation:* The ear specimen is divided into equal linear segments. The average percent of an inflammatory cell infiltrate of all the segments is scored 0, 1, 2, 3, or 4. A score of 0 is assigned when the inflammatory cells in all segments are not beyond that of normal background level. A score of 1, 2, 3, or 4 is assigned when the average percent for the segments is <25, 25–50, 51–75, or >75%, respectively. *Intensity of inflammation:* The intensity of the inflammatory infiltrate in the dermis is assessed as being of a loose or dense nature. A score of 0 is assigned when inflammatory cells in the dermis are not beyond the normal

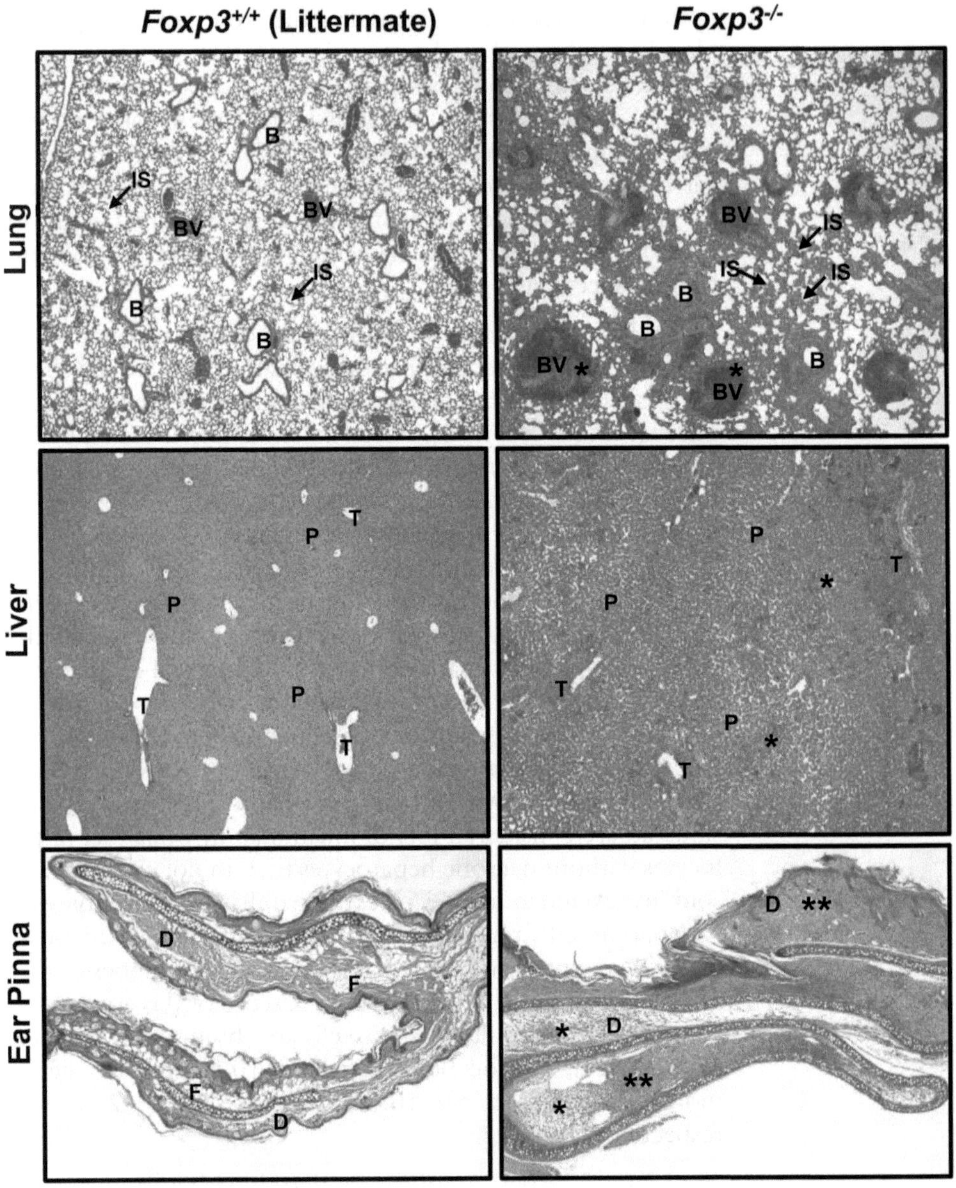

Fig. 3. Microscopic H&E illustrations of *Foxp3*⁺ and *Foxp3*⁻ littermate mice of the lung, liver and ear pinna. (*Lung*) The *Foxp3*⁺ mouse does not have inflammatory cells in or around either the bronchioles (B) or blood vessels (BV), and the interstitial septae (IS) are narrow, thin, and lack inflammatory cell infiltrates. In the *Foxp3*⁻ mouse, inflammatory cell infiltrates (*) surround bronchioles (B) and pulmonary blood vessels (PV) and focally thicken the interstitial septae (IS). (*Liver*) The portal tracts (T) and liver lobular parenchyma (P) of the *Foxp3*⁺ mouse lack inflammatory infiltrates. Inflammatory cells fill some portal tracts (T) of the *Foxp3*⁻ mouse and they infiltrate the periportal hepatocytes broadening the portal tracts consistent with interface hepatitis. Foci of inflammatory cells (*) are randomly scattered through the liver lobular parenchyma of the *Foxp3*⁻ mouse. (*Ear pinna*) The dermis (D) and fatty (F) tissue of the *Foxp3*⁺ are void of inflammatory cells. Inflammatory cell infiltrates are present in a loose (*) and dense (**) pattern. The dermis is thicker and the fat tissue is obscured with inflammatory cell infiltrates in the *Foxp3*⁻ mouse.

background level. The intensity of the inflammation is considered loose when the majority of the inflammatory cells do not abut one another. The inflammation intensity is considered dense when the majority of the inflammatory cells abut one another. When all the inflammation is of the loose nature, a score of 1 is assigned. When there is a mixture of loose and dense inflammatory cell infiltrates, a score of two is assigned when the loose form is dominant; A score of three is assigned when the dense form is dominant; A score of 4 is assigned when all of the inflammation is of a dense nature.

Histological scoring parameters:

1. Lung
   (a) Peribronchiolar inflammation.

   Score 0: Minimal or no inflammation.

   Score 1: <10% of bronchioles with inflammation.

   Score 2: 10–50% of bronchioles with inflammation.

   Score 3: >50% of bronchioles with inflammation.

   (b) Perivascular inflammation.

   Score 0: Minimal or no inflammation.

   Score 1: <10% of blood vessel with inflammation.

   Score 2: 10–50% of blood vessel with inflammation.

   Score 3: >50% of blood vessel with inflammation.

   (c) Interstitium.

   Score 0: Minimal or no inflammation.

   Score 1: <10% of interstitium with inflammation.

   Score 2: 10–50% of interstitium with inflammation.

   Score 3: >50% of interstitium with inflammation.

2. Liver
   (a) Portal tract inflammation.

   Score 0: Portal tracts with minimal or no inflammation.

   Score 1: <25% of tracts with inflammation.

   Score 2: 25–75% of tracts with inflammation.

   Score 3: >75% of tracts with inflammation.

   (b) Portal/periseptal interface hepatitis.

   Score 0: Portal tracts with minimal or no inflammation.

   Score 1: A few tracts with a focus of interface hepatitis.

   Score 2: Most tracts with a focus of interface hepatitis.

   Score 3: <50% of tracts with multiple foci of interface hepatitis.

   Score 4: >50% of tracts with multiple foci of interface hepatitis.

(c) Hepatitis lobular inflammation.

Score 0: Minimal or no parenchymal inflammatory cells.
Score 1: One focus or less per 10x field.
Score 2: Two to four foci per 10× field.
Score 3: Five to ten foci per 10× field.
Score 4: More than ten foci per 10× field.

3. Ear pinna

(a) Percent of ear with inflammation.

Score 0: Inflammatory cells not beyond background level.

Score 1: <25% of dermis with inflammation.

Score 2: 25–50% of dermis with inflammation.

Score 3: 51–75% of dermis with inflammation.

Score 4: >75% of dermis with inflammation.

(b) Intensity of inflammation in the ear.

Score 0: Inflammatory cells not beyond background level.

Score 1: Inflammatory cells have loose arrangement.

Score 2: Inflammatory cells have a loose and dense arrangement with the former dominating.

Score 3: Inflammatory cells have a loose and dense arrangement with the latter dominating.

Score 4: Only dense arrangement of inflammatory cells.

4. Notes

1. The optimal manufacturer and lot number of FBS/FCS can vary; therefore, this must be determined empirically. Prior to use in assays, FBS must be heat inactivated for 30 min at 56°C. Following heat inactivation, FBS can be stored at 4°C for up to 1 month.
2. Caution: *Mycobacterium tuberculosis* is an inflammatory reagent. Avoid inhalation, skin or eye exposure. Use gloves, protective eyewear, and a mask when handling the reagent. Refer to the manufacturer's MSDS for more details.
3. Caution: *Bordatella pertussis* toxin is a bacterial virulence factor. Avoid skin or eye exposure, or inhalation. Use gloves, protective eyewear, and a mask when handling the reagent. Refer to the manufacturer's MSDS for more details.
4. Panning is an economical method for B cell depletion and will reduce the sort time required. Panning is done following RBC lysis (step 5). The cells are incubated on sterile nontissue

culture treated plates coated with goat antimouse Ig. Thus, the B cells will bind to the plate and the nonadherent cells (predominantly T cells) are collected and stained for sorting.

5. Antibody conjugates can be altered depending upon laser availability. Optimal antibody concentrations must be determined empirically

6. Typically, there will be three groups of recipient mice: T_{conv} only, T_{conv} + "control" T_{regs}, and T_{conv} + "Experimental" T_{regs}. Ideally, it is important to have at least three mice receive T_{conv} cells only as this group is the control group to which the T_{conv} + T_{regs} groups will be compared. Each mouse will receive either 2×10^6 T_{conv} or 2×10^6 T_{conv} + 0.5×10^6 "control" or "Experimental" T_{regs}. Occasionally, a T_{reg} only group is included, especially if molecules regulating T_{reg} homeostatic control are being studied.

7. It is important to add the cells in 1 ml aliquots and to vortex very well prior to centrifugation to ensure that all samples are treated equally and equivalent cell numbers are achieved in all groups. Deviation from this may result in variability in cell numbers during analysis.

8. For example, if you are injecting four mice, resuspend in 2.0 ml. Also, to account for loss due to cell transfers, etc., it is advisable to add an additional 10% by volume of PBS + 0.1% FBS.

9. It is critical to ensure that all air bubbles have been removed from syringes prior to intravenous injections. Introducing air into the vein can cause lethal lung, heart, or brain failure.

10. Upon isolation of the spleens, it is important that the spleens are kept separate throughout the entire analysis. The total number of T_{conv} cells can be determined on a per mouse basis, thus increasing statistical power.

11. Alternatively, an anti-Fc receptor antibody (e.g., clone 2.4G2) can be used to block Fc receptors.

12. The incidence of colitis upon naive T cell transfer can vary substantially between institutions (<20 100%) due to variations in the intestinal microbiota. Investigators should first perform a pilot experiment to determine incidence in their facility empirically. Typically, the greatest variable is *Helicobacter spp* colonization, but other strains are likely to influence colitis incidence (16, 30, 31). It is recommended to have at least five mice per experimental group. In *Helicobacter*-free mouse facilities or when using specific pathogen free mice, the incidence of colitis can be approximately 40%. In *Helicobacter*-positive mice, the incidence of the disease is greater. Therefore, assuming that there will be at least a 50% incidence of colitis following T_{conv} injection, and there are

three groups of T_{regs} to be assessed for their function (e.g., No T_{reg}, wild type T_{reg}, and experimental T_{reg}), use at least 30 *Rag1*$^{-/-}$ mice per experiment. *Rag1*$^{-/-}$ mice are the most common strain used for these experiments but in theory any lymphopenic strain could be used. However, this would need to be verified empirically.

13. It is important that the weight of the mice is measured as accurately as possible. Body weight of the mice should be taken every week on the same day and preferably at the same time using the same weighing balance during the entire course of the study to avoid any artifacts. Use a reference of known weight (such as a plastic lid or a tube filled with a set volume of water) to ensure the balance is accurate each time before and after the mice are weighed.
14. The time window for optimal T_{reg} adoptive transfer is small (1–2 days) so careful planning is critical. The availability and flexibility of your institutional sorting facilities should be considered in this planning process. Planning should be done in advance as much as possible. Typically, the mice will not reach 5% weight loss at the same time so several sorts over consecutive days are often required (~3–4 days, 3–4 weeks from T_{conv} cells injection). Do not set up different groups of T_{regs} on different days but rather a few from each group of T_{regs} on each day.
15. It is important to inject the mice as soon as they lose 5% of their body weight to minimize differences in the severity of disease between the mice used for T_{reg} injection.
16. Typically, tissue sections are stained with H&E to determine cellular infiltration and tissue destruction. However, Alcian blue/ PAS staining (to distinguish between acidic and neutral mucin in the colonic tissue and aid in goblet cell identification), anti-CD3, and Foxp3 staining (to detect T_{conv} and T_{regs}) may also be used on the tissue sections.
17. Most IACUC protocols require the euthanasia of mice that lose over 10% of their body weight.
18. Significant decrease in disease symptoms has been observed after injection of 1×10^6 natural T_{regs} (19). Higher (up to 4×10^6) or lower (0.5×10^6) numbers of T_{regs} can also be used.
19. The length of emulsion formation varies significantly and might be affected by the salt concentration and purity of the peptide. Cooling the syringes filled with CFA and peptide to 4°C helps in emulsion formation. The quality of the emulsion can be tested by dropping a small amount on the surface of a beaker of water. If the emulsion is good, it will remain intact on the water surface. If not, the oil component will rapidly

spread out over the entire water surface and the drop will disintegrate.

20. Mice usually develop symptoms between days 10 and 18, and the experiment can be terminated 30–40 days post immunization.
21. Since the scoring system is subjective, the scoring should be performed in a blinded manner (i.e., the identity of the experimental groups should be hidden from the person who is scoring the mice). Additionally, the same person should score the mice throughout the experiment to prevent person to person variability in scoring.
22. Most institutions require euthanasia of a mouse after several days at a score of 4. At the point of euthanasia, the mouse is scored as 5. Contact your IACUC committee for protocol approval and institutional EAE guidelines.
23. Usually, injection of 100 μg of MOG35-55 results in a peak average score of 2.5. The peptide activity can vary between batches, so an optimal concentration of peptide should be determined based on disease symptoms. Ideally, an HPLC grade purified peptide should be used. If you have difficulty inducing EAE, check that the mice are handled gently, and that they are not housed under stressful conditions. Additionally, the mice should be rested for at least 7 days after arrival to your animal facility. If necessary, increase the amount of peptide antigen.
24. B16 cells should be frozen at $4–5 \times 10^6$/ml in 1 ml aliquots of 10% DMSO in FBS. The B16 thawing protocol is based upon a highly concentrated B16 cell frozen stock ($4–5 \times 10^6$/ml/vial) and should be used only as a guide. The proper dilution volume must be determined empirically, based upon the density of the frozen stock and the growth characteristics of the cells postthaw. B16 cells takes approximately 18 h to double in culture.
25. One T175 flask at 75–80% confluence will provide enough cells to challenge approximately 25 mice.
26. It is not necessary to include the CD45RB antibody in the purification of $CD4^+$ T cells for B16 tumor experiments. Instead, splenocytes should be stained with anti-CD8 antibody in addition to CD4 and CD25 to facilitate purification of $CD8^+$ T cells.
27. It is important to note that most IACUC guidelines require that all cell lines must be tested for murine virus contaminants prior to injecting into mice.
28. Isofluorane (1-chloro-2, 2,2-trifluoroethyl difluoromethyl) is a general inhalation anesthetic drug. Institutional approval is

required for use and proper training should be completed prior to use. Although isofluorane has the largest margin of safety of all potent halogenated agents, care should be taken to avoid excessive exposure.

29. Following isofluorane anesthesia, animals have reduced body temperature, which can cause distress. To aid in recovery, mouse cages can be placed on a heating pad or under a heat lamp. Institutional guidelines may require additional monitoring, so it is imperative to check with your institution for instructions.
30. In most cases, IACUC approval must be obtained to photograph animals.
31. For cell count consistency and accurate comparison between treated and untreated groups, we routinely take 6 lymph nodes: 2 inguinal, 2 axillary, and 2 cervical. Additional lymph nodes can be taken if so desired.
32. Caution: Irritant and suspected carcinogen. Refer to the manufacturer's MSDS for more details.

Acknowledgments

We thank Terrence Geiger and Hongbo Chi for advice and critical discussion and Samir Burns for technical guidance regarding EAE experiments. We are also grateful to Mary Jo Turk for advice and technical guidance regarding the B16 tumor model. We thank Karen Forbes, Tara Moore, Jessica Magwood, and Amy Krause for maintenance and breeding of mouse colonies, Andrea Szymczak-Workman for IBD histological analysis, Richard Cross, Greig Lennon and Stephanie Morgan for FACS, the St Jude VPC Laboratory for histological analyses, the staff of the Shared Animal Resource Center at St Jude for the animal husbandry, Matthew Smeltzer for advice on statistical analysis and the Hartwell Center for Biotechnology and Bioinformatics at St Jude for MOG synthesis and purification. LWC is supported by an Individual NIH NRSA (F32 AI072816). MB is supported by a Juvenile Diabetes Research Foundation International postdoctoral fellowship (3-2009-594). DAAV is supported by the National Institutes of Health (NIH) (AI39480, AI52199, AI072239), Juvenile Diabetes Research Foundation International (1-2004-141 [The Robert and Janice Compton Research Grant, In Honor of Elizabeth S. Compton] and 1-2006-847), a Cancer Center Support CORE grant (CA21765) and the American Lebanese Syrian Associated Charities (ALSAC).

References

1. Thornton, A. M., and Shevach, E. M. (1998) CD4+CD25+ immunoregulatory T cells suppress polyclonal T cell activation in vitro by inhibiting interleukin 2 production, *The Journal of Experimental Medicine 188*, 287–296.
2. Asseman, C., Mauze, S., Leach, M. W., Coffman, R. L., and Powrie, F. (1999) An essential role for interleukin 10 in the function of regulatory T cells that inhibit intestinal inflammation, *The Journal of Experimental Medicine 190*, 995–1004.
3. Dieckmann, D., Plottner, H., Berchtold, S., Berger, T., and Schuler, G. (2001) Ex vivo isolation and characterization of CD4(+) CD25(+) T cells with regulatory properties from human blood, *The Journal of Experimental Medicine 193*, 1303–1310.
4. Belkaid, Y. (2007) Regulatory T cells and infection: a dangerous necessity, *Nature Reviews 7*, 875–888.
5. Tang, Q., and Bluestone, J. A. (2008) The Foxp3+ regulatory T cell: a jack of all trades, master of regulation, *Nature Immunology 9*, 239–244.
6. Bettini, M., and Vignali, D. A. (2009) Regulatory T cells and inhibitory cytokines in autoimmunity, *Current Opinion in Immunology 21*, 612–618.
7. Fontenot, J. D., Rasmussen, J. P., Williams, L. M., Dooley, J. L., Farr, A. G., and Rudensky, A. Y. (2005) Regulatory T cell lineage specification by the forkhead transcription factor foxp3, *Immunity 22*, 329-341.
8. Collison, L. W., Workman, C. J., Kuo, T. T., Boyd, K., Wang, Y., Vignali, K. M., Cross, R., Sehy, D., Blumberg, R. S., and Vignali, D. A. (2007) The inhibitory cytokine IL-35 contributes to regulatory T-cell function, *Nature 450*, 566–569.
9. Workman, C. J., and Vignali, D. A. (2005) Negative regulation of T cell homeostasis by lymphocyte activation gene-3 (CD223), *Journal of Immunology 174*, 688–695.
10. Goldrath, A. W., Bogatzki, L. Y., and Bevan, M. J. (2000) Naive T cells transiently acquire a memory-like phenotype during homeostasis-driven proliferation, *The Journal of Experimental Medicine 192*, 557–564.
11. Cho, B. K., Rao, V. P., Ge, Q., Eisen, H. N., and Chen, J. (2000) Homeostasis-stimulated proliferation drives naive T cells to differentiate directly into memory T cells, *The Journal of Experimental Medicine 192*, 549–556.
12. Allez, M., and Mayer, L. (2004) Regulatory T cells: peace keepers in the gut, *Inflammatory Bowel Diseases 10*, 666–676.
13. Maloy, K. J., Salaun, L., Cahill, R., Dougan, G., Saunders, N. J., and Powrie, F. (2003) CD4+CD25+ T(R) cells suppress innate immune pathology through cytokine-dependent mechanisms, *The Journal of Experimental Medicine 197*, 111–119.
14. Powrie, F., Leach, M. W., Mauze, S., Menon, S., Caddle, L. B., and Coffman, R. L. (1994) Inhibition of Th1 responses prevents inflammatory bowel disease in scid mice reconstituted with CD45RBhi CD4+ T cells, *Immunity 1*, 553–562.
15. Mottet, C., Uhlig, H. H., and Powrie, F. (2003) Cutting edge: cure of colitis by CD4+CD25+ regulatory T cells, *Journal of Immunology 170*, 3939–3943.
16. Cahill, R. J., Foltz, C. J., Fox, J. G., Dangler, C. A., Powrie, F., and Schauer, D. B. (1997) Inflammatory bowel disease: an immunity-mediated condition triggered by bacterial infection with Helicobacter hepaticus, *Infection and Immunity 65*, 3126–3131.
17. Yen, D., Cheung, J., Scheerens, H., Poulet, F., McClanahan, T., McKenzie, B., Kleinschek, M. A., Owyang, A., Mattson, J., Blumenschein, W., Murphy, E., Sathe, M., Cua, D. J., Kastelein, R. A., and Rennick, D. (2006) IL-23 is essential for T cell-mediated colitis and promotes inflammation via IL-17 and IL-6, *The Journal of Clinical Investigation 116*, 1310–1316.
18. Kohm, A. P., Carpentier, P. A., Anger, H. A., and Miller, S. D. (2002) Cutting edge: CD4+CD25+ regulatory T cells suppress antigen-specific autoreactive immune responses and central nervous system inflammation during active experimental autoimmune encephalomyelitis, *Journal of Immunology 169*, 4712–4716.
19. Selvaraj, R. K., and Geiger, T. L. (2008) Mitigation of experimental allergic encephalomyelitis by TGF-beta induced Foxp3+ regulatory T lymphocytes through the induction of anergy and infectious tolerance, *Journal of Immunology 180*, 2830–2838.
20. Jager, A., Dardalhon, V., Sobel, R. A., Bettelli, E., and Kuchroo, V. K. (2009) Th1, Th17, and Th9 effector cells induce experimental autoimmune encephalomyelitis with different pathological phenotypes, *Journal of Immunology 183:* 7169–7177.

21. O'Connor, R. A., Prendergast, C. T., Sabatos, C. A., Lau, C. W., Leech, M. D., Wraith, D. C., and Anderton, S. M. (2008) Cutting edge: Th1 cells facilitate the entry of Th17 cells to the central nervous system during experimental autoimmune encephalomyelitis, *Journal of Immunology 181*, 3750–3754.
22. Langrish, C. L., Chen, Y., Blumenschein, W. M., Mattson, J., Basham, B., Sedgwick, J. D., McClanahan, T., Kastelein, R. A., and Cua, D. J. (2005) IL-23 drives a pathogenic T cell population that induces autoimmune inflammation, *The Journal Of Experimental Medicine 201*, 233–240.
23. Seliger, B., Wollscheid, U., Momburg, F., Blankenstein, T., and Huber, C. (2001) Characterization of the major histocompatibility complex class I deficiencies in B16 melanoma cells, *Cancer Research 61*, 1095–1099.
24. Fidler, I. J. (1973) Selection of successive tumour lines for metastasis, *Nature: New Biology 242*, 148–149.
25. Poste, G., Doll, J., Hart, I. R., and Fidler, I. J. (1980) In vitro selection of murine B16 melanoma variants with enhanced tissue-invasive properties, *Cancer Research 40*, 1636–1644.
26. Turk, M. J., Guevara-Patino, J. A., Rizzuto, G. A., Engelhorn, M. E., Sakaguchi, S., and Houghton, A. N. (2004) Concomitant tumor immunity to a poorly immunogenic melanoma is prevented by regulatory T cells, *The Journal of Experimental Medicine 200*, 771–782.
27. Zhang, P., Cote, A. L., de Vries, V. C., Usherwood, E. J., and Turk, M. J. (2007) Induction of postsurgical tumor immunity and T-cell memory by a poorly immunogenic tumor, *Cancer Research 67*, 6468–6476.
28. Brunkow, M. E., Jeffery, E. W., Hjerrild, K. A., Paeper, B., Clark, L. B., Yasayko, S. A., Wilkinson, J. E., Galas, D., Ziegler, S. F., and Ramsdell, F. (2001) Disruption of a new forkhead/winged-helix protein, scurfin, results in the fatal lymphoproliferative disorder of the scurfy mouse, *Nature Genetics 27*, 68–73.
29. Fontenot, J. D., Gavin, M. A., and Rudensky, A. Y. (2003) Foxp3 programs the development and function of CD4+CD25+ regulatory T cells, *Nature Immunology 4*, 330–336.
30. Ivanov, I. I., Atarashi, K., Manel, N., Brodie, E. L., Shima, T., Karaoz, U., Wei, D., Goldfarb, K. C., Santee, C. A., Lynch, S. V., Tanoue, T., Imaoka, A., Itoh, K., Takeda, K., Umesaki, Y., Honda, K., and Littman, D. R. (2009) Induction of intestinal Th17 cells by segmented filamentous bacteria, *Cell 139*, 485–498.
31. Gaboriau-Routhiau, V., Rakotobe, S., Lecuyer, E., Mulder, I., Lan, A., Bridonneau, C., Rochet, V., Pisi, A., De Paepe, M., Brandi, G., Eberl, G., Snel, J., Kelly, D., and Cerf-Bensussan, N. (2009) The key role of segmented filamentous bacteria in the coordinated maturation of gut helper T cell responses, *Immunity 31*, 677–689.

Chapter 10

In Vivo Depletion of FoxP3$^+$ Tregs Using the DEREG Mouse Model

Katharina Lahl and Tim Sparwasser

Abstract

In recent years, researchers have increasingly focused on the modulation of regulatory T cell (Treg) function to interfere with the outcome of virtually every type of immune response. For a long time, specific in vivo targeting of Tregs was precluded due to the lack of appropriate markers. Only after the discovery of Foxp3 as a Treg-specific transcription factor, was the development of Treg-specific mouse models feasible. We generated DEREG mice (DEpletion of REGulatory T cells), a BAC (bacterial artificial chromosome) transgenic mouse line, which allows direct in vivo analysis and depletion of this exceedingly important cell type. Our DEREG mice carry a DTR-eGFP transgene under the control of an additional Foxp3 promoter, thereby allowing specific depletion of Treg by application of diphtheria toxin at any desired point of time during an ongoing immune response.

This chapter will elaborate the advantages and disadvantages of employing different genetic approaches and discuss further parameters used in the studies focusing on employment of diphtheria toxin and its degree of general toxicity in mice. Additionally, we will address the question: to which extent DEREG mice are suitable for studying the effect of long-term Treg depletion during specific immune responses.

Key words: DEREG, DT mediated Treg depletion, Long-term depletion, Depletion efficacy

1. Introduction

When Sakaguchi et al. rediscovered regulatory T cells in 1995 (1), it became broadly accepted that these cells are indispensable for balancing immune responses, and researchers agreed on their crucial impact in maintaining peripheral tolerance and regulating immunity (2, 3). However, genetic models have not been available until recently to fully understand their importance and address mechanisms of tolerance induction. Previous attempts

George Kassiotis and Adrian Liston (eds.), *Regulatory T Cells: Methods and Protocols*, Methods in Molecular Biology, vol. 707, DOI 10.1007/978-1-61737-979-6_10,

have been made to study the role of Treg mainly by using depleting anti-CD25 antibodies. The fact that CD25 is also up-regulated on activated T cells and presence of CD25 negative Treg subpopulation (4–7) along with a study claiming that the abovementioned anti-CD25 antibody sheds the epitope rather than depleting the cells (8), limits the interpretation of the data (9). Additionally, it has been shown that the anti-CD25 antibody clone, namely PC61 used for in vivo depletion experiments, lingers in the system for several days and results in steric hindrance for fluorochrome conjugated anti-CD25 antibody (used later for analysis), a technical issue that further impedes reliable analysis of generated data (10).

A relatively new approach to fully deplete certain cell types within mice has been published in 2001 (11). Here, the authors achieve the ablation of cells by genetic introduction of primate diphtheria toxin receptor (DTR) targeted to a specific tissue by usage of a specific promoter. Diphtheria toxin, a *Corynebacterium diphtheriae* derived enterotoxin, efficiently blocks protein synthesis in mammalian cells and causes rapid cell death via apoptosis. Rodent cells are at least 10^3–10^5 times less susceptible to DT-induced cell death due to their low affinity DT receptor, as compared to the simian or human counterpart (12, 13). Thus, transgenic expression of the high affinity DT receptor version in a specific murine cell type allows for their specific ablation upon DT injection and provides a powerful tool to address the role of the targeted cell type upon inducible depletion. Fusion of the primate DTR with eGFP further allows for tracking of the targeted cells and monitoring their position and efficacy of depletion, as has been published in a conventional mouse model targeting dendritic cells (14).

In the past, requirement of a specific promoter to actively target a certain cell type using the DT approach, delayed the usage of this model to analyze Treg function. Only after the discovery of Foxp3 as a Treg-specific transcription factor (15, 16), it became feasible to introduce faithful DTR expression only in these cells. Generally, two targeting strategies can be employed to introduce transgene expression into cells: Transgenic approaches and knock-in technologies, with the latter being by far more time- and labor intensive. A major limitation in using transgenic methods has been the requirement of knowledge about the full promoter region of the desired target gene. Further, lack of specificity has often been an issue due to the lack of epigenetic elements, which have been shown to be occasionally located up to 50 kb up- or downstream of the actual genetic sequence. Random integration of the transgene upon pronucleus injection can also lead to differential expression of the transgene due to its location within the genome (homo- or heterochromatin, adjacent promoters, lack of insulators, etc.). In order to overcome these problems, BAC transgenesis (bacterial artificial chromosome) has been developed as a more refined mode to create transgenic mice combining the advantages of conventional transgenic approaches, such as speed and ease of use, with those of standard knock-in technologies as usage of uncharacterized promoters and complete regulons (17–19). Here, large genomic fragments containing the gene of interest are introduced into the genome, assuring the presence of potential enhancers and silencers, cis-elements that define expression domains (20) and protecting the transgene via insulators within the surrounding sequence, one reason why BAC transgenic mice are often referred to as pseudo knock-in mice. This technology can be widely used not only for the constitutive and conditional ablation of cells but also for other approaches such as expression mapping and identification of murine regulatory regions, cell fate mapping and identification of progenitor cells, conditional inactivation of genes, or the humanization of mice (reviewed in ref. (19)). Of note, determination of a frameshift mutation within "scurfin" (Foxp3) in the natural scurfy mouse mutant, Brunkow et al. proved correlation of the disrupted scurfin expression with the development of multi-organ autoimmune disease by introducing a BAC containing the wild type (WT) coding region (21).

Various methods have been developed to insert genes or delete DNA on BACs in vitro. In our experience, transfection of the BAC-containing E.coli with a vector consisting of the transgene, flanked by homologue boxes to the target region and containing RecA, has been shown to be most efficient (22). Many different genetically targeted mice carrying the DTR coding sequence in order to be expressed by specific cell types have been created in the recent years, spanning approaches to deplete many different immunologic cell types, such as macrophages (23, 24), natural killer cells (25), conventional dendritic cells (14), and langerhans cells (26, 27). In one mouse model, the DTR coding region follows a floxed stop codon knocked into the ubiquitously active ROSA26 locus, leading to DTR expression only in cre recombinase expressing cells (28). These mice could theoretically be used to deplete all specific cell types for which specific cre expressing mice are available in a double transgenic approach. One bottleneck of this approach in Treg depletion, however, is the inefficient thymocyte depletion due to weak expression of the *Gt(ROSA)26Sor* promoter in these cells (28).

By creating a BAC transgenic mouse model carrying the DTR-eGFP fusion protein under the control of the promoter driving expression of the Treg specific transcription factor Foxp3, we could show that depletion of Tregs in neonatal mice leads to the development of severe autoimmune disease causing death within a short timeframe (29). In parallel, Kim et al. published another, knock-in-based mouse model, in which depletion of

Tregs led to the development of autoimmunity, similar to our results (30). Disease development in their study did not depend on the age as demonstrated using adult mice.

The phenotypic differences between both mouse models can probably be best explained by using the different targeting approaches. Recently, another BAC technology-based mouse model was published, closely resembling the depletion effect of our mice. Here, the transgenic BAC was injected directly into NOD mice, a mouse model of autoimmune diabetes. Treg depletion led to rapid onset of the specific disease, but not to a strong systemic, multi-organ autoimmune pathology (31). Knock-in technology might still be more efficient and lead to expression of the transgene in 100% of all positive cells, whereas BAC transgenesis might be incomplete in terms of transgene expression. As a consequence, depletion in DEREG mice leads to clearance of $Foxp3^+$ Tregs to only 95–98% (29). Potentially this is not sufficient for disease development in adult mice, presumably due to the absence of lymphopenia-induced cell proliferation as can be observed in neonates (32) in addition to other developed regulatory mechanisms. Delayed expression of the transgene in BAC transgenic mice, upon Treg renewal after depletion, could be another explanation, in which case reinjection of the toxin does not lead to complete depletion.

Using our BAC transgenic-based mouse model, we could further investigate the role of Foxp3 as such in Tregs. Since these mice carry an additional Foxp3 promoter driving the transgene instead of Foxp3, crossing DEREG mice to scurfy mice allowed investigation of Foxp3 deficient "wanna-be" Tregs based on their GFP expression. Using this approach, we could demonstrate that those cells are not per se pathogenic (33).

It has been recently suggested that Foxp3 expression is not restricted to the Treg lineage, but that epithelial cells also express Foxp3, which could be the prime cause of death upon DT injection independently of Treg depletion in both models. However, careful analysis of both the transgene and endogenous Foxp3 expression, as well as reconstitution assays, proved the specificity of Treg depletion without affecting the epithelial tissue (39).

Taken together, DT injection in DEREG mice allows for highly specific depletion of Treg at any desired time point during an ongoing immune response. Depletion may not be fully complete, and long-term depletion is not feasible in these mice due to the outgrowth of transgene-negative $Foxp3^+$ Tregs over a time course exceeding 2 weeks (Fig. 1), nonetheless usage of these mice in various models has shown substantial effect of Treg depletion on the immunologic outcome already as a result of short-term depletion and thus underlines the exceptionally high suppressive potency of this cell type. In vitro suppression assay,

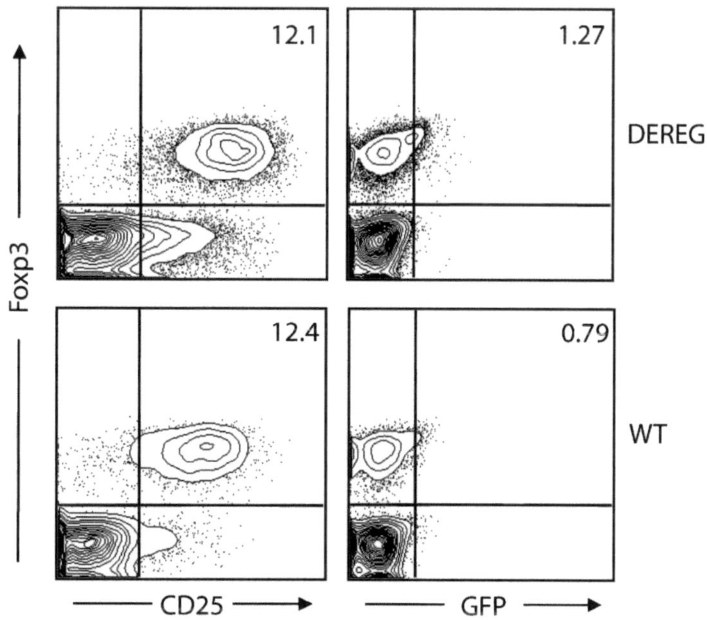

Fig. 1. Recovered Tregs after multiple rounds of DT application are negative for the transgene. DEREG and WT mice were treated with DT for 3 consecutive weeks as described in material and methods. One day after the third round of depletion, mice were sacrificed and lymph node cells were stained in order to assess Foxp3 vs. transgene (GFP) expression. Most recovered Tregs in DEREG mice are GFP negative. Plots show live-gated CD4+ cells.

cell proliferation measurement, and delayed type hypersensitivity are methods proposed in this chapter to test for the effects on Treg function following long-term depletion regimens. However, we could already show in our model that Treg ablation during the priming phase significantly enhances CD8+ T cell responses by day 7 and depletion during vaccination led to better clearance of *Listeria monocytogenes*, upon infection even several weeks later (34). How long this effect persists and if it reflects true amelioration of memory needs to be further investigated. As has been shown earlier, one important characteristic of DT mediated cell death is that it follows the apoptotic pathway rather than necrosis (26). Enhanced vaccination efficiency was not mediated by dying cells which might provide adjuvant effect, as mice deficient in the adaptor molecule ASC (required for inflammasome activation) (35) showed comparable CD8 responses upon Treg depletion (unpublished data). Given that our mice do not succumb to catastrophic autoimmune disease after Treg depletion in an adult stage, this model is highly suitable also to address the role of Tregs in settings other than full multi-organ autoimmune disease, such as tumor immunology, allergy, transplantation, infection, or antigen-induced autoimmune diseases.

2. Materials

2.1. In Vivo Depletion and Monitoring of Tregs in DEREG Mice

1. Diphtheria toxin (Merck, unnicked from *C. diphtheriae*, catalog number 322326): 1 μg in 100 μl phosphate buffered saline (Invitrogen, Carlsbad, CA).
2. Bleeding: Heparinized capillary tubes (VWR), anesthesia: Medetomidin(Domitor®),Midazolam(Midazolamratiopharm®), Fentanyl (Fentanyl®), Atipamezol (Antisedan®), Flumazenil (Anexate®), Naloxon (Naloxon®).
3. Antibodies for FACS analysis: Ethidium monazide (EMA, Invitrogen, Carlsbad, CA), αCD16/32 (Fc block, clone 93, eBioscience), αCD4-PerCP (clone RM4-5, BD biosciences, San Jose, CA), αCD25-APC (clone PC61.5, eBioscience), αFoxp3-PE (FJK-16s, eBioscience).
4. Buffers for FACS analysis: 1% BSA (Sigma Aldrich, St. Louis, MO) in PBS, fix/perm solution for Foxp3 staining (eBioscience, San Diego, CA).
5. V-bottom 96-well plates (Fisher, catalog number 07-200-108).

2.2. Assessment of CD8⁺ T Cell Responses After Priming in the Absence of Tregs

1. DT treatment and bleeding: see above.
2. Sensitization: ovalbumin, grade V (Sigma Aldrich), CpG ODN1886 (Cooley Pharmaceutics, Palo Alto, CA).
3. Antibodies and dyes: EMA (EMA, Invitrogen, Carlsbad, CA), αCD16/32 (Fc block, clone 93, eBioscience), αCD8α pacific blue (clone 53-6.7, eBioscience), αCD62L PE (clone MEL-14, eBioscience), IFNγ APC (clone XMG1.2, eBioscience).
4. Restimulation and staining buffers: $OVA_{257-264}$ (SIINFEKL) peptide (Peprotech, Rocky Hill, NJ), Brefeldin A (eBioscience), Fixation buffer (eBioscience), Permeabilization buffer (eBioscience).

2.3. Analysis of Recovered Tregs After Depletion

2.3.1. Proliferation Measurements

αKi67-PE (clone B56, BD biosciences), αIL-2-APC (clone JES6-5H4, eBioscience), αFoxp3-APC (FJK-16s, eBioscience), 1% BSA in PBS, EMA, fix/perm solution.

2.3.2. In Vitro Suppression Assay

1. Pooled lymph nodes from at least three mice per group.
2. CD4 negative depletion kit (Dynal, Invitrogen, Carlsbad, CA).
3. αCD25-PE (clone PC61.5, eBioscience, San Diego, CA).
4. αPE microbeads, αCD90 microbeads and MS columns (Miltenyi Biotech, Heidelberg, Germany).
5. Unconjugated αCD3 antibody (clone 145-2C11, eBioscience).

6. 96-well round bottom cell culture plates (Costar).
7. RPMI1640 medium (PAA, Linz, Austria) containing 10% FCS (Hyclone, Waltham, MA), 1% penicillin/streptomycin (PAA, Linz, Austria), 1% l-glutamin (PAA, Linz, Austria), and β-mercaptoethanol (Invitrogen, Carlsbad, CA).

2.3.3. Delayed Type Hypersensitivity

General delayed type hypersensitivity induction: for materials and general method see ref. (36).

Purification and transfer of Tregs:

1. CD4 negative depletion kit (Dynal, Invitrogen).
2. αCD25 PE (clone PC61.5, eBioscience, San Diego, CA), αPE microbeads, and MS columns (Miltenyi Biotech, Heidelberg, Germany).
3. SubQ 1 ml syringes (BD, Franklin Lakes, NJ).

2.4. Bone Marrow Chimeras

1. Syringes and needles, PBS, erythrocyte lysis buffer (Sigma, St. Louis, MO).
2. αCD90 magnetic beads and magnet (Miltenyi Biotech, Heidelberg, Germany).
3. αCD45.1-APC (clone A20, eBioscience).

3. Methods

3.1. In Vivo Depletion of Tregs in DEREG Mice

1. 1 mg of diphtheria toxin is reconstituted in 1 ml of sterile PBS and aliquoted into 100 μl stocks. DT is relatively unstable, requiring snap freezing on dry ice upon aliquoting to maintain its activity. Each vial can further be diluted with sterile PBS to give 10 ml of 1 μg DT/100 μl PBS working stock. Generally, this concentration of DT application works for most DT batches, but each batch should be tested for in vivo activity to obtain best results with as little toxin as possible (see Notes 1 and 2).
2. Administration of 1 μg DT i.p. for two consecutive days leads to almost complete depletion of Foxp3+ cells by day 3. Longer depletion is not recommended since efficiency does not improve due to absence of transgene expression in remaining cells. For neonatal mice, 100 ng in a volume of 25 μl has been used to deplete Tregs on days 1 and 8. This was sufficient for effective depletion of Tregs for at least 10 days (see Note 3).
3. Absence of Fop3+ Treg is monitored in peripheral blood on day 3 by retro-orbital bleeding of anesthetized mice. As little as 100 μl peripheral blood is sufficient for analysis in case multiple bleeding is required. Fully antagonisable injectable anesthesia has been the most appropriate method because it

allows anesthetizing the mice for the exact length of procedure intended. Anesthetic solution containing 0.5 mg/kg mouse Medetomidin, 5 mg/kg mouse Midazolam, and 0.05 mg/kg mouse Fentanyl is injected i.p. After bleeding retro-orbital (this bleeding method is particularly favorable in cases of multiple bleeding procedures), anesthetics will be antagonized by i.p. injection of 2.5 mg/kg mouse Atipamezol, 0.5 mg/kg Flumazenil, and 1.2 mg/kg mouse Naloxon (method adapted from TU Munich). Tregs are generally absent for only few days, starting to re-arise by day 5 in peripheral blood and immunological organs and Treg levels are completely back to normal by day 14.

4. FACS analysis is performed in V-bottom 96-well plates; EMA staining (1 mM in a volume of 50 μl FACS-buffer containing 0.5 μg Fc block) for 30 min on ice under light is followed by surface staining for 20 min on ice in the dark (50 μl FACS buffer containing 0.5 μl αCD4-PerCP and 0.25 μl αCD25-APC). Fixation and permeabilization are performed using 200 μl fix/perm solution for 30 min. This step can be prolonged up to overnight; however, best GFP signals are detected after shorter incubation. Intracellular staining of Foxp3 is performed in FACS buffer (0.25 μl in 50 μl staining volume). Two washing steps in 200 μl FACS buffer follow each staining or fixation step, centrifugation is performed at 1,300 rpm for 3 min. If necessary, samples can be stored at 4°C for up to 3 days prior to FACS analysis, although direct acquisition is recommended, especially when tandem conjugates are used for staining (see Note 4).

3.2. Assessment of CD8⁺ T Cell Responses After Priming in the Absence of Tregs

1. Mice are sensitized by dorsal subcutaneous injection with a mixture of 10 μg OVA protein and 10 nM CpG ODN1886 in PBS followed by DT injections on the next 2 days (see above).
2. On day 7, mice are anesthetized and bled for analysis or sacrificed if long-term monitoring of CTL responses is not required. Erythrocytes are lysed, the remaining blood cells counted and are plated into 96-well round bottom wells in a volume of 100 μl complete RPMI medium.
3. 2×10^6 cells can be easily restimulated and stained per well; generally all cells are used for peripheral blood leukocytes.
4. 100 μl of a two times concentrated stock solution containing 2 μl/ml Brefeldin A and 2 μM $OVA_{257\text{-}264}$ (SIINFEKL) peptide in complete RPMI1640 is added to the cells and the plates are incubated for 6 h at 37°C.
5. Staining is performed as described above for Foxp3 with following exceptions: αCD8 pacific blue staining is used instead of αCD4 staining (1:200), fixation buffer is used instead of

fix/perm, followed by intracellular staining for IFNγ instead of Foxp3 diluted 1:100 in permeabilization buffer (see Note 5).

3.3. Functional Analysis of Recovered Tregs

3.3.1. Proliferation Measurements

1. After three rounds of Treg depletion (two injections per week on consecutive days for 3 weeks, analysis on the day after the second DT injection in third week), LN from DEREG and WT mice are removed, filtered through 70 μM sieves and washed with PBS containing 1% BSA.
2. Staining for Ki67 (Ki67 PE, 1:20) is performed in order to measure the proliferation status of the cells. An isotype control is included in the Ki67 staining kit from BD. The staining protocol follows the same steps and utilizes the same buffers as the Foxp3 staining (see above, here: Foxp3 APC, dilution 1:50).
3. For results and interpretation see Note 6.

3.3.2. In Vitro Suppression Assay

1. Following the same depletion regimen as for measurements of proliferation, peripheral lymph nodes are removed from three mice per group to assess their in vitro suppressive capacity.
2. CD4+ cells are enriched by negative depletion kit in order to purify untouched CD4+ T cells following the manufacturer's recommendations (see Note 7).
3. Isolation is followed by incubation with αCD25-PE antibody (1:100). After washing, αPE microbeads are added as described in the product specifications and cells are applied to Miltenyi MS columns. The flow-through is kept as responder T cells, whereas the positive fraction reflects the Treg compartment.
4. Responder T cells are stained with 5 μM CFSE prior to culture to assess proliferation. For this, cells are washed twice in PBS and then stained with CFSE in PBS for 10 min at 37°C (residual FCS quenches the CSFE and thus needs to be removed prior to staining). After incubation, 10 ml of complete RPMI (containing 10% FCS) is added and cells are placed on ice for 10 min. After one additional washing step in RPMI/FCS, cells are ready to be plated. This staining protocol leads to an intensity that is distinguishable in brightness from the potentially remaining GFP signal in Treg. Allogenic markers can be used in addition if available.
5. Both cell types are adjusted to 1×10^6 cells per ml and cells are seeded into 96-well plate in varying ratios of 20:1, 10:1, 5:1, and 1:1 responder T cells:Treg in a final volume of 100 μl 1×10^6 antigen-presenting cells (irradiated splenic cells that have been depleted of CD90+ cells by magnetic bead separation) and 0.2 μg αCD3 are added in a volume of 50 μl each.
6. After 4 days of incubation, cells are harvested, stained for CD4 and allogenic markers, and CFSE dilution in responder T cells is assessed by flow cytometry.

3.3.3. Delayed Type Hypersensitivity

Generally, the published method is used (36)

1. To investigate regulatory function of rebounded Treg, the above-mentioned DT injection regimen is applied.
2. Mice are sensitized 1 day before third weeks' depletion. In case where magnetic bead-based sorted WT Treg are adoptively transferred to prove diminished immuno-regulation after multiple depletion, 1×10^6 of those cells in PBS are injected i.v. on the day of sensitization. Tregs are purified following the same protocol as used for the in vitro suppression assay.
3. Challenge is performed in concordance with the original protocol 6 days after sensitization and footpad thickening is assessed 24 h later (for directions of accurate measurements of footpad swelling follow instructions in the original protocol (36)).

3.4. Bone Marrow Chimeras for Origin Determination

1. DEREG mice are lethally irradiated (10.3 Gy) and reconstituted with 8×10^6 CD45.2 DEREG cells and 2×10^6 CD45.1 wt cells from bone marrow. Donor bone marrow cells are harvested from indicated mice by removing tibiae and femurs and flushing the bone marrow out of the bones using PBS. After erythrocytes are lysed using 1 ml lysis buffer per donor mouse for 1 min, cells are incubated with magnetic beads directly conjugated with αCD90 for 20 min and purified using Miltenyi columns in order to deplete mature T cells. Cells are counted, adjusted to 2×10^8 cells per ml, and mixed at a ratio of 5:1 DEREG(CD45.2): wt(CD45.1). 200 μl of the mixed cell suspension is injected i.v. 4 h after irradiation.
2. To allow for complete reconstitution of the immune system, mice are left for 8 weeks.
3. Prior to the first round of depletion, mice are bled to monitor the basic ratio of CD45.2 over CD45.1.
4. From then on, mice are depleted for 2 consecutive days every week and bled the day after second DT injection. Cells are stained for CD4, CD45.1, and Foxp3.
5. For results and interpretation see Note 8.

4. Notes

1. It is crucial to carefully titrate the DT batch since batches vary with regard to activity and purity. In contrast to mouse models targeting other cell types like dendritic cells (14, 26), depletion of Tregs requires approximately 50 times more DT, leading to a much higher potential of toxic, transgene unrelated, side effects (lower steady-state protein synthesis in

Tregs as compared to accessory cells might be one explanation). As an example, using a DT batch from one vendor with the same final concentration of DT as for a batch from another vendor, mice showed massive wasting and died within 3 weeks due to toxic effects (Fig. 2). Careful analysis of inner organs by histology revealed liver congestion with dilated sinusoids filled with red blood cells consistent with congestive heart failure as a putative cause of death in these mice. Signs of autoimmunity could not be observed.

2. Tregs can be depleted in DEREG cell cultures by using 100 ng DT per ml medium.
3. Onset of autoimmunity in neonatal DEREG mice upon depletion is a fine balance. The tendency for mice to succumb to the disease seems to be highly dependent not only on timing and DT doses but also on environmental factors. It might be helpful to avoid changing cages.
4. Whenever possible, Foxp3 APC staining should be avoided as much brighter signal is obtained when PE-conjugated version of the antibody is used.
5. Foxp3 staining and intracellular cytokine staining may be combined when using the Foxp3 staining protocol. Foxp3 staining does not work using the permeabilization buffer for intracellular cytokine staining. Since the GFP protein is fused

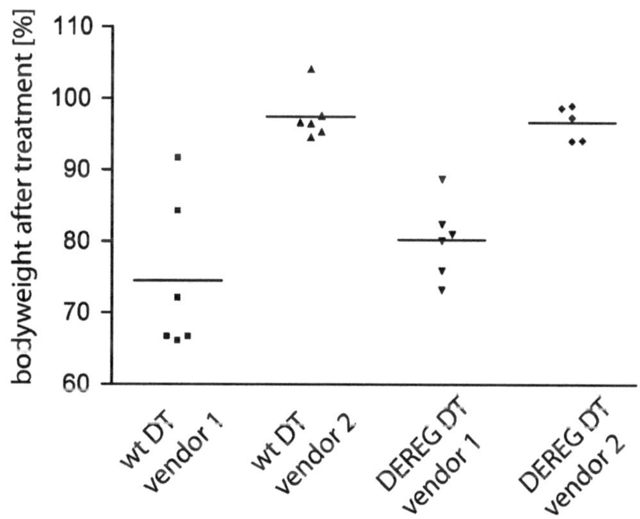

Fig. 2. Each DT batch requires careful titration and evaluation for biological activity when used in concentrations as high as being required for Treg depletion. DEREG and WT mice were injected with 1 μg DT (from two different vendors to compare biologic activity and toxicity) i.p. every 48 h. One day after the fourth DT injection, mice were sacrificed because of significant weight loss in two out of four groups. The graph shows the severe drop in body weight in both DEREG and control group as a consequence of injection of DT from one batch, but not from the other.

to the DTR, GFP will not be washed out and can still be assessed after permeabilization of the cells.

6. For long-term ablation of Tregs, the DEREG model is not well suited. As has been shown before, Tregs rebound fast after depletion. Interestingly, the recovered Tregs are GFP negative and thus cannot be depleted using DT anymore due to a general lack of transgene expression (Fig. 1 shows CD4 gated cells after DT injection twice a week for 3 weeks), showing that inefficient depletion is not a consequence of generation of neutralizing antibodies upon repetitive DT injections. This issue has been addressed more thoroughly by Buch et al. in a mouse model of cre-inducible DTR expression (28). However, preliminary data suggest the absence of Treg function during recovery of the population. Sorted Tregs based on CD25 expression were not efficient in suppressing allogenic effector T cell proliferation in an in vitro suppression assay (data not shown). To further analyze effects on Treg function after multiple depletions, Ki67 stainings were performed on both endogenous Tregs and effector T cells ex vivo after the third round of depletion, showing that both populations proliferated extensively (Fig. 3). Based on in vitro

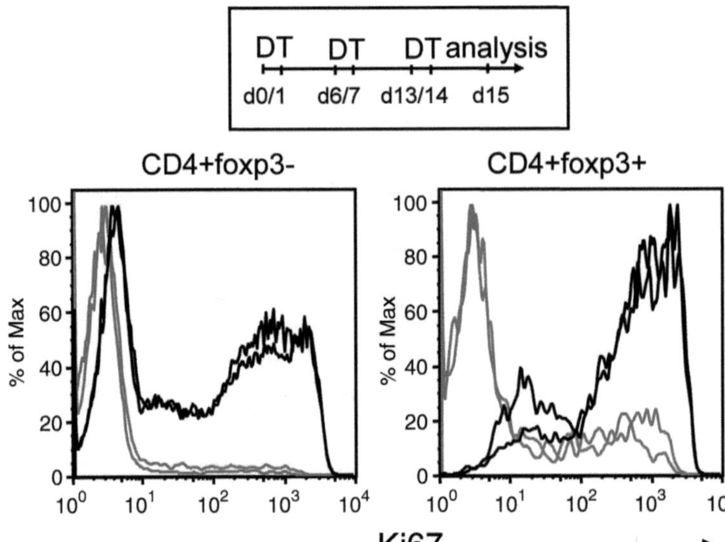

Fig. 3. Tregs as well as effector T cells proliferate extensively after multiple rounds of Treg depletion. After 3 weeks of depletion, lymph node cells from DEREG and WT mice were purified and stained for the proliferation marker Ki67. For gating, cells were stained with CD4 pacific blue and Foxp3 APC. Each line represents one individual mouse. DEREG-derived cells are shown in black and WT cells in grey. Histograms in the left panel show Foxp3⁻ gated cells (effector T cells) and histograms in the right column represent Foxp3⁺ gated cells (Tregs). Plots are representative for 3 individual experiments, each containing 2 mice per group.

data, it has been suggested that Tregs are nonfunctional during extensive proliferation (37). To test for functional relevance and to exclude hyperactivation of effector T cells after multiple depletions, delayed type hypersensitivity reactions were induced during the third round of depletion (Fig. 4). Footpad swelling could be detected in Foxp3^{+}GFP^{-} Treg-containing multiple depleted DEREG mice, but not in non-depleted controls. Further, adoptive transfer of natural resting Tregs from untreated DEREG mice rescued tolerance in the former group. These data suggest that Tregs cannot be depleted in long-term experiments, but can be rendered functionally impaired in DEREG mice. No development of general autoimmunity could be detected.

7. In vitro suppression assays work best with column purified Tregs. FACS-based sorting of Tregs results in diminished functional activity.

8. A major concern using transgenic mice when compared to knock-in mouse models is the potentially incomplete or unspecific expression of the transgene. BAC transgenesis

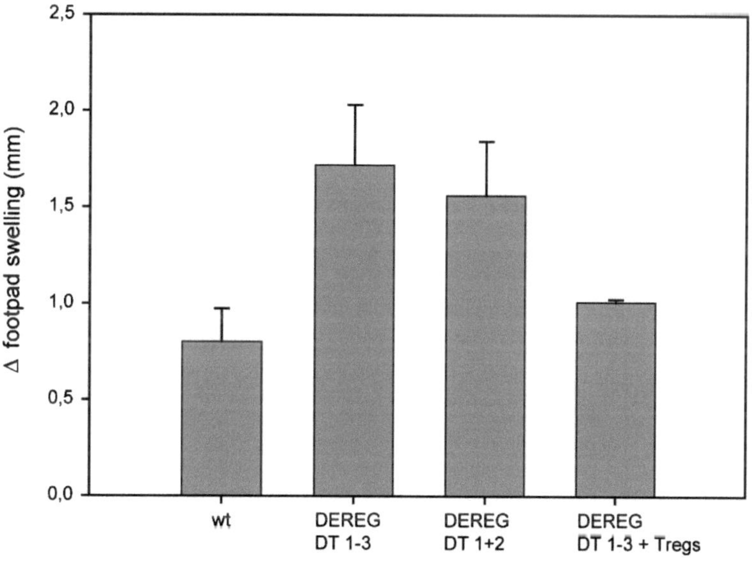

Fig. 4. Recovered Foxp3^{+}GFP^{-} Tregs after multiple DT injections do not rescue mice from the development of delayed type hypersensitivity. Mice were depleted for 2 weeks, twice each week, prior to sensitization. Sensitization was performed in week 3, 1 day before the third depletion round and mice were challenged 6 days later. The diagram shows footpad thickening of differentially treated mice, one group being depleted every week, one only before sensitization and one control group being replenished with WT Tregs prior to sensitization. No major differences in footpad swelling could be observed between the WT and the Treg replenished group. Both Treg depleted groups showed enhanced swelling, irrespective of mice being DT treated during sensitization or not. This confirms our hypothesis of diminished regulatory activity exerted by the outgrowing transgene negative Treg population following multiple depletions in the DEREG mice.

reflects an approach to avoid these problems by insertion of a large genomic fragment, containing all putative epigenetic regulators of the targeted promoter. We have shown that transgene expression is specific for Foxp3$^+$ cells in DEREG mice. However, we could not exclude the existence of a minor population of GFP$^-$Foxp3$^+$ Tregs. In light of the previous set of experiments using multiple rounds of depletion, we asked whether outgrowing GFP$^-$Foxp3$^+$ cell populations might arise from generally GFP$^-$ precursors. To address this, we set up mixed bone marrow chimeras using CD45.2 DEREG bone marrow and CD45.1 WT bone marrow in a ratio of 5:1 (Fig. 5). We used DEREG mice as host to account for potentially incomplete irradiation of endogenous Tregs, as these cells have been shown to be relatively resistant toward irradiation (38). Depletion of Tregs from DEREG mice led to a drop in Treg numbers according to the ratio of injected precursors. After 3 weeks of depletion, Treg numbers were recovered to normal in concordance with the earlier studies. Interestingly, Tregs were almost exclusively CD45.1$^+$, arguing for a scenario in which GFP$^-$ cells within the endogenous

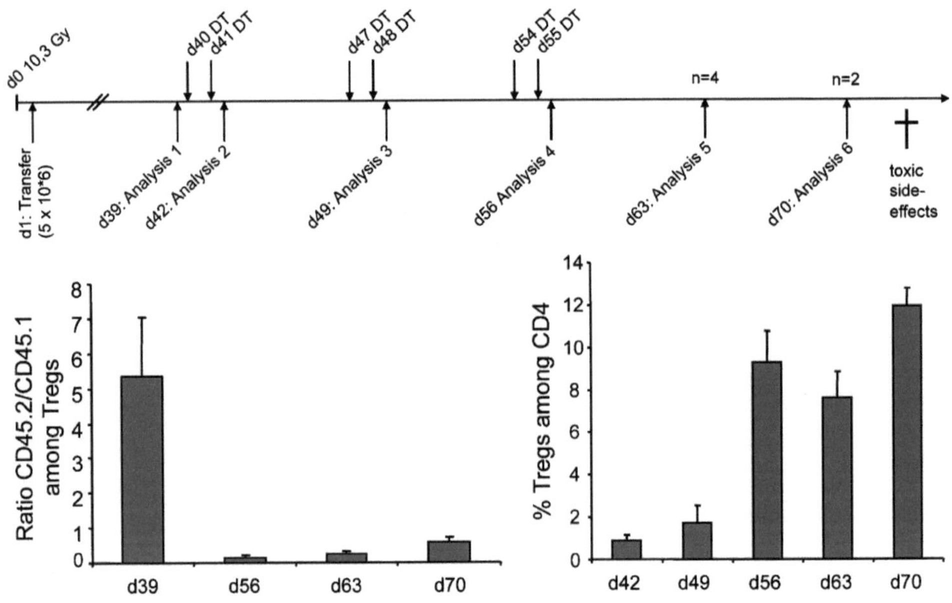

Fig. 5. Recovered Tregs after multiple rounds of depletion are likely to arise from a minor, GFP$^-$ subpopulation. Irradiated DEREG mice were reconstituted with 80% CD45.2 DEREG bone marrow and 20% CD45.1 wt bone marrow. The upper scheme shows the depletion regimen and time points of evaluation. As expected, the ratio of CD45.2 to CD45.1 was around 5 before administration of DT. Injection of the toxin led to depletion of DEREG derived Treg but not of WT derived Tregs, and percentages of DEREG derived Treg when compared to WT derived Tregs did not recover over time, probably due to the efficient fill-up of the niche by the WT derived cells. In this model of incomplete depletion, all remaining Tregs after toxin administration must have differentiated from the transgene negative donor with almost no in vivo conversion of DEREG derived effector T cells into transgene negative Tregs or outgrowth of DEREG derived transgene negative Tregs.

Treg population might heavily proliferate after GFP+ Treg cell depletion in order to refill the niche.

Taken together, our DEREG mouse model is well suited to study short-term immune responses in the absence of efficient Treg function. This model can be employed to dissect effects of Tregs on priming or challenge of the immune system, or to study cell–cell interactions in detail. Indeed few Tregs in these mice are not depleted due to insufficiency or absence of transgene expression which could be the explanation to why our DEREG mice do not develop massive autoimmune disease in an adult stage. This in turn allows use of DEREG mice in other models than full-blown autoimmunity in which Treg function matters.

References

1. Sakaguchi, S. et al. (1995) Immunologic self-tolerance maintained by activated T cells expressing IL-2 receptor alpha-chains (CD25). Breakdown of a single mechanism of self-tolerance causes various autoimmune diseases. *J Immunol* 155 (3), 1151–1164
2. Sakaguchi, S. (2004) Naturally arising CD4+ regulatory t cells for immunologic self-tolerance and negative control of immune responses. *Annu Rev Immunol* 22, 531–562
3. Belkaid, Y. and Rouse, B.T. (2005) Natural regulatory T cells in infectious disease. *Nat Immunol* 6 (4), 353–360
4. Fontenot, J.D. et al. (2005) Regulatory T cell lineage specification by the forkhead transcription factor foxp3. *Immunity* 22 (3), 329–341
5. Wan, Y.Y. and Flavell, R.A. (2005) Identifying Foxp3-expressing suppressor T cells with a bicistronic reporter. *Proc Natl Acad Sci U S A* 102 (14), 5126–5131
6. Stephens, L.A. and Mason, D. (2000) CD25 is a marker for CD4+ thymocytes that prevent autoimmune diabetes in rats, but peripheral T cells with this function are found in both CD25+ and CD25− subpopulations. *J Immunol* 165 (6), 3105–3110
7. Lehmann, J. et al. (2002) Expression of the integrin alpha Ebeta 7 identifies unique subsets of CD25+ as well as CD25− regulatory T cells. *Proc Natl Acad Sci U S A* 99 (20), 13031–13036
8. Kohm, A.P. et al. (2006) Cutting edge: anti-CD25 monoclonal antibody injection results in the functional inactivation, not depletion, of CD4+CD25+ T regulatory cells. *J Immunol* 176 (6), 3301–3305
9. Couper, K.N. et al. (2009) Anti-CD25 antibody-mediated depletion of effector T cell populations enhances susceptibility of mice to acute but not chronic Toxoplasma gondii infection. *J Immunol* 182 (7), 3985–3994
10. Couper, K.N. et al. (2007) Incomplete depletion and rapid regeneration of Foxp3+ regulatory T cells following anti-CD25 treatment in malaria-infected mice. *J Immunol* 178 (7), 4136–4146
11. Saito, M. et al. (2001) Diphtheria toxin receptor-mediated conditional and targeted cell ablation in transgenic mice. *Nat Biotechnol* 19 (8), 746–750
12. Naglich, J.G. et al. (1992) Expression cloning of a diphtheria toxin receptor: identity with a heparin-binding EGF-like growth factor precursor. *Cell* 69 (6), 1051–1061
13. Mitamura, T. et al. (1997) Structure-function analysis of the diphtheria toxin receptor toxin binding site by site-directed mutagenesis. *J Biol Chem* 272 (43), 27084–27090
14. Jung, S. et al. (2002) In vivo depletion of CD11c(+) dendritic cells abrogates priming of CD8(+) T cells by exogenous cell-associated antigens. *Immunity* 17 (2), 211–220
15. Fontenot, J.D. et al. (2003) Foxp3 programs the development and function of CD4+CD25+ regulatory T cells. *Nat Immunol* 4 (4), 330–336
16. Hori, S. et al. (2003) Control of regulatory T cell development by the transcription factor Foxp3. *Science* 299 (5609), 1057–1061
17. Yang, X.W. et al. (1997) Homologous recombination based modification in Escherichia coli and germline transmission in transgenic mice of a bacterial artificial chromosome. *Nat Biotechnol* 15 (9), 859–865

18. Heintz, N. (2001) BAC to the future: the use of bac transgenic mice for neuroscience research. *Nat Rev Neurosci* 2 (12), 861–870
19. Sparwasser, T. and Eberl, G. (2007) BAC to immunology – bacterial artificial chromosome-mediated transgenesis for targeting of immune cells. *Immunology* 121 (3), 308–313
20. McKnight, R.A. et al. (1992) Matrix-attachment regions can impart position-independent regulation of a tissue-specific gene in transgenic mice. *Proc Natl Acad Sci U S A* 89 (15), 6943–6947
21. Brunkow, M.E. et al. (2001) Disruption of a new forkhead/winged-helix protein, scurfin, results in the fatal lymphoproliferative disorder of the scurfy mouse. *Nat Genet* 27 (1), 68–73
22. Sparwasser, T. et al. (2004) General method for the modification of different BAC types and the rapid generation of BAC transgenic mice. *Genesis* 38 (1), 39–50
23. Duffield, J.S. et al. (2005) Selective depletion of macrophages reveals distinct, opposing roles during liver injury and repair. *J Clin Invest* 115 (1), 56–65
24. Cailhier, J.F. et al. (2005) Conditional macrophage ablation demonstrates that resident macrophages initiate acute peritoneal inflammation. *J Immunol* 174 (4), 2336–2342
25. Walzer, T. et al. (2007) Identification, activation, and selective in vivo ablation of mouse NK cells via NKp46. *Proc Natl Acad Sci U S A* 104 (9), 3384–3389
26. Bennett, C.L. et al. (2005) Inducible ablation of mouse Langerhans cells diminishes but fails to abrogate contact hypersensitivity. *J Cell Biol* 169 (4), 569–576
27. Kissenpfennig, A. et al. (2005) Dynamics and function of Langerhans cells in vivo: dermal dendritic cells colonize lymph node areas distinct from slower migrating Langerhans cells. *Immunity* 22 (5), 643–654
28. Buch, T. et al. (2005) A Cre-inducible diphtheria toxin receptor mediates cell lineage ablation after toxin administration. *Nat Methods* 2 (6), 419–426
29. Lahl, K. et al. (2007) Selective depletion of Foxp3+ regulatory T cells induces a scurfy-like disease. *J Exp Med* 204 (1), 57–63
30. Kim, J.M. et al. (2007) Regulatory T cells prevent catastrophic autoimmunity throughout the lifespan of mice. *Nat Immunol* 8 (2), 191–197
31. Feuerer, M. et al. (2009) How punctual ablation of regulatory T cells unleashes an autoimmune lesion within the pancreatic islets. *Immunity* 31 (4), 654–664
32. McHugh, R.S. and Shevach, E.M. (2002) Cutting edge: depletion of CD4+CD25+ regulatory T cells is necessary, but not sufficient, for induction of organ-specific autoimmune disease. *J Immunol* 168 (12), 5979–5983
33. Lahl, K. et al. (2009) Nonfunctional regulatory T cells and defective control of Th2 cytokine production in natural scurfy mutant mice. *J Immunol* 183 (9), 5662–5672
34. Heit, A. et al. (2008) Circumvention of regulatory CD4(+) T cell activity during cross-priming strongly enhances T cell-mediated immunity. *Eur J Immunol* 38 (6), 1585–1597
35. Mariathasan, S. et al. (2004) Differential activation of the inflammasome by caspase-1 adaptors ASC and Ipaf. *Nature* 430 (6996), 213–218
36. Luo, Y. and Dorf, M.E. (2001) Delayed-type hypersensitivity. *Curr Protoc Immunol* Chapter 4, Unit 4.5
37. Thornton, A.M. and Shevach, E.M. (1998) CD4+CD25+ immunoregulatory T cells suppress polyclonal T cell activation in vitro by inhibiting interleukin 2 production. *J Exp Med* 188 (2), 287–296
38. Komatsu, N. and Hori, S. (2007) Full restoration of peripheral Foxp3+ regulatory T cell pool by radioresistant host cells in scurfy bone marrow chimeras. *Proc Natl Acad Sci U S A* 104 (21), 8959–8964
39. Kim, J. et al. (2009) Cutting edge: depletion of Foxp3+ cells leads to induction of autoimmunity by specific ablation of regulatory T cells in genetically targeted mice. *J Immunol* 183(12), 7631–7763

Chapter 11

Antigen-Specific Induction of Regulatory T Cells In Vivo and In Vitro

Carolin Daniel, Hidde Ploegh, and Harald von Boehmer

Abstract

The peripheral induction of Foxp3-expressing regulatory T cells outside the thymus is required in order to maintain local homeostasis in distinct microenvironments such as the gut. Extrathymic induction of Treg may also be exploited to prevent unwanted immune responses. Here, we discuss the methodology allowing for the stable de novo generation of Tregs specific for foreign antigens in peripheral lymphoid tissue via subimmunogenic peptide delivery using either peptide contained in fusion antibodies directed against the DEC205 endocytotic receptor on steady-state dendritic cells or the implantation of peptide-delivering osmotic mini-pumps. Furthermore, we also address methods in order to achieve TGFβ-dependent Treg conversion in vitro, thereby mainly focusing on the role of retinoic acid (RA) to enhance TGFβ-dependent conversion into Tregs.

Key words: Foxp3, Regulatory T cells (Tregs), Conversion, Antigen, DEC205, Sortagging

1. Introduction

A variety of mechanisms have been identified by which the immune system acquires tolerance to self. Studies performed over the last decades have defined deletion of immature thymocytes (negative selection) before acquiring functional maturity in the thymic cortex and medulla as one important mechanism (1–4) that is supported by other mechanisms allowing for ectopic presentation of tissue antigens in the thymus (5–8). However, central tolerance mechanisms can fail (9, 10) permitting the escape of some self-reactive cells into the periphery. These cells probably bear TCRs that recognize weak self epitopes and thus peripheral tolerance mechanisms are needed that deal with such autoreactive cells. Peripheral tolerance mechanisms comprise deletion, reversible

George Kassiotis and Adrian Liston (eds.), *Regulatory T Cells: Methods and Protocols*, Methods in Molecular Biology, vol. 707, DOI 10.1007/978-1-61737-979-6_11, © Springer Science+Business Media, LLC 2011

anergy (11, 12) as well as dominant mechanisms that are imposed by regulatory T cells (Treg). The identification and characterization of Treg definitively confirmed the existence of dominant tolerance (13–15). In order to maintain local homeostasis, the organism requires the peripheral generation of Tregs in distinct microenvironments, such as the gut or at those sites chronically exposed to tumors or microbes. Tregs generated in the peripheral tissue outside of the thymus persist as resting cells at an intermitotic stage, independent of further supply of their agonist ligand that induced their formation. This is an important feature of the immune system, as it allows for the prospective induction of Tregs that suppress unwanted immunity. When Treg encounter their agonist TCR ligand, Tregs home to antigen-draining lymph nodes where they can undergo considerable expansion (16, 17). The specificity of Treg-mediated suppression results from the co-recruitment of Treg and other T cells in antigen-draining lymph nodes, thus Tregs of one particular specificity are able to suppress a variety of effector cells with different specificity when being colocalized at the same antigen-presenting cell (18, 19). Trafficking and migration to tissues and secondary lymphoid organs at sites of inflammation are required for Treg suppressive function (20) and allows for suppression of Th1 and Th17 responses.

Transcriptional profiling of Foxp3-expressing Tregs compared to naïve or activated T cells showed a distinct number of differentially expressed genes comprising some genes normally upregulated in activated T cells, such as *IL2ra* (CD25), *Ctla4* (CTLA4), and *Tnfrsf18* (glucocorticoid-induced TNF receptor; GITR), thus representing signature target genes for Foxp3 (21, 22).

We have developed protocols that allow the de novo generation of Tregs specific for foreign antigens in peripheral lymphoid tissue by delivering minute doses of peptide contained in fusion antibodies directed against the DEC205 endocytic receptor on steady-state dendritic cells (DCs) (17) or via the implantation of osmotic mini-pumps (23). These two approaches are both suitable methods to induce antigen-specific Tregs, provided that the delivery process does not result in DC activation. To assure the generation of Tregs under conditions of subimmunogenic antigen presentation, the peptide dose is not critical when the infusion method is used, but, the amount of the same antigen is crucial with the DEC205 delivery protocol, resulting from the fact that peptide infusion generally leads to a quick elimination of peptides, while with DEC205 delivery one creates an antigen deposit that critically influences T-cell activation.

Here, we will describe in detail the two approaches of peptide delivery allowing the stable induction of antigen-specific $CD4^+CD25^+$

suppressor T cells in vivo. In addition, we will discuss methods to achieve TGFβ-dependent Treg conversion in vitro, mainly focusing on the impact of retinoic acid (RA) to enhance TGFβ-dependent Treg conversion.

2. Materials

2.1. De Novo Production of Antigen-Specific Suppressor Cells by Peptide Infusion Via Mini-Osmotic Pumps

1. Mice: RAG2$^{-/-}$ Marilyn mice expressing a transgenic TCR for MHC class II restricted HY-peptide were a generous gift from P. Matzinger (National Institutes of Health, Bethesda, MD), Foxp3-GFP-expressing Marilyn mice were obtained by crossing Marilyn mice with Foxp3-GFP reporter mice (kindly provided by Dr. Rudensky). Mice were bred in the Dana-Farber Cancer Institute animal facility under specific pathogen-free conditions. Animal care and all procedures were in accordance with the guidelines of the Animal Care and Use Committee of the Dana Farber Cancer Institute.
2. Surgical instruments: Micro dissecting forceps, scissors, delicate hemostatic forceps as well as Reflex 9 Wound Clip Applier (all from Roboz, MD), Reflex 7 mm Wound Clips (Roboz, MD).
3. Osmotic micro-pump (Alzet, Model 1002, mean pumping rate 0.25 μl/h, mean fill volume 100 μl, infusion time 14 days, Durect corporation).
4. Insulin syringe (1 ml), BD.
5. Peptide of interest, here we used the IA-^{b} restricted *DbY* encoded HY-peptide (NAGFNSNRANSSRSS), synthesized by New England Peptide LLC.
6. Phosphate buffered saline (PBS, GIBCO).
7. Xylazine/Ketaject (Phoenix Pharmaceuticals) for anesthesia, ethanol, sterile gloves.

2.2. De Novo Production of Antigen-Specific Suppressor Cells by DEC205 Delivery

2.2.1. Sortagging of DEC-205 Antibody

2.2.1.1. Expression of Sortagged-DEC205 Antibody in CHOs Cells

1. CHO cells, Invitrogen.
2. Freestyle CHO Expression medium, Invitrogen.
3. Freestyle MAX Reagent, Invitrogen.
4. OptiPro SFM, Invitrogen.
5. L-Glutamine-200 mM (100×), liquid, Invitrogen.

2.2.1.2. Purification and Elution of Sortagged-DEC205 Antibody

1. HiTrap Protein G HP 1 or 5 ml columns (Amersham Biosciences, designed for use with a syringe, peristaltic pump or chromatography system) or Protein G sepharose beads.
2. Binding buffer: 20 mM sodium phosphate, pH 7.0.
3. Elution buffer: 0.1 M glycine-HCl, pH 2.7.
4. HiTrap Desalting, HiPreg 26/10 Desalting, or PD-10 Desalting Columns (Amersham Biosciences).

2.2.1.3. Sortase Reaction

1. Soluble *Staphylococcus aureus* sortase A was expressed and purified as described (24).
2. 10× Sortase reaction buffer: 10 mM $CaCl_2$, 50 mM Tris, pH 7.5, 150 mM NaCl.
3. Oligycine Nucleophile (peptide of interest containing 2–5 glycines at the N-terminus).
4. LPETG substrate (anti DEC205-ab containing the LPETG sequence at the COOH-terminus of the heavy chain of DEC205).

2.2.1.4. Test of Antigen Processing and Presentation of Sortagged-DEC205 Antibody In Vitro

1. MACS CD11c microbeads (Miltenyi Biotec).
2. Cell-culture media: RPMI 1640 (Gibco).
3. 5,6-carboxy fluorescein succinimidyl ester (CFSE, Molecular Probes) or ^{3}H-thymidine (Perkin Elmer).
4. Penicillin-Streptomycin, liquid, GIBCO.
5. FCS, GIBCO.

2.2.1.5. In Vivo Generation of Antigen-Specific Suppressor Cells Via Sortagged-DEC205 Antibody

1. 1 ml syringes, BD.
2. Mice (here, as described above, RAG2$^{-/-}$ Foxp3-GFP Marilyn mice).
3. Anti-DEC205 ab containing the peptide of interest, here anti-DEC205-HY antibody.

2.3. Analysis of Foxp3 Expression

1. PBS (Gibco).
2. EDTA (Boston Bioproducts).
3. FACS-buffer: 1× Hank's balanced salt solution supplemented with 5% FCS (vol/vol) and 10 mM HEPES buffer solution.
4. MACS-separation buffer: PBS supplemented with 0.5% BSA (wt/vol) and 2 mM EDTA.
5. Biotin-anti-CD4 antibody (RM4-5, BD).
6. PerCPCy5.5 anti-CD25 antibody (PC61, BD).
7. MACS Streptavidin Microbeads (Miltenyi Biotec).
8. Streptavidin Fluorochrome (e.g., Pacific Blue, BD).

9. MACS LS columns and MACS separator (all Miltenyi Biotec).
10. 5 ml polystyrene round-bottom tube with cell-strainer cap (e.g., BD Falcon).

2.4. In Vitro Conversion of Naïve into Regulatory T Cells Via TGFβ and Retinoic Acid

1. Foxp3-GFP reporter mice were kindly provided by Dr. Rudensky.
2. mAbs specific for CD4 (RM4-5), CD25 (PC61), CD44 (IM7), CD62L (MEL-14), BD, used as biotin, FITC, PE, PerCP-Cy5.5, allophycocyanin (APC), or Pacific Blue conjugates.
3. All-trans retinoic acid (Sigma).
4. Human Transforming Growth Factor β1, TGF-β1 (R&D systems).
5. Anti-CD3, anti-CD28 antibodies, BD.
6. Recombinant murine IL-2, Peprotech.
7. Cell-culture media, RPMI1640 (GIBCO).
8. Supplements for cell culture, GIBCO, see above.

3. Methods

3.1. Induction of HY-Specific Tregs in Female Foxp3-GFP-Marilyn Mice by Peptide Infusion

To induce antigen-specific tolerance, female 6–8 week-old Foxp3-GFP Marilyn mice are implanted subcutaneously with osmotic mini-pumps infusing daily 10 μg of HY-peptide or PBS for a time period of 14 days (pumping rate 0.25 μl/h).

3.1.1. Filling of the Micro-Osmotic Pumps

1. The empty pump needs to be weighed and filling of the pump is accomplished with a small syringe (1 ml) and the provided blunt-tipped, 27 gauge filling tube.
2. Filling of the pumps needs to be performed slowly in order to avoid the introduction of air bubbles. With the flow moderator removed, hold the pump in an upright position, insert the filling tube, and fill in the peptide solution in PBS until solution appears on the outlet. Here used, HY-peptide in PBS allowing for the infusion of 10 μg of HY-peptide at a pumping rate of 0.25 μl/h. Control pumps are filled with PBS only.
3. Wipe off excess solution and insert the flow moderator until the white flange is flush with the top of the pump.
4. The filled pump is weighed and the difference in weights obtained in steps 1 and 4 will give the net weight of the solution loaded. For most aqueous solutions, the weight in milligrams is approximately the same as the volume in microliters.

3.1.2. Subcutaneous Implantation of Mini-Osmotic Pumps

1. The mice are anaesthetized by intraperitoneal application of xylazine/Ketaject (Phoenix Pharmaceuticals, at a dose of 40/200 mg/kg).
2. Hair of the mice is shaved on the back; skin is cleaned with alcohol and painted with iodine.
3. Using micro-dissecting scissors a half inch mid-scapular skin incision is made slightly posterior to the scapulae.
4. A subcutaneous pocket is created using a hemostat and the pump is inserted into it with the flow moderator pointing away from the incision.
5. The wound is closed using a wound clip applier with 7 mm clips. Clips are removed 7 days post surgery.
6. In vivo conversion into Tregs in spleen, mesenteric and peripheral lymph nodes is analyzed by multi-color FACS analysis 14 days after implantation of the mini-osmotic pumps.

3.2. De Novo Production of Antigen-Specific Suppressor Cells by DEC205 Delivery

3.2.1. Sortagging of DEC-205 Antibody

We have described the production of anti-DEC205-peptide fusion antibodies in detail elsewhere (17, 25). In order to facilitate this approach in accordance with higher yields and less time needed for antibody production and therefore to allow straightforward installation of peptides covalently linked to the DEC-205 antibody we made use of sortase-mediated transpeptidation, or sortagging, a versatile orthogonal protein labeling method (24).

Bacterial sortases are thiol-containing enzymes that covalently attach proteins to the bacterial cell wall (*26*). The covalent linkage is directed by a sorting signal, LPETG, which is recognized by sortase A from *S. aureus*. Sortase A then cleaves the peptide bond between threonine and glycine. The carboxyl group of threonine is amide linked to the amino group of a pentaglycine nucleophile on the peptide of interest, which is in vivo provided by a cell wall precursor (*27*). Synthetic peptides containing 2–5 glycines at the N-terminus readily serve as nucleophiles in this reaction, and so allow the installation of any peptide of interest, provided it is preceded by a short run of glycines at the N-terminus.

Constructs encoding the anti-DEC205 Ig heavy chain with the LPETG tag as well as for the Ig heavy chain GL117 with LPETG tag (isotype control) were used (Ploegh et al, unpublished).

3.2.2. Expression of Sortag-DEC205 Antibody in CHOs Cells

1. Chinese Hamster Ovary (CHO) cells are used for transfection, expression, and production of sortaggable-DEC205-antibody. The CHO-S cell line is adapted to serum-free suspension growth in FreeStyle CHO expression medium, a serum- and protein-free medium which does not need to be changed after transfection.

2. Establish CHO cell culture and incubate cells in a 37°C incubator containing a humidified atmosphere of 8% CO_2 in air on an orbital shaker platform rotating at 125 rpm.
3. Subculture cells 24–48 h after thawing by seeding at 0.3×10^6 viable cells/ml in pre-warmed CHO expression medium supplemented with 8 mM L-Glutamine in 125 or 250 ml Erlenmeyer flasks.
4. Transfection of cells is done using the cationic lipid-based FreeStyle MAX reagent. Use cells at a density of 1×10^6 cells/ml. Passage cells ~24 h before transfection. It is important for high transfection results that viability of cells is over 95%.
5. For a cell volume of 30 ml use 20 μg of DNA encoding for the DEC205 Ig heavy chain as well as 20 μg of DNA encoding for the DEC205 Ig light chain. Dilute with OptiProSFM to 0.6 ml. Pipette plasmid DNA into a 15-ml conical tube. In a separate tube, dilute 40 μl of MaxReagent by adding 0.6 ml OptiProSFM and mix gently by inverting the tube (avoid vortexing). Immediately add diluted MAX reagent to diluted DNA and mix gently. Incubate for 10 min at room temperature to allow complexes to form.
6. Slowly add 1.2 ml of DNA-FreeStyle MAX reagent complex in the 125 ml flask containing cells while slowly swirling the flask. It is important to add the MAX DNA complex drop by drop.
7. Transfected cells are incubated at 37°C, 8% CO_2, on an orbital shaker for 6–7 days.

3.2.3. Purification and Elution of Sortaggable-DEC205 Antibody

1. Pour transfected CHO cells in 2–3 50 ml conical tubes, centrifuge at 6,500 rpm for 10 min, filter supernatant through 2–3 PVDF membrane tips attached to 60 ml syringes into 2 or 3 new 50 ml conical tubes.
2. The sample should be adjusted to the composition of the binding buffer either by diluting with binding buffer (for details refer to Subheading 2) or by buffer exchange using HiTrap Desalting, HiPrep 26/10 Desalting, or PD-10 Desalting Columns.
3. Prepare collection tubes by adding 60–200 μl 1 M Tris–HCl, pH 9.0 per ml of fraction to be collected.
4. Fill the syringe or pump tubing with binding buffer. Remove the stopper and connect the column to the syringe with the provided adaptor or pump tubing, drop to drop to avoid introducing air into the column.
5. Remove the snap-off end at the column outlet. Wash the column with ten column volumes of binding buffer at 5 ml/min for 5 ml column.

6. For the application of the sample, we use a syringe fitted to the luer adaptor or by pumping it onto the column.
7. Wash with 5–10 volumes of binding buffer.
8. Elute with 2–5 column volumes of elution buffer.
9. Check pH of all fractions and correct to pH7.4 as acidic or basic pH will denature the antibody.

3.2.4. Sortase Reaction

To perform the sortagging reaction, as exemplified here for a 100-μl reaction, one needs:

1. LPETG substrate, in our case the sortaggable-anti-DEC205-LPETG antibody.
2. Add 10 μl of 10× sortase reaction buffer (see Note 1).
3. Add peptide of interest, in our case GGG-HY-peptide (the glycines need to be appended to the N-terminus of the peptide) at a concentration of 1 mM.
4. Add sortase A at 50 μM.
5. Shake at 37°C overnight.
6. For removal of excess His-6-tagged sortase A we use NI-NTA beads. Wash NI-NTA beads once with water and once with TBS. Add 25 μl beads per 100 μl reaction and shake at 4°C overnight.
7. Centrifuge at 5,000 rpm at 4°C for 10 min and pipette of supernatant completely. Aliquot and store sortagged-anti-DEC205-HY at −80°C.

3.2.5. Test of Antigen Processing and Presentation of Sortag-DEC205 Antibody In Vitro

In order to confirm that the antigenic peptide delivered in form of the sortag-anti-DEC205 antibody is properly processed and presented, anti-DEC205 CD11c$^+$ DCs can be tested for their capacity to induce proliferative responses in antigen-specific CD4$^+$ T cells in vitro.

1. Purify CD11c$^+$DCs and incubate them with various amounts of anti-DEC205 antibody or control antibody for 30 min at 37°C. To remove unbound antibodies, wash cells twice.
2. Test the capacity of pulsed DCs to induce proliferation of antigen-specific CD4$^+$ T cells by setting up co-cultures with naïve CD4$^+$ T cells from TCR transgenic mice in 200 μl RPMI1640 medium supplemented with 10% (vol/vol) FCS in 96-well round-bottomed plates. As positive control use 10 μg/ml of specific peptide (here used HY-peptide) and add to some cultures.
3. Analysis of proliferation of antigen-specific T cells can be performed either by incorporation of ^{3}H-thymidine added for the last 12 h of a 70-h culture period followed by scintillation counting or by flow cytometric measurement of CFSE dilution. For CFSE labeling of T cells for in vitro proliferation,

incubate $<10^7$ cells in 1 ml of 0.1% BSA in PBS (wt/vol) and add 1 μl of 1 mM CFSE in DMSO. Incubate for 10 min at 37°C in the dark.

4. Wash cells once and resuspend cells in cell culture medium before adding to the culture. In case more than 10^7 cells are labeled in one reaction, scale up the labeling buffer volume in 1 ml increments as well as the amount of CFSE in 1-μl increments accordingly.
5. Flow cytometric analysis of CFSE dilution allows for the simultaneous analysis of the expression of activation markers (e.g., CD25 as an early T-cell activation marker). Example data are demonstrated in Fig. 1.

3.2.6. In Vivo Generation of Antigen-Specific Suppressor Cells Via Sortagged-DEC205 Antibody

1. To perform in vivo conversion of naïve T cells, here in our example into HY-specific Tregs via targeting the HY-peptide to steady-state DCs, inject 6–8-week old female Foxp3 GFP Marilyn mice intraperitoneally with a single dose of either anti DEC205-HY antibody or control antibody.

   Inject individual mice with titrated amounts of antibodies covering a spectrum of antibody concentrations ranging from 0.001 to 0.1 μg (see Note 2).
2. In vivo conversion into Tregs in spleen, mesenteric, and peripheral lymph nodes is analyzed by multi-color FACS analysis 14 days after injection of sortag-DEC205 antibodies. Example data are shown in Fig. 1.

3.3. Analysis of Foxp3 Expression Following In Vivo Conversion of Naïve T Cells into Tregs

1. Single cell suspensions from spleens, mesenteric, and peripheral (axial and inguinal) lymph nodes are prepared.
2. All staining reactions are preceded by a 10-min incubation with a blockade mixture made of 2.4G2 supernatant (Fc-block) and 10% rat and mouse sera (Jackson ImmunoResearch Laboratories).
3. Magnetic bead separation for the enrichment of induced suppressor cells:

   It is recommended that $CD4^+$ cells from the spleen and lymph nodes of individual mice are purified using a magnetic bead-based enrichment step.

   Single cell suspensions of spleens and lymph nodes are resuspended (~10^8 cells) in 1 ml FACS buffer (for recipe see Subheading 2). Add fluorochrome-conjugated Abs (e.g., PerCP Cy5.5 anti-CD25 and APC anti-Thy1.2) and biotin-anti-CD4 in appropriate amounts previously determined by titration experiments, and incubate in the dark on ice for 30 min.

   Wash cells once with >10 ml FACS buffer and aspirate supernatant completely. Resuspend cell pellet in 250 μl FACS

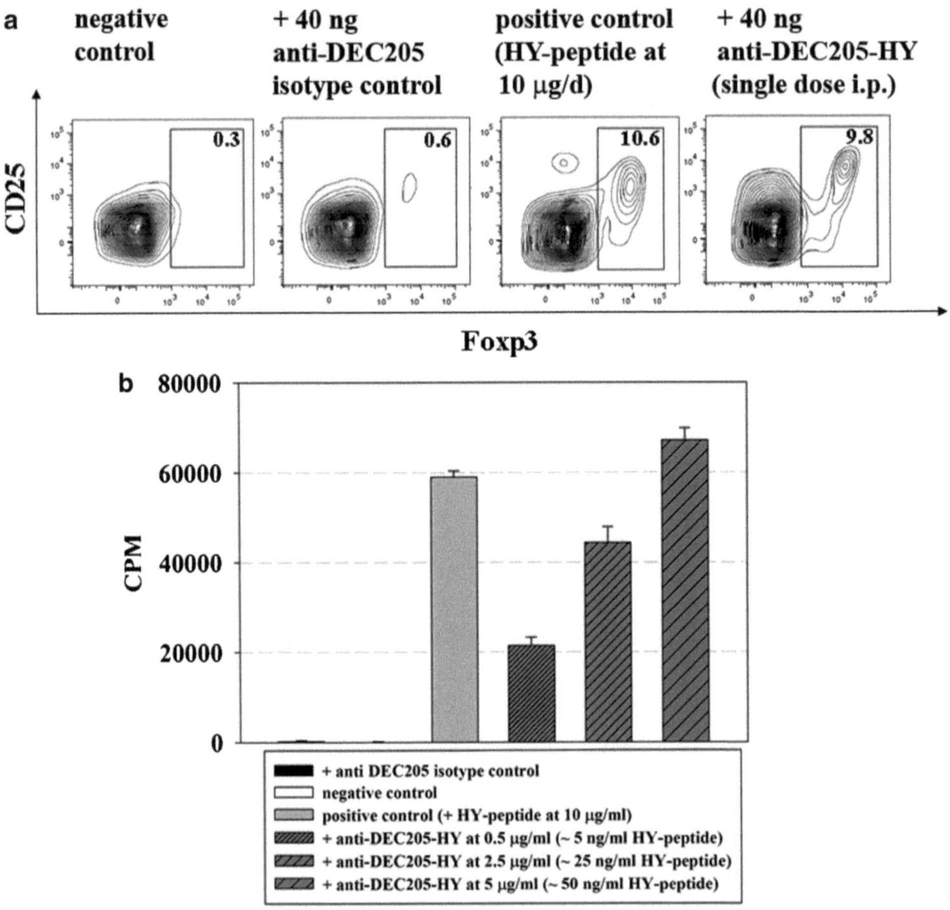

Fig. 1. (**a**) Induction of HY-specific Tregs in female Foxp3-GFP-Marilyn mice by peptide infusion or anti-DEC205-HY-delivery. Female 6–8-week-old Foxp3-GFP-Marilyn mice are subcutaneously implanted with mini-osmotic pumps infusing HY-peptide (10 μg/day) or injected with a single dose of 40 ng titrated anti-DEC205-HY antibody. After 14 days, in vivo conversion into Tregs in spleen, mesenteric, and peripheral lymph nodes is analyzed by multi-color FACS analysis. Numbers indicate the percentages of $CD4^+CD25^+Foxp3^+$ cells. (**b**) Anti-DEC205-HY delivery of antigen to dendritic cells in vitro. Purified splenic $CD11c^+$ DCs were incubated with the indicated amounts of fusion antibodies. After unbound antibodies were removed by washing, pulsed DCs were co-cultered with HY-specific $CD4^+$ T cells at a 1:1 ratio. Analysis of proliferation of antigen-specific T cells is performed by incorporation of ^{3}H-thymidine added for the last 12 h of 70 h culture period followed by scintillation counting. The anti-DEC205-HY antibody induces strong activation and proliferation of HY-specific $CD4^+$ T cells. DCs incubated with the same amount of isotype control ab failed to mediate a proliferative T-cell response. Free HY-peptide served as a positive control.

buffer per 10^8 cells and add 25 μl of streptavidin microbeads. This cell-to-bead ratio has been optimized in our laboratory and differs from the recommendation of the manufacturer (see Note 3). Incubate in the dark for 15 min on ice; add an appropriate amount of streptavidin-fluorochrome (e.g., pacific blue) and incubate for an additional 10 min.

Wash cells once with >10 ml FACS buffer and resuspend cell pellet in 3 ml degassed MACS separation buffer (for recipe refer to Subheading 2). Filter sample through a 40-mm mesh or filter to eliminate cell clumps (e.g., 5 ml polystyrene round-bottom tube with cell strainer cab). Apply cell suspension to the prepared MACS separation LS column, as indicated by the manufacturer. In order to achieve maximum purity of the $CD4^+$ fraction retained on the column, wash column twice with 7 ml MACS buffer. The eluted fraction is usually composed of >95% $CD4^+$ cells.

4. Enumeration of cells and acquisition are performed by using FACSAria and FACSDiva software (Becton Dickinson). Single-cell data analyses are done by the use of the FlowJo software (Tree Star).

3.4. In Vitro Conversion of Naïve into Regulatory T Cells Via TGFβ and Retinoic Acid

It has been reported that RA enhances Treg cell conversion by inhibiting the secretion of cytokines that interfere with conversion. A more recent analysis of carefully separated T cell subsets concluded that RA elicits its effect via "contaminating" activated $CD44^{hi}$ cells that secrete cytokines in response to antigenic stimulation, while these cytokines in turn prevent the conversion of naïve T cells into Tregs (28). Moreover, in a recent exchange of letters, it was proposed that RA directly affects conversion of naïve T cells, possibly via the inhibition of cytokine secretion by naïve T cells (29, 30).

We have investigated the role of RA in the Treg conversion process in more detail by analyzing the contribution of $CD44^{hi}$ cells, titrating costimulating CD28 antibodies as well as cytokines in order to optimize protocols for in vitro conversion of naïve T cells into Tregs.

The results show that RA can interfere with the negative effect of costimulation and certain cytokines on naïve T cells, in addition to directly inhibiting cytokine secretion. Furthermore, RA can enhance Treg cell conversion of naïve T cells in the absence of secreted inhibitory cytokines.

1. For in vitro conversion assays, Foxp3-GFP reporter mice were used. T cells are purified from spleen and lymph nodes of 6–8-week old mice. Highly purified naïve $CD4^+$ T cells are FACS sorted (FACS Aria cell sorter, BD) as $CD4^+CD44^{lo}CD62L^{hi}CD25^-Foxp3\text{-}GFP^-$ T cells.
2. Naïve $CD4^+$ T cells are activated with plate-bound anti-CD3 alone or together with anti-CD28 antibodies at a concentration of 5 μg/ml in the presence of 100 U/ml recombinant murine IL-2. T cells are then cultured in 96-well flat bottom plates at a concentration of 0.5×10^5 cells per well for 3 days.
3. Naïve T cells are treated with recombinant TGFβ at a concentration of 1 ng/ml and/or RA at 2.5 nM.
   The effect of RA on the conversion process is clearly diminished but not abolished when cells are cultured with CD3 and CD28 antibodies at a 1:1 ratio. The magnitude of the RA

enhancing impact is dependent on the degree of costimulation, since it is reduced when only CD3 antibodies are used. Excess costimulation (ratio CD28:CD3 2:1) significantly decreases the conversion rate in the absence of RA. Addition of RA at 2.5 nM allows for full reversal of this decrease. Thus, the direct effect of RA on the conversion of naïve T cells can be best seen under conditions of enhanced costimulation.

4. After 3 days of culture cells are examined by FACS for expression of GFP.

4. Notes

1. It is important that the sortase reaction is performed using buffers that contain no phosphate.
2. Anti-DEC205 antibodies need to be carefully titrated to allow for efficient conversion under subimmunogenic conditions thus avoiding DC activation.
3. Magnetic bead separation for the enrichment of induced Tregs:

   Note that an optimized cell-to-bead ratio is used for the streptavidin microbeads which differs from that recommended by the manufacturer.

Acknowledgments

These studies were supported by NIH grant NIH-AI-53102 to Harald von Boehmer. Carolin Daniel was supported by a Leopoldina research fellowship (BMBF-LPD 9901/8-184) and by LOEWE (LiFF) program of the Federal State of Hessen, Germany.

References

1. Burnet FM. The Clonal Selection Theory. 200. Cambridge Press, London, 1959.
2. Kappler JW, Roehm N, Marrack P. T cell tolerance by clonal elimination in the thymus. Cell 1987;49(2):273–280.
3. Kisielow P, Bluthmann H, Staerz UD, Steinmetz M, von BH. Tolerance in T-cell-receptor transgenic mice involves deletion of nonmature CD4+8+ thymocytes. Nature 1988;333(6175):742–746.
4. Lederberg J. Genes and antibodies. Science 1959;129(3364):1649–1653.
5. Albert ML, Pearce SF, Francisco LM et al. Immature dendritic cells phagocytose apoptotic cells via alphavbeta5 and CD36, and cross-present antigens to cytotoxic T lymphocytes. J Exp Med 1998;188(7): 1359–1368.
6. Albert ML, Sauter B, Bhardwaj N. Dendritic cells acquire antigen from apoptotic cells and induce class I-restricted CTLs. Nature 1998;392(6671):86–89.
7. Antonia SJ, Geiger T, Miller J, Flavell RA. Mechanisms of immune tolerance induction

through the thymic expression of a peripheral tissue-specific protein. Int Immunol 1995;7(5):715–725.

8. Jolicoeur C, Hanahan D, Smith KM. T-cell tolerance toward a transgenic beta-cell antigen and transcription of endogenous pancreatic genes in thymus. Proc Natl Acad Sci U S A 1994;91(14):6707–6711.
9. Kishimoto H, Sprent J. A defect in central tolerance in NOD mice. Nat Immunol 2001;2(11):1025–1031.
10. Liston A, Hardy K, Pittelkow Y et al. Impairment of organ-specific T cell negative selection by diabetes susceptibility genes: genomic analysis by mRNA profiling. Genome Biol 2007;8(1):R12.
11. Rocha B, von BH. Peripheral selection of the T cell repertoire. Science 1991;251(4998): 1225–1228.
12. Rocha B, Tanchot C, von BH. Clonal anergy blocks in vivo growth of mature T cells and can be reversed in the absence of antigen. J Exp Med 1993;177(5):1517–1521.
13. Fontenot JD, Gavin MA, Rudensky AY. Foxp3 programs the development and function of CD4+CD25+ regulatory T cells. Nat Immunol 2003;4(4):330–336.
14. Sakaguchi S. Naturally arising Foxp3-expressing CD25+CD4+ regulatory T cells in immunological tolerance to self and non-self. Nat Immunol 2005;6(4):345–352.
15. Sakaguchi S, Yamaguchi T, Nomura T, Ono M. Regulatory T cells and immune tolerance. Cell 2008;133(5):775–787.
16. Apostolou I, Sarukhan A, Klein L, von BH. Origin of regulatory T cells with known specificity for antigen. Nat Immunol 2002;3(8):756–763.
17. Kretschmer K, Apostolou I, Hawiger D, Khazaie K, Nussenzweig MC, von BH. Inducing and expanding regulatory T cell populations by foreign antigen. Nat Immunol 2005;6(12):1219–1227.
18. Jaeckel E, von BH, Manns MP. Antigen-specific FoxP3-transduced T cells can control established type 1 diabetes. Diabetes 2005;54(2):306–310.
19. Tarbell KV, Yamazaki S, Olson K, Toy P, Steinman RM. CD25+ CD4+ T cells, expanded with dendritic cells presenting a single autoantigenic peptide, suppress autoimmune diabetes. J Exp Med 2004;199(11): 1467–1477.
20. Rudensky AY, Campbell DJ. In vivo sites and cellular mechanisms of T reg cell-mediated suppression. J Exp Med 2006;203(3):489–492.
21. Marson A, Kretschmer K, Frampton GM et al. Foxp3 occupancy and regulation of key target genes during T-cell stimulation. Nature 2007;445(7130):931–935.
22. Chen C, Rowell EA, Thomas RM, Hancock WW, Wells AD. Transcriptional regulation by Foxp3 is associated with direct promoter occupancy and modulation of histone acetylation. J Biol Chem 2006;281(48): 36828–36834.
23. Verginis P, McLaughlin KA, Wucherpfennig KW, von BH, Apostolou I. Induction of antigen-specific regulatory T cells in wild-type mice: visualization and targets of suppression. Proc Natl Acad Sci U S A 2008;105(9): 3479–3484.
24. Popp MW, Antos JM, Grotenbreg GM, Spooner E, Ploegh HL. Sortagging: a versatile method for protein labeling. Nat Chem Biol 2007;3(11):707–708.
25. Kretschmer K, Heng TS, von BH. De novo production of antigen-specific suppressor cells in vivo. Nat Protoc 2006;1(2):653–661.
26. Ton-That H, Liu G, Mazmanian SK, Faull KF, Schneewind O. Purification and characterization of sortase, the transpeptidase that cleaves surface proteins of Staphylococcus aureus at the LPXTG motif. Proc Natl Acad Sci U S A 1999;96(22):12424–12429.
27. Marraffini LA, Schneewind O. Targeting proteins to the cell wall of sporulating Bacillus anthracis. Mol Microbiol 2006;62(5):1402–1417.
28. Hill JA, Hall JA, Sun CM et al. Retinoic acid enhances Foxp3 induction indirectly by relieving inhibition from CD4+CD44hi Cells. Immunity 2008;29(5):758–770.
29. Hill JA, Hall JA, Sun CM et al. Response to letter from Mucida et al. Immunity 2009;30:472–473.
30. Mucida D, Pino-Lagos K, Kim G et al. Retinoic acid can directly promote TGF-beta-mediated Foxp3(+) Treg cell conversion of naive T cells. Immunity 2009;30(4):471–472.

Chapter 12

In Vitro Expansion of Alloantigen-Specific Regulatory T Cells and Their Use in Prevention of Allograft Rejection

Clémence Nouzé, Lise Pasquet, and Joost P.M. van Meerwijk

Abstract

Regulatory T lymphocytes expressing CD4, high levels of CD25, and the transcription factor Foxp3 play a crucial role in the control of immune responses to self and nonself antigens. In contrast to immunosuppressive drugs currently used to treat immunopathology, these cells act in a very specific manner. Consequently, their clinical potential in the treatment of autoimmune disorders, inflammatory diseases, graft-versus-host disease, and allograft rejection is currently extensively studied in experimental animal models as well as in clinical trials. We have previously shown that appropriately in vitro stimulated $CD4^{+}CD25^{high}$ regulatory T cells can be used to prevent rejection of bone marrow, skin, and heart allografts in the Mouse. We here describe the protocols used in our laboratory to isolate mouse regulatory T cells, to stimulate them in vitro in order to enrich in cells specific for donor-antigens, and to transplant bone marrow under cover of regulatory T cells. Thus, generated hematopoietic chimeras may subsequently be transplanted with solid tissues and organs from the same donor.

Key words: Immunology, Immunoregulation, Regulatory T lymphocyte, Transplantation, Hematopoietic chimerism, Mouse, Allograft rejection, Immunosuppression

1. Introduction

Regulatory T lymphocytes (Treg) play a central and nonredundant role in the control of immune responses (1). One of the best-characterized regulatory T cell populations expresses the coreceptor CD4, high levels of the IL-2Rα chain CD25, and the forkhead/winged helix transcription factor Foxp3 (2). Absence of these cells because of mutations in the FOXP3 gene leads to the syndrome Immunodysfunction Polyendocrynopathy Enteropathy X-linked (IPEX) (3). This observation clearly demonstrates the crucial role of Treg in prevention of autoimmune

George Kassiotis and Adrian Liston (eds.), *Regulatory T Cells: Methods and Protocols*, Methods in Molecular Biology, vol. 707, DOI 10.1007/978-1-61737-979-6_12,

pathology and also strongly suggests that these cells play an important role in the control of immune responses to nonself antigens. From experimental animal studies and clinical research, we now know that Treg control immune responses not only to self-antigens but also to tumors (4), to pathogens (5), and to the fetus (6).

Given the fundamental physiological role of Treg in control of immune responses, the use of these cells for therapeutic purposes appears very tempting. In contrast to immunosuppressive drugs, Treg act in an antigen-specific manner (7), and their clinical use should therefore avoid the severe side effects of currently used drugs. The observation that Treg control maternal immune responses to paternal antigens of the fetus (6) suggested that these cells may be very efficient in preventing immune responses to antigens expressed by allografts. We have tested this hypothesis in, initially, a bone marrow transplantation model in the Mouse (7, 8). Host-derived Treg were isolated by selecting $CD4^{+}CD25^{high}$ splenocytes. To enrich this population in cells specific for donor antigens, we cultured them in presence of donor spleen-derived antigen-presenting cells. We also added high levels of IL-2 to break the in vitro anergic state of Treg (9). These cultured Treg were subsequently injected in preconditioned hosts that were simultaneously transplanted with donor bone marrow. Thus, the allograft was efficiently protected from rejection by the host's immune system. We showed that protection was durable and donor-specific. When the generated hematopoietic chimeras were subsequently grafted with skin or heart from the same donor, the latter allografts were fully protected from acute and chronic rejection (10). Importantly, prevention of chronic rejection required that the injected Treg were specific for donor antigens directly presented by donor APC and indirectly by host APC. The latter observation indicated that protection from solid allograft rejection was due to the injected Treg and not (solely) to the previously induced hematopoietic chimerism. It also has important implications for the in vitro Treg culture protocol.

We here describe the detailed protocols for isolation of splenic Treg, their in vitro culture, and allogeneic bone marrow transplantation under cover of Treg in the Mouse. The protocol has allowed for permanent acceptance of bone marrow allografts in all of the numerous semi- or fully allogeneic host/donor combinations we tested. The generated hematopoietic chimeras can subsequently be transplanted with skin or heart allografts from the same donor using specialized protocols previously described (11, 12).

2. Materials

2.1. Isolation of Splenic Treg

1. Mice: Any strain of inbred mouse can be used. These mice are commercially available from several suppliers. We always use specific pathogen free (SPF) animals.
2. RPMI 1640 medium (Eurobio, Les Ulis, France) supplemented with 10% heat-inactivated fetal calf serum (FCS), 2 mM L-glutamine, penicillin, streptomycin, 10 mM Hepes, 50 μM 2-mercaptoethanol (2-ME), 1 mM nonessential amino acids, 1 mM sodium pyruvate.
3. Lympholyte-M (Cedarlane laboratories, Hornby, ON, Canada).
4. MACS Buffer: phosphate buffered saline (PBS), supplemented with 3% BSA (Bovine Serum Albumin) and 0.5 mM EDTA. Sterilize by filtration on a 0.2-μM membrane filter (e.g., Millipore, Billerica, MA). Store at 4–8°C.
5. Mouse CD4 Cell Negative Isolation Kit (Dynal Biotech, Oslo, Norway).
6. Hybridoma supernatants: hybridomas are cultured in complete medium with 5% FCS. When more than 90% of the cells are dead, supernatants are harvested by centrifugation and subsequent filtration on a 0.4-μM membrane filter.
7. MicroBeads coated with anti-PE antibody (Miltenyi Biotec, Paris, France).
8. MS columns and MiniMACS separator (Miltenyi Biotec, Paris, France).
9. Fluorochrome-conjugated antibody to mouse antigens: CD25-PE (PC61), CD4-APC (L3T4) (eBiosciences, San Diego, CA; BD Pharmingen, San Jose, CA).

2.2. Flow Cytometry

1. ACK buffer: 10 mM $KHCO_3$, 155 mM NH_4Cl, 0.1 mM Na_2EDTA in H_2O, pH 7.2–7.4. Membrane-filter the solution (0.2 μM) and store at 4°C. Refresh ACK buffer at least every 3 weeks.
2. FACS buffer: PBS, supplemented with 2.5% FCS, filtered on a 0.2-μm membrane filter.
3. Appropriate mouse fluorochrome-conjugated antibodies (e.g., from eBiosciences or BD Pharmingen).
4. Flow cytometer: e.g., LSR II (BD Biosciences, San Jose, CA).
5. Analysis: FlowJo software (Tree Star, Ashland, OR).

2.3. Treg Culture

1. Tissue potter.
2. Lympholyte-M (Cedarlane Laboratories).
3. Tissue Culture Plate, 96 well, U-Bottom.
4. RPMI 1640 medium (Eurobio) supplemented with 10% heat-inactivated FCS, 2 mM l-glutamine, penicillin, streptomycin, 10 mM Hepes, 50 μM 2-mercaptoethanol (2-ME), 1 mM nonessential amino acids, 1 mM sodium pyruvate.
5. IL-2: filtered supernatant of EL4.IL-2 cells (American Type Culture Collection [ATCC], Manassas, VA) stimulated during 24 h with 10 ng/ml of phorbol myristate acetate [PMA]. IL-2-concentration is determined by ELISA.

2.4. Bone Marrow Chimeras

1. Mice: Any strain of inbred mouse can be used as donors and hosts. These mice are commercially available from several suppliers. We use male or female 8–10-week-old SPF animals.
2. Cs^{134} γ-ray research irradiator.
3. Hybridoma supernatants: Anti-Thy1 antibody (AT83 for Thy1.2, HO22.11 for Thy1.1, ATCC) prepared as described in Subheading 2.1, item 4.
4. Rabbit complement (Saxon Europe, Suffolk, UK).
5. Antibiotics: 0.4% pediatric suspension of Bactrim (Roche, Basel, Switzerland) in the drinking water.

3. Methods

The following protocols are established for one spleen, usually allowing for isolation of 0.3 to 1×10^6 Treg. After in vitro culture, typically a 20-fold increase in Treg cell numbers is observed. For generation of ten hematopoietic chimeras, we typically use five host-type spleens for isolation of Treg, three donor-type spleens to be used as source of antigen-presenting cells, and three to four bone marrow donors.

Since the protocols heavily depend on primary cell cultures, particular attention needs to be paid to avoid contamination. Use, as much as possible, laminar flow hoods and, for interventions on dead or live animals, clean procedures.

3.1. Preparation of a Total Host-Type Splenocyte Suspension

1. Euthanize the host-type mouse by cervical dislocation, clean the left flank with 70% alcohol, make an incision with sterile scissors, and carefully remove the spleen using sterile forceps. Transport the spleen in ice-cold complete medium.
2. Make a raw splenocyte suspension in complete medium by careful mechanical disruption of the spleen in a potter. Centrifuge at $345 \times g$ for 5 min at 4°C.

3. Wash cells by resuspending the cell pellet in 10 ml complete medium. Centrifuge.
4. Pass cells through sterile cotton-wool in a syringe.
5. Resuspend cells in 8 ml of complete medium and carefully deposit them on 2 ml of Lympholyte-M in a 15-ml tube (see Note 1).
6. Centrifuge at 1,118 × *g* for 15 min at room temperature (RT) without brake.
7. Recover the leukocyte layer between the Lympholyte-M and the medium (see Note 2).
8. Wash cells twice in complete medium, resuspend cells at 3×10^7 cells/ml in complete medium.
9. When the prepared splenocytes are stained with anti-CD4, anti-CD25, and anti-Foxp3 antibodies and analyzed by Flow cytometry, results similar to those shown in Fig. 1a should be obtained.

3.2. Enrichment of CD4+ T Cells

1. Incubate the prepared splenocytes on ice with saturating concentrations of the following hybridoma supernatants: anti-CD8 (53.6.7), anti FcγRII/III (2.4G2), and anti-MHC class II (M5) for 30 min. Agitate every 10 min.
2. Centrifuge at 345 × *g* for 5 min at 4–8°C.
3. Resuspend cells in 1 ml of complete medium.
4. Add 40 μl of the antibody cocktail provided in the Dynal CD4 cell negative isolation kit.
5. Mix well and incubate for 10 min on ice.
6. Wash cells by adding 9 ml of complete medium, centrifuge at 345 × *g* for 5 min at 4°C.
7. Resuspend splenocytes in 2 ml of complete medium.
8. Wash (3×) 250 μl of the anti-rat IgG-coated Dynabeads provided in the kit (see Note 3) by adding 10 ml of complete medium. Then, place the tube in a Dynal magnet for 1 min and discard the medium.
9. Add the cells to the washed beads and incubate for 30 min on ice, inverse tube regularly to resuspend cells and beads.
10. Add ice-cold complete medium up to 10 ml, place the tube in the Dynal magnet for 1 min and transfer the cell suspension to a new tube.
11. Place the new tube in the magnet for 1 min and transfer the cell suspension to another tube, centrifuge cells at 345 × *g* at 4°C.
12. Wash cells once, resuspend them in 10 ml complete medium and centrifuge. Resuspend the cells at 3.10^7 cells/ml.

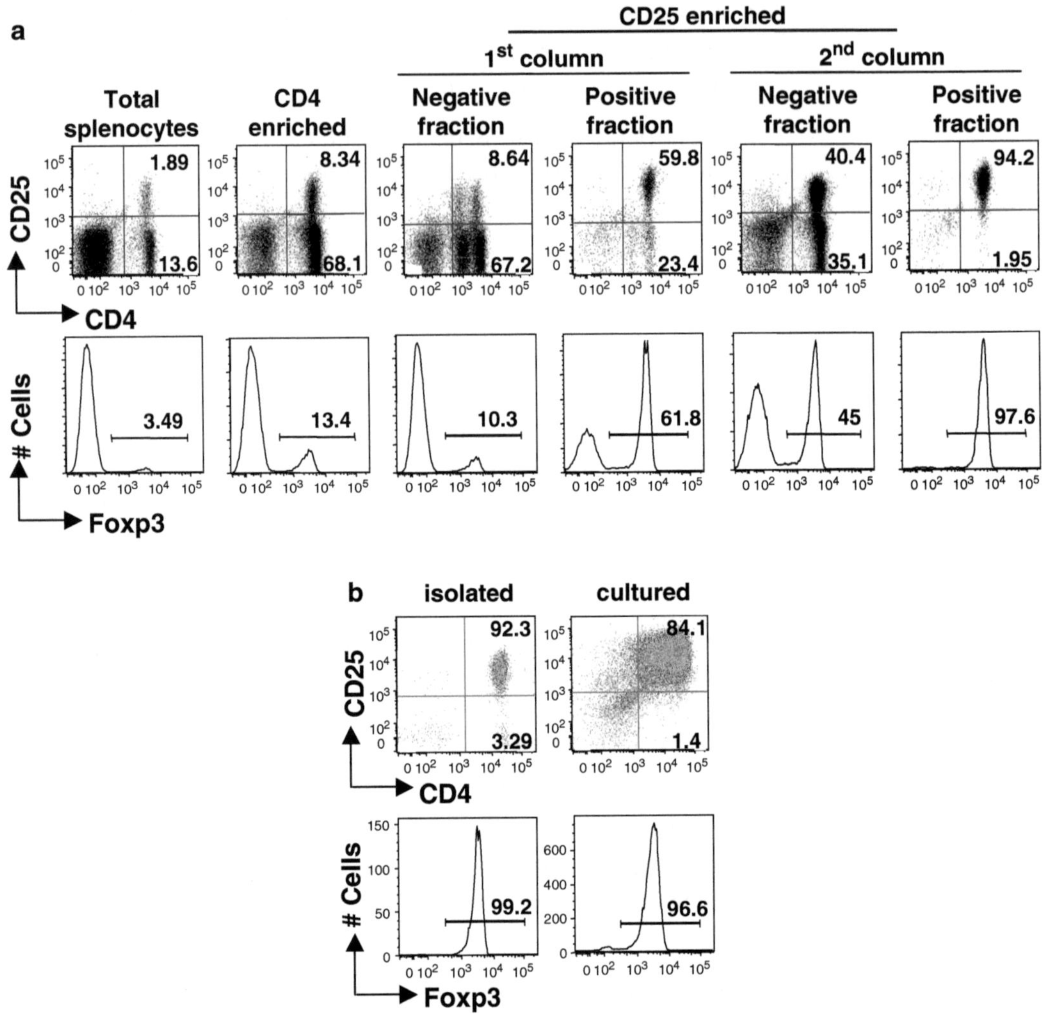

Fig. 1. Purification and culture of mouse regulatory T cells. (**a**) Total mouse splenocytes from mice transgenic for a bacterial artificial chromosome containing an EGFP-encoding sequence under control of the *Foxp3* promoter (13) were prepared as described in Subheading 3.1, stained with antibodies to CD4 and CD25, and analyzed by flow cytometry. Life cells are gated on forward and side scatter, and CD4/CD25 distribution (*upper panel*) and EGFP fluorescence (Foxp3) (*lower panel*) shown. Cell-samples from subsequent steps in the isolation procedure were analyzed similarly: CD4-enriched (Subheading 3.2), CD25⁺ cells magnetic bead sorted once (3.3.15) or twice (3.3.16). "Negative fraction" corresponds to the flow-through of the column, "positive fraction" to the cells retained on the magnetic column. (**b**) $CD4^{+}CD25^{high}$ cells thus isolated (*left hand panels*) from Foxp3-IRES-EGFP mutant mice (generously provided by Dr. Bernard Malissen, Marseille, France) (14) were cultured as described in Subheading 3.4 (*right hand panels*) and analyzed similarly. Numbers indicate percentages of cells within indicated gates.

13. When the prepared cells are stained with anti-CD4, anti-CD25, and anti-Foxp3 antibodies and analyzed by Flow cytometry, results similar to those shown in Fig. 1a should be obtained.

3.3. Isolation of $CD25^{high}$ Cells

1. Add saturating amounts of anti-mouse CD25-PE to the CD4-enriched cells.

2. Carefully mix suspension and incubate for 20 min in the dark on ice.
3. Wash cells twice in MACS buffer (see Note 4). Centrifuge at 345 × *g*, 4°C.
4. Resuspend cell pellet in 80 μl MACS buffer per 10^7 cells.
5. Add 5 μl of anti-PE Miltenyi microbeads per 10^7 total cells, mix well.
6. Incubate 20 min at 4°C.
7. Centrifuge at 345 × *g* for 5 min at 4°C.
8. Resuspend up to 10^8 cells in ice-cold 500 μl of MACS buffer.
9. Place Miltenyi MS column in the MiniMACS separator.
10. Prepare column by rinsing it with 500 μl of MACS buffer.
11. Apply cells suspension on the column.
12. Collect flow-through in a tube and add, 4 times, 500 μl of ice-cold buffer to the column. Collect total effluent.
13. Remove column from separator and place it on a collection tube.
14. Pass 1 ml of ice-cold MACS buffer and flush out the labeled fraction by softly applying the plunger.
15. Repeat this magnetic separation (steps 7–14) with a new column to increase the purity.
16. Check the purity of the different fractions by flow cytometry. We typically obtain results similar to those shown in Fig. 1a.

3.4. In Vitro Expansion of Alloantigen-Specific Treg

1. Prepare a suspension of donor-type total splenocytes (3×10^7 cells/ml) as described in Subheading 3.1, step 1–3. Then, the cells are γ-irradiated (17.5 Gy), passed through sterile cotton wool in a syringe, counted, and washed once more (see Note 5).
2. Coculture-purified regulatory T cells (2,000/well) and allogenic-irradiated splenocytes (2.5×10^5/well) in 100 μl/well complete RPMI medium complemented with 100 U/ml IL-2 in 96-well round-bottom plates at 37°C, 5% CO_2. Fill as many wells as the number of isolated $CD4^+CD25^{high}$ cells allows.
3. At day 7, add 100 μl of fresh medium (complete RPMI with 100 U/ml IL-2) and culture cells for another 7 days.
4. Harvest and pool cells from all wells, wash twice, and resuspend at 10^7 cells/ml in complete medium.
5. Analyze cultured cells by flow cytometry for expression of CD4, CD25, and Foxp3. Results typically obtained in our laboratory are shown in Fig. 1b.
6. Just prior to injection, pellet the cells and resuspend them in ice-cold PBS at 2×10^6 cells/50 μl.

3.5. Allogenic Bone Marrow Graft

1. γ-irradiate (5 Gy) hosts 1 day before bone marrow transplantation.
2. Add antibiotic to the drinking water during the complete duration of the experiment.
3. Collect tibias and femurs from donor mice in complete medium.
4. Carefully cut off the ends of the bones with scissors, keep them with forceps and thoroughly flush them with complete medium using a 26-G needle.
5. Carefully pipette the collected cells in complete medium to dissociate clumps. Wash cell suspension with complete medium (see Note 6).
6. Resuspend bone marrow cells in RPMI with 1% FCS (no other additives) at 10^7 nucleated cells/ml.
7. Add appropriate concentrations of anti-Thy1 antibody-containing hybridoma supernatant and rabbit complement (see Note 7).
8. Incubate 1 h at 37°C in a water bath. Fill the tube with ice-cold complete medium containing 10% FCS, centrifuge cells.
9. Wash cells twice more in complete RPMI, count, and resuspend them at 10^7 cells/150 μl PBS.
10. Intravenously coinject 150 μl (=10^7) bone marrow cells and 50 μl (=2.10^6) Treg into host mice irradiated 1 day earlier.

3.6. Determination of Allograft Acceptance

1. Collect blood samples in a tube containing 5–10 μl 500 mM EDTA.
2. Wash cells 3× with 500 μl of ice-cold FACS buffer, centrifuge at 220×*g* for 5 min at 4°C.
3. Resuspend the pellet in 500 μl of ACK buffer.
4. Incubate 10 min at RT.
5. Stop the reaction by adding 500 μl of FACS buffer.
6. Centrifuge at 220×*g* for 5 min at 4°C (see Note 8).
7. Wash cells once more with FACS buffer.
8. Resuspend the pellet in 100 μl of 2.4G2 (anti-FcγR) hybridoma supernatant.
9. Incubate 20 min on ice.
10. Add antibodies to donor and host MHC class I (or other appropriate marker) and incubate 20 min on ice.
11. Wash cells and analyze them by flow cytometry.
12. Results routinely obtained in our laboratory are shown in Fig. 2.

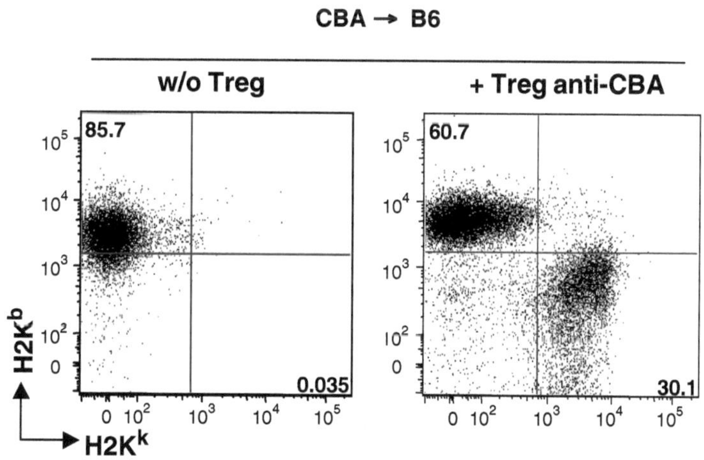

Fig. 2. In vitro cultured regulatory T cells prevent rejection of bone marrow allografts. CBA (H-2^k) bone marrow was transplanted into B6 (H-2^b) mice without further treatment (*left hand panel*) or under cover of in vitro cultured regulatory T cells as described in Subheading 3.5 (*right hand panel*). Three weeks later, blood samples were taken and analyzed by flow cytometry as described in Subheading 3.6.

4. Notes

1. Lympholyte-M has to be conserved at 4°C but needs to be used at RT. Make sure that the cells suspension, the centrifuge, and the buckets are at RT. The technique can be realized by two different manners:
   (a) The cell suspension is slowly deposited on the Lympholyte-M.
   (b) The Lympholyte-M is slowly deposited under the cell suspension using a Pasteur pipette.
2. After centrifugation with the Lympholyte-M, carefully recover the white interface layer using a Pasteur pipette.
3. Careful prewashing of Dynabeads is a very crucial step since the solution in which they are conserved is toxic.
4. The MACS buffer should always be used cold to avoid non-specific retention of cells on the magnetic column.
5. For bone marrow transplantation, use donor-type splenocytes to stimulate Treg. If, after induction of hematopoietic chimerism, transplantation of solid tissues or organs is envisaged, use (donor x host)F1 splenocytes to enrich Treg specific not only for directly but also for indirectly presented alloantigens (10).
6. To recover a maximum of cells, flushing of bones must be continued until bones are fully white.

7. Appropriate concentrations of anti-Thy1 antibody and rabbit complement need to be determined by complement lysis.
8. After the first incubation with ACK buffer, the cell pellet may still be red showing that some erythrocytes are left. Do not hesitate to repeat ACK-mediated lysis to remove all erythrocytes allowing for better analysis by flow cytometry.

References

1. Sakaguchi, S., Yamaguchi, T., Nomura, T., and Ono, M. (2008) Regulatory T cells and immune tolerance, *Cell* **133**, 775–787.
2. Tang, Q., and Bluestone, J. A. (2008) The Foxp3+ regulatory T cell: a jack of all trades, master of regulation, *Nat. Immunol.* **9**, 239–244.
3. Ziegler, S. F. (2006) FOXP3: of mice and men, *Annu. Rev. Immunol.* **24**, 209–226.
4. Beyer, M., and Schultze, J. L. (2006) Regulatory T cells in cancer, *Blood* **108**, 804–811.
5. Belkaid, Y., Blank, R. B., and Suffia, I. (2006) Natural regulatory T cells and parasites: a common quest for host homeostasis, *Immunol. Rev.* **212**, 287–300.
6. Aluvihare, V. R., Kallikourdis, M., and Betz, A. G. (2004) Regulatory T cells mediate maternal tolerance to the fetus, *Nat. Immunol.* **5**, 266–271.
7. Joffre, O., Gorsse, N., Romagnoli, P., Hudrisier, D., and van Meerwijk, J. P. M. (2004) Induction of antigen-specific tolerance to bone marrow allografts with CD4+CD25+ T lymphocytes, *Blood* **103**, 4216–4221.
8. Joffre, O., and van Meerwijk, J. P. M. (2006) CD4+CD25+ regulatory T lymphocytes in bone marrow transplantation, *Sem. Immunol.* **18**, 128–135.
9. Itoh, M., Takahashi, T., Sakaguchi, N., Kuniyasu, Y., Shimizu, J., Otsuka, F., and Sakaguchi, S. (1999) Thymus and autoimmunity: production of CD25+CD4+ naturally anergic and suppressive T cells as a key function of the thymus in maintaining immunologic self-tolerance, *J. Immunol.* **162**, 5317–5326.
10. Joffre, O., Santolaria, T., Calise, D., Al Saati, T., Hudrisier, D., Romagnoli, P., and van Meerwijk, J. P. M. (2008) Prevention of acute and chronic allograft rejection with CD4+CD25+Foxp3+ regulatory T lymphocytes, *Nat. Med.* **14**, 88–92.
11. Coudert, J. D., Coureau, C., and Guery, J. C. (2002) Preventing NK cell activation by donor dendritic cells enhances allospecific CD4 T cell priming and promotes Th type 2 responses to transplantation antigens, *J. Immunol.* **169**, 2979–2987.
12. Corry, R. J., Winn, H. J., and Russell, P. S. (1973) Primarily vascularized allografts of hearts in mice. The role of H-2D, H-2K, and non-H-2 antigens in rejection, *Transplantation* **16**, 343–350.
13. Lahl, K., Loddenkemper, C., Drouin, C., Freyer, J., Arnason, J., Eberl, G., Hamann, A., Wagner, H., Huehn, J., and Sparwasser, T. (2007) Selective depletion of Foxp3+ regulatory T cells induces a scurfy-like disease, *J. Exp. Med.* **204**, 57–63.
14. Wang, Y., Kissenpfennig, A., Mingueneau, M., Richelme, S., Perrin, P., Chevrier, S., Genton, C., Lucas, B., DiSanto, J. P., Acha-Orbea, H., Malissen, B., and Malissen, M. (2008) Th2 lymphoproliferative disorder of LatY136F mutant mice unfolds independently of TCR-MHC engagement and is insensitive to the action of Foxp3+ regulatory T cells, *J. Immunol.* **180**, 1565–1575.

Part IV

Human

Chapter 13

Analysis of Human FOXP3⁺ Treg Cells Phenotype and Function

Eva d'Hennezel and Ciriaco A. Piccirillo

Abstract

Naturally occurring regulatory T (nT_{Reg}) cells play a critical role in the establishment of immunological self-tolerance in humans. Currently, the analysis of nT_{Reg} cell function from bulk PBMC has led to discrepancies, largely due to the failure to discriminate T_{Reg} cells from other antigen-experienced $CD4^+$ T cells in states of inflammation. We developed a novel, multiparametric, single-cell strategy approach, which consists of isolating and expanding individual $CD4^+CD25^+$ T cells into clones, in turn allowing us to discriminate *bona fide* T_{Reg} cells from activated, $FOXP3^+$ T_{Eff} cells, which frequently confound bulk $CD25^{High}$ T_{Reg} functional assays. This approach enabled us to compare their phenotype and function at the single-cell level and to uncover the functional heterogeneity that exists among the $CD4^+FOXP3^+$ T_{Reg} cell population in human PBMC.

Key words: Regulatory T cells, FOXP3, IL-2, Single-cell sorting, Cloning, Suppression, Anergy

1. Introduction

Naturally occurring $CD4^+$ regulatory T cells (nT_{Reg}) arise during normal thymic lymphocyte development, and represent 1–10% of $CD4^+$ T cells in humans and mice (1). They are characterized by the expression of high levels of the IL-2R alpha (α) chain ($CD25^{High}$) and the FOXP3 transcription factor. They can suppress the activation of autologous T cells, are hyporesponsive to in vitro TCR-induced proliferation (anergic) in the absence of IL-2, and secrete low levels of inflammatory cytokines (1–4). $CD4^+$ T_{Reg} cells can also differentiate extrathymically from non-regulatory precursors upon immune activation in particular immunological settings (5), and are termed induced T_{Reg} (iT_{Reg}) cells. They display an array of phenotypes and functions that can

George Kassiotis and Adrian Liston (eds.), *Regulatory T Cells: Methods and Protocols*, Methods in Molecular Biology, vol. 707, DOI 10.1007/978-1-61737-979-6_13, © Springer Science+Business Media, LLC 2011

differ from nT_{Reg} cells, although they often express FOXP3 and may operate via similar suppression mechanisms.

Currently, isolating $CD4^{+}CD25^{High/Bright}$ T cells is the most common strategy to assess the phenotype and function of nT_{Reg} cells from human blood or tissue (4, 6, 7). However, markers such as CD25, FOXP3, and CD127, are also readily upregulated by all activated T cell subsets in chronic inflammatory states (8–12). Thus, the $CD4^{+}CD25^{High/Bright}$ T cell pool in normal PBMC is enriched for regulatory function, but represents a heterogeneous population which includes nT_{Reg} cells but also a variety of other $CD4^{+}$ T cells with a spectrum of antigen experiences, phenotypes, and functional profiles.

We developed a novel, multiparametric approach to dissect the human $CD25^{High}$ pool down to the single-cell level, and, in turn, allowing us to uncover the functional heterogeneity contained in this population (13). Our approach consists of correlating the expression of known T_{Reg} markers with the suppressive, proliferative, cytokine-producing potential in in vitro expanded primary cell lines for $CD4^{+}$ T cell subsets from PBMC of healthy subjects (13). The expanded T cells recapitulate the phenotype and function of cells directly ex vivo, and this approach has proven to be very valuable in the phenotypic and functional characterization of $CD4^{+}FOXP3^{+}$ T_{Reg} cells in health and states of disease (13, 14).

2. Materials

2.1. Cell Preparation

1. Ficoll-Paque Plus (GE-Healthcare). Store at 4°C. Warm up to room temperature before use.
2. Dulbecco's Phosphate Buffer Saline (DPBS) (Gibco). Store at 4°C after opening. Warm up to room temperature before use.
3. Roswell Park Memorial Institute Medium 1640 (RPMI), supplemented with 10% fetal bovine serum (FBS), qualified and certified, HEPES (10 mM), MEM NEAA (1×), sodium pyruvate (1 mM), penicillin (100 UI/ml), streptomycin (100 mg/ml) (all form Gibco) thereafter referred to as cRPMI.
4. Hank's Balanced Saline Solution (HBSS) with calcium and magnesium (Gibco).
5. Variable-speed pipette aid.
6. 3 ml Transfer pipette, sterile.
7. 9" Pasteur pipette, sterile.
8. A 2 ml soft rubber bulb fitting Pasteur pipettes (Fisher scientific).

2.2. Cell Culture

1. Feeder cells: human allogeneic PBMCs, freshly isolated or frozen, preferably from at least two different healthy donors, 3×10^7 cells.
2. Recombinant human interleukin-2 (rhIL-2), resuspended in DPBS at 10^4 UI/ml and stored as 1 ml aliquots at −20°C. Thaw on ice prior to use. Aliquots can be kept at 4°C for up to 10 days.
3. Affinity-purified, low-endotoxin, antihuman CD3 antibody, clone OKT3, at 1 mg/ml (BDBiosciences).
4. RPMI, supplemented with 10% FBS, qualified and certified, HEPES (10 mM), MEM NEAA (1X), sodium pyruvate (1 mM), penicillin (100 UI/ml), streptomycin (100 mg/ml) (all form Gibco) thereafter referred to as cRPMI.
5. Multichannel pipettor with range 50–300 μl.
6. Gamma-irradiator.
7. 96-Well polystyrene cell culture plates for suspension cultures, round bottom.
8. A plate-holder allowing working with 96-well plates tilted at a 45° angle.

2.3. Cell Staining and Sorting

1. Staining buffer: DPBS supplemented with 2% FBS.
2. Fluorochrome-conjugated antibodies against human CD4, CD25, CD14, CD56, and CD8. The fluorochrome combination should be chosen so that all five markers can be examined simultaneously, for example CD4-FITC, CD25-APC, CD14-PE, CD56-PE-Cy7, CD8-PercP (BDBioscience).
3. FACSAria (BDBiosciences) or MoFlo (Beckman Coulter) FACS-sorter, with plate-carrier module.

2.4. Microscaled Functional Assays

Cell numbers and amounts are provided for 100 clones.

1. Feeder cells: human allogeneic PBMCs, freshly isolated or frozen, preferably from at least two different healthy donors, 5×10^7 cells.
2. Target cells: human allogeneic PBMCs, freshly isolated or frozen, also allogeneic to the feeder cells, 3×10^7 cells.
3. RPMI, supplemented with 10% FBS, qualified and certified, HEPES (10 mM), MEM NEAA (1×), sodium pyruvate (1 mM), penicillin (100 UI/ml), streptomycin (100 mg/ml) (all form Gibco) thereafter referred to as cRPMI.
4. Carboxyfluorescein Succinimidyl Ester (CFSE) (Sigma).
5. Recombinant human interleukin-2 (rhIL-2).
6. Affinity-purified, low endotoxin, anti-CD3 antibody, clone OKT3 (BDBiosciences).

7. Solution of ^{3}H-thymidine with an activity of 6.7 mCi/nmol (Perkin Elmer).
8. Automated 96-well plate harvester (Tomtec, NewHaven, CT).
9. Trilux Scintillation counter (Wallac/Perkin Elmer, Waltham, MA).
10. Phorbol 12-myristate 13-acetate (PMA), resuspended in EtOH at a concentration of 250 mg/ml, stored as 100 μl aliquots at –20°C.
11. Ionomycin calcium salt, resuspended in EtOH at a concentration of 2 mg/ml, stored as 100 μl aliquots at –20°C.
12. Golgi-Stop (BDBiosciences).
13. Staining buffer: DPBS (Gibco) supplemented with 2% FBS, qualified, certified (Gibco).
14. Intracellular staining kit from eBioscience (see Note 1).
15. Anti-FOXP3 antibody, clone 236A/E7, from eBioscience (see Note 2).
16. Fluorochrome-conjugated anti-IFNγ, anti-IL-10, anti-IL-17, anti-CD4, anti-CD25, anti-CD127, anti-IL-2.

3. Methods

The following methods describe how to isolate individual $CD4^{+}$ T_{Reg} cells from PBMCs and expand them in vitro to obtain primary clonal lines. These clones can then each be subjected to several phenotypic and functional tests in parallel. The data collected in this way can then be correlated to define various relevant regulatory T cell subsets and/or to monitor these characterized subsets in the context of disease.

Unless otherwise specified, every procedure described here should be performed in sterile conditions, in a biosafety cabinet.

3.1. Isolation of PBMCs from Blood

1. Blood sample is to be collected in the presence of an anticoagulating agent, such as K_2EDTA or Heparin, for instance using BD-Vacutainer lavender-cap or green-cap blood collection tubes. Should the processing of the sample be postponed to more than 1 h after collection, the sample should be kept at about 4–10°C until processing, which should occur as soon as possible, and preferably no later than 5 h after collection.

   The blood sample for single-cell cloning can be as small as 2 ml from a normal adult. In the following instructions, we will exemplify the processing of a 10 cc sample.
2. A centrifuge with swinging buckets and adapters for 50 cc tubes is kept at 20°C. The Ficoll and DPBS are brought to room temperature.

3. The blood sample is diluted with 15 ml of DPBS.
4. 12.5 ml of the Ficoll solution is dispensed into a 50 cc tube.
5. The following step consists in overlaying the blood solution onto the Ficoll, so as to form a density gradient. The quality of the interface generated in this gradient will directly affect the quality of PBMC separation, yield, and purity. The following directions should be observed with care. Additional information, as well as graphic instructions can be obtained from the Instruction Manual of the Ficoll-Paque.
   1. Hold the open Ficoll tube in one hand, and tilt it gently towards the other hand to the point where the Ficoll is about 1 cm away from the rim of the tube.
   2. Using a pipette-aid and a 25-ml pipette tip, aspirate the totality of the 25 ml of blood solution (about 2 drops/s), and transfer the blood onto the Ficoll as slowly as possible, letting the blood flow onto the part of the wall of the tube comprised between the surface of the Ficoll, and the rim. This prevents the flow of blood to directly land onto the interface, which would create fluid disturbances.
   3. When the volume of transferred blood solution is about 7 ml, the speed of flow can be slightly increased to 1 ml/s. While continuing to transfer the blood solution, the tube can be slowly tilted back to a standing position. Transfer the remaining sample volume to completion.
6. As soon as the gradient has been prepared, the tube is very carefully carried to the centrifuge, so as to not disturb the fragile gradient interface.
7. The centrifuge will be set so that the brake is disabled, or set to the minimum. Failing to do so will unmistakably lead to a disruption of the separating gradient at the time of deceleration, and to the failure of the whole separation procedure (see Note 3).
8. Centrifuge the preparation at a speed of 700 × *g* for 30 min at 20°C.
9. The tube is taken out of the centrifuge very carefully (see Note 4).

   At this step, the tube presents four distinct phases:
   - A dark red blood cell pellet of about 10 ml
   - A rather turbid colorless phase of about 7.5 ml
   - A white opaque interface of about 1 mm, containing the PBMCs
   - A yellow transparent supernatant of about 20 ml, containing the serum and platelets
10. The supernatant is removed cautiously with a transfer pipette, down to about 1 cm above the PBMC interface, and discarded

(see Note 5). Keep the extremity of the pipette close to the surface and always make sure that the pipette produces an aspirating displacement while collecting the supernatant. Any ejection of the volume while still into the tube will disturb the interface.

11. Adjust the rubber bulb onto the Pasteur pipette, and holding the pipette firmly, collect as much of the remaining supernatant as possible, and discard it. Stop the aspiration at the interface without aspirating it.
12. Using the same pipette, gently collect the interface, and transfer it to a new 50 cc tube. By bringing the tube to eye level, ensure a complete collection: the collection is complete when no trace of the interface remains visible. The collection volume should be about 7.5–10 ml.
13. Fill up the collection tube containing the collected interface to 50 ml with cRPMI. This removes the Ficoll from the cells, and provides them with a recovery-friendly environment.
14. Centrifuge at $450 \times g$ for 10 min at 10°C.
15. Decant the supernatant into a new tube. Gently disrupt the pellet, and resuspend in 35 ml of HBSS. The following two steps will further devoid the sample of platelets.
16. Centrifuge supernatant and sample at $300 \times g$ for 13 min at 4°C.
17. Discard the supernatants by inverting the tubes and keeping it up-side down for about 2 s, making sure the last drop falls off. Loosen the pellets and resuspend them in 35 ml of HBSS.
18. Centrifuge at $300 \times g$ for 13 min at 4°C.
19. Discard the supernatants by inverting the tubes, loosen the pellets, resuspend them in cRPMI, and pool them.
20. Count the cells and adjust the concentration to 10^8 cells/ml, centrifuging if necessary. The cell recovery from a 10 cc blood sample of a healthy adult is expected to be between 10 and 15 million cells.
21. Keep aside 10^5 cells for each of the staining controls, and add to the sample the appropriate concentration of CD4, CD25, CD8, CD56, CD14, or respective isotype control antibodies. The amount of antibody used should be standardized/optimized for each specific antibody and individual cell type to be examined.
22. Incubate at 4°C for 30 min, then wash sample and controls with cRPMI.
23. Resuspend the cell preparation at $5–10 \times 10^6$ cells/ml of cRPMI (see Note 6).

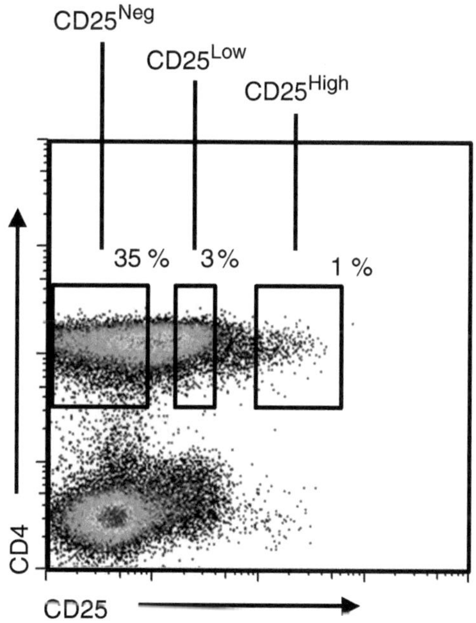

Fig. 1. Gating strategy for cell sorting. A representative gating strategy for single-cell sorting is shown. PBMCs were recovered from the blood of a healthy adult. Cells were stained with FITC-conjugated anti-CD4, and PE-conjugated anti-CD25. Cells are gated on live lymphocytes by FSC-H/SSC-H discrimination.

24. After setting the FACS sorter for voltages and compensation, the gating strategy should encompass an FSC-A vs. SSC-A gate delineating the live lymphocytes, FSC-W and FSC-H gates excluding the doublets, then gates excluding $CD8^+$, $CD56^+$, and $CD14^+$ cells. The resulting subset should display CD4 vs. CD25 expression, on which the sorting gates will be set as shown in Fig. 1.

3.2. Preparation of the Culture Plates

1. If the feeder cells are obtained freshly, proceed to step 2. If the feeder cells were frozen, follow these directives to optimize recovery:
   a. Bring 45 ml of cRPMI to 37°C. This assumes a maximum of three cell vials of 1 ml each.
   b. Allow the cell vials to thaw in a 37°C water-bath, just until it forms a loose ice cube.
   c. Promptly pour the content of the vials into the warm RPMI. With a 1 ml micropipette, make sure to recover the complete content of the vial.
   d. Resuspend the suspension uniformly by inverting the tube gently once, and then centrifuge at 350 × *g* for 8 min at room temperature.

e. Decant, loosen the pellet, and count the cells.

f. Plate the cells in a 25 cm^2 cell culture vial in 20 ml of cRPMI.

g. Incubate for 2 h in a 37°C.

h. Transfer the cells into a 50 cc tube. In order to recover the adhering cells which will have stuck to the bottom of the vial, carefully scrape the vial using a Teflon® cell scraper. It is important to not deplete adhering cells from the feeder cell sample, as adhering cells are largely comprised of antigen-presenting cells.

2. Resuspend the cells at 5×10^6 cells/ml of cRPMI.
3. Irradiate the cells suspension at 3,000 rads with a gamma-irradiator.
4. Centrifuge the suspension at $350 \times g$ for 8 min at 4°C. This step allows for the elimination of free radicals, toxic to the cells, which are created in the medium by the gamma-irradiation.
5. Resuspend the 3×10^7 feeder cells in 200 ml of cRPMI.
6. Prepare the stimulation medium: to the feeder cell suspension, add 200 U/ml of rhIL-2 (800 μl), and 30 ng/ml of anti-CD3 antibody (6 μl).
7. Dispense 200 μl/well of this stimulation medium into 10 round bottomed 96-well plates.
8. The cells are directly single-cell sorted into each well. Two plates will be seeded with $CD4^+CD25^{Neg}$ cells, two with $CD4^+CD25^{Low}$ cells, and 4 with the $CD4^+CD25^{High}$ cells. One plate will remain devoid of clones, as a control for potential undesired growth arising from the feeders.

3.3. Maintenance of the Culture for Clonal Expansion

Cloning cultures follow a cycle of 10 days, after which a new stimulation is needed. During each stimulation cycle, high levels of IL-2 need to be maintained by periodically adding it freshly to the culture.

Replenishing IL-2

1. On the fourth day of culture (i.e., 96 h after seeding), remove 95 μl from each well using a multichannel pipettor. Using a plate holder to maintain the plate at a stable 45° angle can make this operation easier. Make sure to not touch or resuspend the pellet of cells at the bottom of the well.
2. Prepare the feeding solution: to 100 ml of cRPMI, add 400 μl of IL-2 solution.
3. Dispense 100 μl of the feeding solution in each well.
4. Repeat this operation after 3 more days of culture.

Restimulation

5. On the tenth day, examine each individual well for growth using an inverted microscope.
6. While most wells will present with what seems to be dead/dying cells, and debris, some will present a significant amount of viable, blasting T cells, displaying either a rounded or typical activated (i.e., pear-like) morphology. The clonability can vary greatly from one individual to the next, however, the number of clones in the $CD25^{-}$ and $CD25^{Low}$ plates should be greater than in the $CD25^{High}$ plates. Note that several of the wells which seem negative do carry a growing clone, which has yet to overgrow the remains of the feeder cells.
7. Mark and record the wells containing clones with more than 10^4 cells. The cellularity of the clones cannot be individually counted, and should be assessed approximately by sight.
8. Each overconfluent clone will be split into as many wells as is necessary to obtain less than 10^4 cells/well. This requires a good level of organization, as this will be repeated several times for each clone before the end of the culture.
9. It is advised to create a "daughter" plate for each original cloning plate (the "mother" plate). The bordering row of wells for each daughter plate will be filled with 300 μl of sterile DPBS, to protect the proximal wells from evaporation. A referencing system is needed to identify and track individual clones, for instance the coordinates on the mother plate. One column of 6 wells will also be reserved for each clone, so as to keep as much as possible the wells of a same clone in close proximity. Finally, when splitting a clone, it is advised to transfer the totality of the clone to the daughter plate, as opposed to keeping a fraction of the culture in the initial well on the mother plate, which could complicate downstream studies (see Note 7).
10. For the clones split into more than 2 wells, adjust the final volume to 100 μl/well with fresh cRPMI.
11. From each of the other wells (positive or not), the top 95 μl of the medium will be removed.
12. Thaw or isolate allogenic PBMCs. 3×10^4 cells will be needed for each well.
13. Irradiate these feeder cells at 3,000 rads.
14. After centrifuging, resuspend at 3×10^5 cells/ml of cRPMI.
15. Prepare the restimulation medium: to the feeder cell suspension, add 400 U/ml (8 μl/ml) of IL-2 solution, and 60 ng/ml of anti-CD3. This will produce a final concentration identical to the initial conditions of seeding.
16. Dispense 100 μl of restimulation medium in each well.

Passaging, and replenishing IL-2

17. On the fourth day after restimulation, observe each well for growth with an inverted microscope. Mark or record each positive clone, and their approximate level of confluence. One convenient way to do so is to directly assess in which multiple the clone needs to be split, and to mark it on the lid of the plate with a permanent marker.
18. Split the clones as needed, following the same directions as for the day of restimulation. Adjust the final volume of the wells resulting from splitting to 100 μl with fresh cRPMI.
19. Remove the top 95 μl of culture medium from all the wells which are either negative or do not need splitting. At this stage, the feeder cells dispensed for restimulation are still abundant, and may lead to false negatives.
20. Prepare the feeding solution: to 150 ml of cRPMI, add 200 U/ml of IL-2 solution (600 μl).
21. Dispense 100 μl of the feeding solution in each well.
22. Repeat this operation after 3 more days of culture, passaging the cultures as needed.

After a total of 20 days of culture, the clones are ready for harvest, and individual phenotypic and functional testing.

3.4. Preparation of the Clones for Functional Testing

In order to test individual clones functionally, they need to be harvested and counted appropriately prior to testing. This process is long and necessitates careful preparation.

1. Prepare the day before:
   - 1 l of cRPMI
   - A complete list of the clones that will be harvested
   - Sterile, capped 5 ml culture tubes ("FACS" tubes), at least one for each clone, labeled with its reference number
2. On the day of harvest, for each individual clone, resuspend each well by pipetting twice up and down with a 200 μl micropipettor set on 100 μl. Transfer each well into the collection tube labeled for this clone. Wash each well with 100 μl of fresh cRPMI, which is also transferred to the collection tube. Transfer the tube on ice, and proceed to the next clone.
3. Once all the clones are harvested, centrifuge them at 350 × *g* for 6 min at 4°C.
4. Decant each tube by inverting the tube, letting every drop fall off, and blotting the tube on sterile gauze. In order to prevent dryness of the resulting pellet, it is strongly recommended to decant 10–15 clones at a time, and immediately resuspended.

5. Loosen the pellets by gentle tapping, and resuspend in 200 μl cRPMI. For clones requiring more than one tube, pool the content of the tubes and resuspend in a total of 1 ml.
6. Count each clone with a hemocytometer and evaluate viability by the method of Trypan Blue exclusion. We recommend mixing 1 μl of Trypan Blue with 9 μl of cell suspension. Clones are required to be of at least 50,000 cells in order to carry out both phenotypic and functional testing.
7. Adjust each clone to a final concentration of 150–200,000 cells/ml of cRPMI.
8. Keep the cells at 4°C until ready to plate in functional and phenotypical assays.

3.5. Phenotypic Analysis Assays

The phenotype of each clone for various surface and intracellular markers can be assessed after clonal expansion. Surface markers can be examined directly after harvest, whereas intracellular markers, such as cytokines, typically require a step of restimulation in vitro prior to staining.

Surface markers

Examining surface markers on clones can be performed on as little as 25,000 cells, provided a few precautions are observed (see Note 8).

1. Collect 150 μl from each of the clones, and transfer it to individual wells of a round bottomed or V-bottomed 96-well plate. It is recommended to label the lid of the plate with the reference of clone for each well.
2. Centrifuge at 400 × *g* for 5 min at 4°C.
3. Decant and blot dry onto paper towel.
4. Holding the plate closed firmly; loosen the pellets by vortexing the plate a few seconds.
5. Prepare a staining solution: combine fluorochrome-conjugated antibodies directed against desired surface markers in titrated amounts, adding staining buffer to a total of 30 μl of staining solution per well. We recommend staining of CD4, in order to verify that no other T cell gave rise to the clone. A few markers of interest include CD25, CD127, HLA-DR, and GITR.
6. Dispense 30 μl/well, incubate for 30 min at 4°C.
7. Prepare the fixation buffer according to manufacturer's instruction; dispense 50 μl/well
8. Incubate at 4°C for 15 min.
9. Wash by adding 200 μl of staining buffer per well, centrifuge, decant, and resuspend as above.

10. Prepare the FOXP3 staining solution: prepare 1× permeabilization buffer according to manufacturer's instruction, 50 μl/well. Add anti-FOXP3 antibody at 1:10 dilution.
11. Dispense 50 μl/well, incubate for 30 min at 4°C.
12. Wash by adding 200 μl of staining buffer per well, centrifuge, decant, and resuspend as above.
13. Repeat the washing step a total of three times.
14. When acquiring the samples, use microtitre tubes rather than traditional 5 ml tubes, to reduce the minimal sample volume. If acquiring the samples on a flow-cytometer equipped with a protective sheath and pumping system around the sample probe (such as the FASCCalibur from BDBioscience), it is strongly recommended removing this sheath. Failing to do so will lead to major loss of the sample. Consult the facility manager for assistance.

    Intracellular markers

    The detection of cytokine production, and upregulation of many markers, requires the clones to be restimulated in vitro. In order to be able to correlate the observed phenotype with the functional results, we have chosen to restimulate the clones in a fashion very close to that used for the proliferation assays, i.e., in the presence of feeder cells and soluble anti-CD3, rather than stimulation by pate-bound anti-CD3, or PHA.
15. Isolate or thaw allogenic PBMCs. For each cytokine tested, 8×10^5 feeder cells are needed for each clone.
16. Irradiate the feeders at 3,000 rads.
17. Feeder cells are stained with CFSE in order to be able to gate them out at the time of analysis.
    1. Bring cRPMI to room temperature.
    2. Resuspend the feeders at a concentration of 10^7 cells/ml in cRPMI in a 50 cc tube.
    3. Prepare in an equal volume of cRPMI a dilution of 1:500 of CFSE from the 10 μM stock. Combine this CFSE solution with the cell suspension, and homogenize by inverting the tube gently twice.
    4. Let sit in a 37°C cell incubator for 5 min.
    5. Wash by filling up the tube with cRPMI, and centrifuge at $350 \times g$ for 6 min at room temperature.
    6. Decant and repeat the washing step twice.
    7. Count the cells. A loss of about 20% from the original counts is expected.
18. Count the feeder cells and resuspend them at 10^6 cells/ml of cRPMI.

19. Prepare the restimulation medium: to the suspension of feeder cells, add 500 U/ml (10 μl/ml of stock) of IL-2, and 75 ng/ml of anti-CD3. These conditions, after the final dilution, will reconstitute the same conditions as the expansion stimulation.
20. Dispense 80 μl of the restimulation medium in each well.
21. Dispense 120 μl of each clone. It is recommended to label the lid of the plate with the reference of each clone.
22. The time at which the culture needs to be stopped differs for each cytokine. IL-2 and IFN-γ are readily detectable early after restimulation (24 h), whereas measurable levels of IL-10 are often detectable after 48 h, and IL-17, after 4 days.
23. Prepare the pulsing solution: for one plate, combine PMA at 500 ng/ml, ionomycin at 20 μg/ml, and GolgiStop (20 μl), complete to 1 ml with cRPMI.
24. Four hours prior to end of the culture, dispense 10 μl of pulsing solution in each well (see Note 9).
25. To terminate the culture, centrifuge the plate(s) at 400 × *g* for 6 min at 4°C, decant, blot, and vortex.
26. Add 50 μl of fixing solution per well, and incubate for 15 min at 4°C.
27. Proceed with permeabilization and intracellular staining, according to manufacturer instruction.

The same precautions as for the surface marker samples need to be applied at the time of acquisition.

3.6. Proliferation Assays

The main functional feature of T_{Reg} cells is to suppress the proliferation of T_{Eff} cells. This proliferation can be measured by two methods: incorporation of tritium, or CFSE dilution. In both instances, the target cells are freshly FACS-sorted $CD4^{+}CD25^{Neg}$ cells, which need to be isolated on the day of the harvest.

Proliferation assay by incorporation of tritiated thymidine

1. This assay requires to be set in triplicates. Also, in order to gain insights in suppressor potency, it is recommended to prepare at least two different T_{Reg}:T_{Eff} ratios. Here, we describe the procedure for 1:1 and 1:3 ratios. We also recommend only using the 60 central wells of 96-well plates, and fill the surrounding rows with 300 μl of sterile DPBS. This minimizes the evaporation and ensures that the volume remains constant in all wells.
2. Prepare the stimulation medium: For 10 clones (60 wells), combine 150,000 T_{Eff} cells with 0.36 μl of anti-CD3 and 600,000 irradiated feeder cells in a final volume of 12 ml of cRPMI.

3. Dispense 200 μl/well of the stimulation medium.
4. For each clone, dispense 15 μl of cell suspension in 3 wells (triplicates of 1:1 ratio), and 5 μl in 3 other wells (triplicates of 1:3 ratio).
5. In order to calculate the percent suppression, 12 wells of stimulation medium will be left "blank," i.e., devoid of any clone cells. These will serve as the T_{Eff} alone reference.
6. It is also recommended to prepare a positive control for in vitro suppression, with freshly sorted $CD4^{+}CD25^{High}$ cells.
7. Incubate the plates at 37°C for 4 days.
8. Prepare a dilution of ^{3}H-thymidine of 50 μCi/ml in cRPMI.
9. Carefully remove the top 100 μl of medium from each well.
10. Add 10 μl of ^{3}H-thymidine dilution in each well, and incubate further for 18–24 h.
11. On the fifth day of coculture, the incubation should be stopped, either by freezing the plates at −20°C, or by directly transferring them to a fiberglass filter with an automated cell harvester. If the plates are frozen, make sure that all the wells are well thawed prior to proceeding to harvesting.
12. The counts per minute (CPM) for each well will be assessed by liquid scintillation. This value is directly proportional to the proliferation in each well, and can therefore by directly used to calculate the percent suppression in cocultures vs. T_{Eff} alone, for each clone, at each T_{Eff}:T_{Reg} ratio. For example profiles, see Fig. 2.

Proliferation assay by CFSE dilution

1. This assay requires to be set at least in duplicates. Also, in order to gain insights in suppressor potency, it is recommended to prepare at least two different T_{Reg}:T_{Eff} ratios. Here, we describe the procedure for 1:1 and 1:3 ratios.
2. T_{Eff} cells are stained with CFSE. This protocol is optimized to be as gentle as possible to the cells and not to alter their proliferation potential
    1. Bring cRPMI to room temperature.
    2. Resuspend the T_{Eff} cells at a concentration of 10^{6}–10^{7} cells/ml in cRPMI in a 15 cc tube.
    3. Prepare in an equal volume of cRPMI, a dilution of 1:500 of CFSE from the 10 μM stock. Combine this CFSE solution with the cell suspension, and homogenize by inverting the tube gently twice.
    4. Let it sit in a 37°C cell incubator for 5 min.
    5. Wash by filling up the tube with cRPMI, and centrifuge at 350 × *g* for 6 min at room temperature.

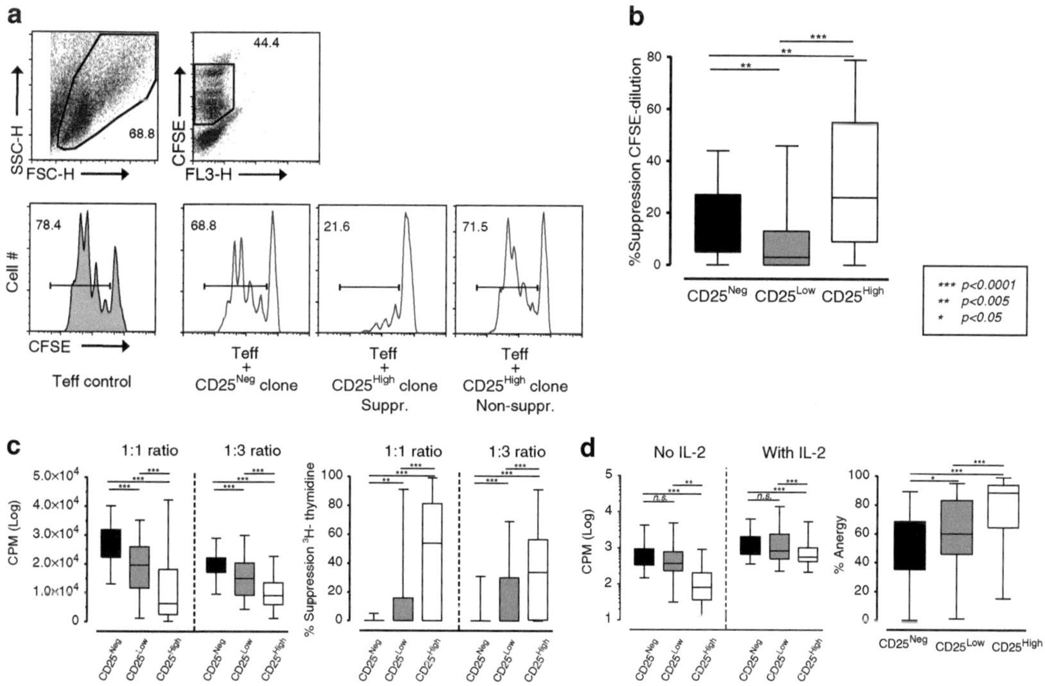

Fig. 2. Functional testing – Suppression and anergy assays. (**a**) Carboxyfluorescein succinimidyl ester (CFSE)-based suppression assay. CFSE-stained TEff cells comprised within the live gate are discriminated from unstained PBMC and clone cells (*upper panel*). Representative coculture results are shown. (**b**) Proliferation is measured as the percentage of $CFSE^{Low}$ cells within the TEff pool. (**c**) Suppression assay by ^{3}H-thymidine incorporation. Proliferation of TEff cells was assessed by ^{3}H-thymidine incorporation (*left*). Suppression is calculated as the ratio of the counts per minute (CPM) values obtained from TEff alone of individual cocultures, at 1:1 and 1:3 ratios (*right*). (**d**) Anergy assay. Proliferation was assessed by ^{3}H-thymididne incorporation in the presence or absence of IL-2 (*left*). Anergy is calculated as the ratio of the CPM values between the two conditions (*right*). ***p <0.0001, **p <0.005, *p <0.05.

6. Decant and repeat the washing step twice.
7. Count the cells. A loss of about 20% from the original counts can be expected.

3. Prepare the stimulation medium: for 10 clones (40 wells), combine 340,000 CFSE-stained T_{Eff} cells with 0.24 μl of anti-CD3 and 1.36×10^6 irradiated feeder cells in a final volume of 6 ml of cRPMI.
4. Dispense 150 μl/well of the stimulation medium.
5. For each clone, dispense 50 μl of cell suspension in 2 wells (duplicates of 1:1 ratio), and 17 μl in 3 other wells (duplicates of 1:3 ratio). The volume in the wells of 1:3 ratio will need to be adjusted to 200 μl by adding 34 μl/well of cRPMI.
6. In order to calculate the percent suppression, 8 wells of stimulation medium will be left "blank," i.e., devoid of any clone cells. These will serve as the T_{Eff} alone reference.

   It is also recommended to prepare a positive control of suppression, with freshly sorted $CD4^+CD25^{High}$ cells.

The necessary control for setting-up the flow-cytometer will also be included.

7. Incubate the plates at 37°C for 4 days.
8. On the fourth day of coculture, the incubation should be stopped. Cytokine secretion of CFSE-labeled target cells can be assessed by intracellular cytokine staining (ICS). If such is the case, the cocultures will be pulsed for the last 4 h, then fixed and permeabilized, and stained as described in Subheading 3.5.
   If CFSE alone will be examined, plates can be immediately acquired on a flow-cytometer, or fixed using a gentle fixation protocol. If ICS is performed, plates will be the samples which are extremely small and caution needs be observed in handling and acquiring them, as detailed in Subheading 3.5, step 14.
9. The feeder PBMCs and the clone cells will be easily excluded at the time of analysis, as they remain unstained. Unproliferated T_{Eff} cells will present a high CFSE fluorescence intensity, whereas proliferating cells will have diluted out the CFSE as they divide, leading to a lower (but positive) CFSE Fluorescence. Proliferation will be calculated as the proportion of the total $CFSE^+$ cells that have an intermediate CFSE fluorescence. Percent suppression can be calculated as the ratio of the proliferation in individual cocultures compared to T_{Eff} alone. For example of profiles, see Fig. 3.

3.7. Anergy Assay

Anergy to TCR induced activation in the absence of IL-2 is a key characteristic of T_{Reg} cells, and can be measured by comparing the proliferation in vitro upon stimulation in the presence or absence of rhIL-2. This proliferation is most easily measured by the incorporation of tritium.

1. This assay requires to be set in triplicates. We recommend only using the 60 central wells of 96-well plates, and fill the surrounding rows with 300 μl of sterile DPBS. This minimizes the evaporation and ensures that the volume remains constant in all wells.
2. Prepare the IL-2 free stimulation medium: for 10 clones (30 wells), combine 300,000 irradiated feeder cells with 0.18 μl of anti-CD3 in a final volume of 6 ml of cRPMI.
3. Prepare the IL-2 stimulation medium: for 10 clones (30 wells), combine 300,000 irradiated feeder cells with 0.18 μl of anti-CD3 and 24 μl of IL-2 in a final volume of 6 ml of cRPMI.
4. Dispense 200 μl/well of the IL-2 free stimulation medium in one half of the plate, and 200 μl/well of the IL-2 stimulation medium in the other half.

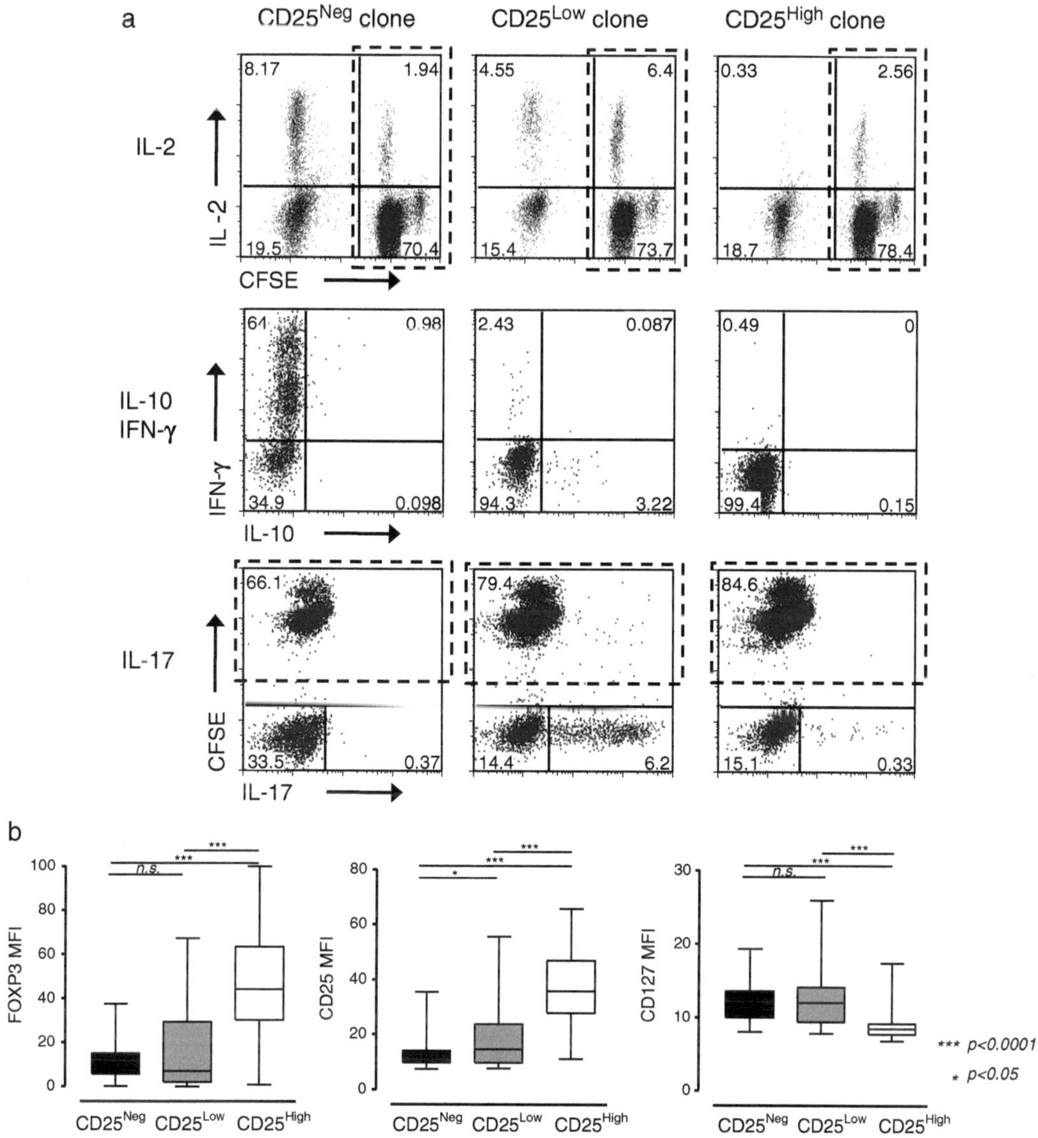

Fig. 3. Phenotypical testing – T_{Reg} markers and cytokines. Intracellular cytokine staining (ICS) was performed on clones after restimulation. Feeder PBMCs can be identified and excluded from analysis as they were stained with CFSE (*dashed boxes*). (**a**) Representative flow cytometry profiles for IL-17, IFN-γ, IL-2, and IL-10 staining on clones. (**b**) Cytokine secretion levels in clones postexpansion. Data obtained in one representative cloning experiment is shown. For IFN-γ and IL-10 stainings, cloned cells are directly shown, as PBMCs were excluded by prior gating. ***$p < 0.0001$, *$p < 0.05$.

5. For each clone, dispense 15 μl of cell suspension in 3 wells of each medium.
6. It is recommended to prepare a positive and a negative control of anergy, with freshly sorted $CD4^+CD25^{High}$ and $CD4^+CD25^-$ cells, respectively.
7. Incubate the plates at 37°C for 4 days.

8. Prepare a dilution of ^{3}H-thymidine of 50 μCi/ml in cRPMI.
9. Carefully remove the top 100 μl of medium from each well.
10. Add 10 μl of ^{3}H-thymidine dilution in each well, and incubate further for 18–24 h.
11. On the fifth day of culture, the incubation should be stopped, either by freezing the plates at −20°C, or by directly transferring them to a fiberglass filter with an automated cell harvester. If the plates are frozen, make sure that all the wells are well thawed prior to proceeding to harvesting.
12. The CPM for each well will be assessed by liquid scintillation. This value is directly proportional to the proliferation in each well, and can therefore by directly used to calculate the relative proliferation for each clone in the presence vs. in the absence, of IL-2.

4. Notes

1. The staining of FOXP3 is particularly sensitive to fixation, and requires gentle fixation protocols. The kit offered by eBioscience offers such conditions. It is not recommended to use kits based on paraformaldehyde, which can be much harsher on both surface stains, and FOXP3 epitopes.
2. The choice of the clone for the anti-FOXP3 antibody should be made with care. Several are available on the market, presenting with various qualities of staining (15). We prefer and recommend the clone 236A/E7 from Ebioscience.
3. The deceleration step needs to be as slow as and as little disruptive as possible. Even when set to a minimum, the deceleration could be too brutal on recent models of centrifuges, particularly table-top. It recommended using full-size, and as old as possible, centrifuges.
4. Ficoll is somewhat toxic to cells; therefore the time the samples spend in contact with this substance should be minimized. Process the sample as promptly as possible. Do not interrupt the isolation until the Ficoll is fully washed off.
5. Removing the supernatant allows to significantly reduce the content of platelets in the sample. This is particularly critical for FACS sorting, where platelets constitute a background noise which reduces the sorting efficiency dramatically.
6. The phenol red added to culture media as a pH indicator could cause some interference (autofluorescence) at the time of sorting. We have not encountered this inconvenience; however, should this be problematic, phenol red-free RPMI can be obtained from Gibco.

7. We advise against transferring bigger clones into flat-bottomed plates, as few of them have reached necessary cellularity, and T cells are extremely sensitive to underconfluence.
8. The usual precautions pertaining to the handling of fluorochromes, such as restricting exposure to light, also need to be observed, and will not be reminded here.
9. PMA and ionomycin are used to exacerbate the production of cytokines. We find that it also produces some level of ectopic expression for other markers, including FOXP3. Hence, FOXP3 expression levels cannot be reliably measured in samples treated with PMA and Ionomycin.

Acknowledgments

The authors wish to thank Evridiki Sgouroudis, Ekaterina Yurchenko, Michal Pyzik and Valerie Hay for discussions and technical support, and Marie-Hélène Lacombe for FACS. We acknowledge the financial support of CIHR grant MOP67211 and a CIHR MOP84041 grant from the New Emerging Team in *Clinical Autoimmunity: Immune Regulation and Biomarker Development in Pediatric and Adult Onset Autoimmune Diseases.* C.A.P holds Canada Research Chair in Regulatory Lymphocytes of the Immune System.

References

1. Levings MK, Allan S, d'Hennezel E, & Piccirillo CA (2006) Functional dynamics of naturally occurring regulatory T cells in health and autoimmunity. *Adv Immunol* 92:119–155.
2. Stephens LA, Mottet C, Mason D, & Powrie F (2001) Human CD4(+)CD25(+) thymocytes and peripheral T cells have immune suppressive activity in vitro. *Eur J Immunol* 31(4):1247–1254.
3. Taams LS, *et al.* (2001) Human anergic/suppressive CD4(+)CD25(+) T cells: a highly differentiated and apoptosis prone population. *Eur J Immunol* 31(4):1122–1131.
4. Baecher-Allan C, Brown JA, Freeman GJ, & Hafler DA (2001) CD4+CD25high regulatory cells in human peripheral blood. *J Immunol* 167(3):1245–1253.
5. Shevach EM (2006) From vanilla to 28 flavors: multiple varieties of T regulatory cells. *Immunity* 25(2):195–201.
6. Baecher-Allan C, Wolf E, & Hafler DA (2005) Functional analysis of highly defined, FACS-isolated populations of human regulatory CD4+CD25+ T cells. *Clin Immunol* 115(1):10–18.
7. Hoffmann P, *et al.* (2006) Isolation of CD4+CD25+ regulatory T cells for clinical trials. *Biol Blood Marrow Transplant* 12(3):267–274.
8. Walker MR, *et al.* (2003) Induction of FoxP3 and acquisition of T regulatory activity by stimulated human CD4+CD25- T cells. *J Clin Invest* 112(9):1437–1443.
9. Walker MR, Carson BD, Nepom GT, Ziegler SF, & Buckner JH (2005) De novo generation of antigen-specific CD4+CD25+ regulatory T cells from human CD4+CD25-cells. *Proc Natl Acad Sci U S A* 102(11): 4103–4108.
10. Yagi H, *et al.* (2004) Crucial role of FOXP3 in the development and function of human CD25+CD4+ regulatory T cells. *Int Immunol* 16(11):1643–1656.
11. Allan SE, *et al.* (2005) The role of 2 FOXP3 isoforms in the generation of human CD4+ Tregs. *J Clin Invest* 115(11):3276–3284.

12. Liu W, *et al.* (2006) CD127 expression inversely correlates with FoxP3 and suppressive function of human CD4(+) T reg cells. *J Exp Med* 203(7):1701–1711.
13. d'Hennezel E, Sgouroudis E, Yurchenko E, Hay V, & Piccirillo CA (2011) Single-cell analysis reveals functional heterogeneity of CD4+FOXP3+ regulatory T cells in human peripheral blood. *Manuscript submitted.*
14. d'Hennezel E, *et al.* (2009) FOXP3 forkhead domain mutation and regulatory T cells in the IPEX syndrome (translated from eng). *N Engl J Med* 361(17):1710–1713.
15. Pillai V & Karandikar NJ (2008) Attack on the clones? Human FOXP3 detection by PCH101, 236A/E7, 206D, and 259D reveals 259D as the outlier with lower sensitivity. *Blood* 111(1):463–464.

Chapter 14

Depletion of Human Regulatory T Cells

Amy C. Hobeika, Michael A. Morse, Takuya Osada, Sharon Peplinski, H. Kim Lyerly, and Timothy M. Clay

Abstract

Regulatory T cells (Treg) have become increasingly relevant in the study of human disease including cancer. Treg cells have been shown to inhibit anti-tumor immune responses, and elevated Treg levels have been associated with certain types of cancer. Similarly, depletion of Tregs by various methods can also enhance anti-tumor immune responses. We have found a prevalence of Treg in cancer patients when compared to normal volunteers. In addition, we have shown that the depletion of Treg using the IL-2 fusion protein denileukin diftitox decreased Treg function and increased antigen-specific T cell response to a cancer vaccine. These results indicate the potential for combining Treg depletion with anti-cancer vaccines to enhance tumor antigen-specific immune responses and the need to explore the dose and schedule of Treg depletion strategies in optimizing vaccine efforts.

Key words: Regulatory T cells, Denileukin diftitox, Antigen specific T cells, Cancer, Tumor antigen

1. Introduction

Regulatory T cells (Treg), defined by their expression of CD4, persistently high expression of the IL-2 receptor component CD25, and intracellular expression of the transcription factor FoxP3 (1), have gained interest in recent years due to their fundamental role in immune homeostasis. While initial studies with Tregs often focused on their involvement in and possible use for inducing tolerance in autoimmune diseases and transplantation, there has also been an increasing interest in blocking their suppressive activity to enhance immunization against both self and non-self antigens. One of the potential uses of anti-Treg strategies is to increase the immune response in the

George Kassiotis and Adrian Liston (eds.), *Regulatory T Cells: Methods and Protocols*, Methods in Molecular Biology, vol. 707, DOI 10.1007/978-1-61737-979-6_14,

field of cancer immunotherapy. Tumor growth is believed to result at least partly from the lack of sufficient immune response to tumor antigens, and elevated Treg levels may account for the poor immune response to cancer (2, 3). Reducing the number of Tregs in the body may boost the immune response to weak tumor antigens.

Several lines of evidence indicate Tregs can suppress host immune responses and induce self-tolerance. Mouse studies show that depletion of Tregs can lead to autoimmune diseases, and can enhance anti-tumor responses (4–7). In vitro studies have shown that for both humans and mice, Tregs suppress proliferation and cytokine production by responder T cells (8–11). Previous studies have indicated elevated levels of Tregs in lung, ovarian, and breast cancer patients, and have correlated Tregs levels in cancer patients with poor prognosis (12–14).

Vaccines to treat or prevent recurrence of cancer have been of considerable interest (12, 15), but one challenge to their efficacy has been an immune suppressive effect of the tumor microenvironment and modulation of T cell expansion by inhibitory cells such as regulatory T cells. Increasing evidence implicates a contribution of Treg to the impaired host immune response against cancer (3, 16). Elevated Treg levels in the peripheral blood, regional lymph nodes, and the tumor microenvironment of cancer patients are associated with reduced survival (2, 13, 17, 18). Depletion of Treg in animal models leads to enhanced anti-tumor immune responses (19–22), and human studies have reported that Treg depletion before immunization enhanced tumor antigen-specific T cell responses (23, 24).

There are multiple strategies for depleting regulatory T cells including anti-CD25 antibodies, cyclophosphamide and the immunotoxin denileukin diftitox. Denileukin diftitox (ONTAK) is a fusion protein of the active domains of diphtheria toxin and IL-2. Denileukin diftitox binds to cells expressing high levels of CD25, whereupon it is internalized, leading to blockade of protein synthesis and cell death (25, 26). Clinically, denileukin diftitox has shown direct antitumor activity against CD25-expressing T cell malignancies (27, 28). Cells expressing the high affinity IL-2 receptor, consisting of three sub-units: β-subunit (CD122), a γ-subunit (CD132), and a α-subunit (CD25), are most susceptible to the effects of denileukin diftitox, and the short half-life of denileukin diftitox (70–80 min) should limit its impact on subsequently activated effector T cells (29). This set of methods describes the evaluation of denileukin diftitox as a targeted therapy for depletion of Treg cells, with a view towards developing a strategy for clinical depletion of Treg prior to immunotherapy of cancer.

2. Materials

2.1. Flow Cytometry

1. Antibodies for staining cocktail: human anti-CD25-FITC, anti-CD3-PerCP, anti-CD4-APC, anti-CD14-APC, anti-CD19-APC and isotype controls (catalog numbers 347643, 347344, 340443, 340436, 340437; BD Biosciences, San Jose, CA).
2. Peptide-MHC Tetramers: PE-HLA-A*0201 CMVpp65 (NLVPMVATV), PE-HLA-A*0201 MART-1 (ELAGIGLTV) and negative control tetramers (Beckman Coulter, Fullerton, CA).
3. FACS washing solution: 1% BSA/PBS (w/v) solution: 5 g Bovine Serum Albumin (BSA, Sigma, St. Louis, MO) completely dissolve in 500 ml PBS (Dulbecco's Phosphate Buffered Saline, Invitrogen, Carlsbad, CA) and filter sterilize. Store at 4°C.
4. BD Falcon Polystyrene round bottom 5 ml snap cap culture tubes (BD Bioscience Discovery Labware, Bedford, MA).
5. 0.5 ml Screw cap conical cryogenic tubes (Bio Plas Inc., San Rafael, CA).
6. eBioscience PE anti-human Foxp3 Staining Set (eBioscience, San Diego, CA).
7. eBioscience 1× RBC Lysis Buffer (eBioscience).
8. BD FACSFlow sheath fluid (BD Biosciences).

2.2. Cell Culture and T Cell Expansion

1. Ficoll-Hypaque™ Plus (Amersham Pharmacia Biotech AB, Upsala, Sweden).
2. Complete Media: RPMI 1640, 10% v/v huAB serum, 25 mM Hepes, 100 U/ml penicillin, 100 μg/ml streptomycin, 2 mM glutamine.
3. Peptide Antigens: CMVpp65(495-503) (NLVPMVATV), MART-1(27-35) (ELAGIGLTV) (synthesized by Invitrogen and HPLC purified to greater than 95% purity).
4. 0.4% Trypan Blue (Invitrogen).
5. BD Falcon 3 ml transfer pipet (BD Biosciences).
6. Deoxyribonuclease (DNase) (Sigma): dissolve 20,000 units into 100 ml HBSS (Hanks Balanced Salt Solution, Sigma) and filter sterilize to make 10× stock solution. Aliquot and store are −20°C.

2.3. Treg Depletion and Proliferation Assay

1. Orthoclone OKT3 human anti-CD3 (Ortho Biotech, Horsham, PA).

2. Carbonate-bicarbonate buffer (pH 9.6): dilute 1 carbonate-bicarbonate capsule (Sigma) in 100 ml deionized water, filter sterilize, and store at 4°C.
3. Denileukin diftitox (ONTAK) (Eisai, Inc., Woodcliff Lake, NJ). Dilute in complete media just prior to use.
4. Human interleukin-2 (IL-2, Proleukin) (Novartis Pharmaceuticals Corporation, East Hanover, NJ). Dilute in complete media just prior to use.
5. (^{3}H]Thymidine (PerkinElmer, Boston, MA): dilute in complete media at 1 μCi/20 μl; use 20 μl/well.
6. PBS (Invitrogen).
7. Scintillation fluid (Betaplate Scintillation Fluid, Wallac, Turku, Finland).

3. Methods

Analysis of Tregs typically involves identification of Treg population by flow cytometric methods and functional analysis by suppression of CD4+CD25− proliferation by CD4+CD25+ cells. Flow cytometry allows for a quantitative description of Tregs by surface staining for CD4 and CD25 and intracellular staining of FoxP3. There are multiple human FoxP3 antibodies available for intracellular staining, and it is important to determine which antibody works best for a given protocol. Flow cytometric acquisition and analysis of Tregs should ideally be performed by an operator with previous experience identifying human Treg cells since human CD4+CD25^{High} expressing cells are not as clearly defined as mouse CD4+CD25+ populations.

Functional analysis of Tregs determined by the suppression of anti-CD3 induced CD4+CD25− T cell proliferation by CD4+CD25+ Tregs is of limited use for many Treg studies. These types of assays, while useful, examine the functional capabilities of a given set of isolated Tregs and determine inhibition of non-specific proliferation (suppression of anti-CD3 stimulated T cells). For Treg depletion studies, our group has been interested in determining not only whether active depletion of Tregs from the total T cell population enhances overall T cell proliferation, but also whether it can benefit antigen specific T cell expansion. We developed assays based on our experience in human immunology studies in cancer and viral diseases to address these issues.

3.1. Flow Cytometry Analysis of Regulatory T Cells

1. Human blood samples are obtained following signed informed consent from cancer patients or healthy human donors according to an institutional review board-approved protocol.

2. Determine the number of blood samples for staining. All antibody cocktails for FACS analysis can be made prior to initiation of staining procedures. Ideally, 0.5 ml V bottom screw top tubes should be used to store pre-made cocktails. The antibody cocktail for Treg staining of a single blood sample should be prepared using the following: 20 μl anti-CD25-FITC, 10 μl anti-CD3-PerCP, 10 μl anti-CD4-APC (see Note 1).
3. Add 200 μl whole blood to 5 ml snap cap polystyrene tubes for positive and negative staining controls, and two experimental tubes for staining for Tregs.
4. Add 20 μl of the Treg cocktail prepared in step 2 to each of the experimental tubes containing blood and mix gently.. Incubate in the dark at room temperature for 30 min.
5. After incubation, add 2 ml of 1× RBC lysis buffer (eBioscience) to each tube. Snap the tube cap on tightly and mix back and forth five times. Incubate for 10 min in the dark at room temperature. Do not allow to incubate longer than 15 min.
6. Add 2 ml PBS to each tube. Centrifuge for 5 min at 1,800 × *g*.
7. Carefully aspirate the supernatant so as not to disturb the cell pellet. Wash the cell pellet by adding 2 ml 1% BSA/PBS, gently mixing, and centrifuge for 5 min at 1,800 × *g*. While centrifuging, prepare the eBioscience Fix/Perm buffer by diluting 1 part Fix/Perm buffer to 3 parts Fix/Perm Diluent (both supplied in eBioscience set).
8. Carefully aspirate the supernatant and gently resuspend the cell pellet with low speed pulse vortexing.
9. Once the pellet is suspended, add 1 ml of the prepared/diluted (step 7) Fix/Perm buffer to each tube. Briefly pulse vortex and incubate at 4°C for 30 min (see Note 2).
10. Wash each tube using 2 ml 1% BSA/PBS and centrifuge for 5 min at 1,800 × *g*. While centrifuging, prepare eBioscience Permeabilization buffer for washes. It is supplied as a 10× solution. Dilute this stock 1:10 in dH_2O to make enough to wash each tube four times with 1 ml buffer.
11. Carefully aspirate the supernatant from the cell pellet and wash each tube with 1 ml Permeabilization buffer. Repeat this step for two total washes. Prepare blocking buffer by making a 2% v/v rat serum (supplied in eBioscience set) in 1× Permeabilization buffer from step 10.
12. Aspirate supernatant and gently resuspend cell pellet with brief pulse vortexing. Add 100 μl blocking buffer to each tube and gently mix. Incubate at 4°C for 15 min.

13. Without washing, add to control Treg tube 15 µl rat anti-IgG isotype (this tube will have surface stain only) and to the other Treg tube 15 µl anti-FoxP3-PE (this tube will contain FoxP3+ surface stain). Mix gently and incubate 30 min at 4°C in the dark.
14. Wash each tube twice using 1 ml of the diluted Permeabilization buffer (from step 10) and centrifuge for 5 min at 1,800 × *g*.
15. Aspirate the final wash and resuspend cell pellets in 200 µl 1% BSA/PBS for analysis by flow cytometry. Acquisition by FACS should ideally be performed within 2–3 h and stained cells stored on ice or at 4°C until acquired.
16. The CD4+CD25+FoxP3+ Treg population is determined by gating on lymphocytes by forward and side scatter and CD3+ T cells, and then the percent of CD4+CD25^{bright} cells that are also greater than 90% FoxP3+.

3.2. Depletion of Treg by Denileukin Diftitox

1. Human blood samples are obtained following signed informed consent from cancer patients or healthy human donors according to an institutional review board-approved protocol.
2. Isolate PBMC from approximately 90 cc of human blood by density gradient centrifugation over Ficoll-Hypaque™ Plus gradient (see Note 3).
3. Culture the PBMC in a 12 well plate at 10^7 cells/well in a total volume 4 ml with complete media alone (RPMI 1640, 10% huAB serum, 25 mM Hepes, 100 U/ml penicillin, 100 µg/ml streptomycin, 2 mM glutamine) or media plus desired concentrations of denileukin diftitox (0.5–8 nM) (see Note 4). Incubate at 37°C for 18–20 h.
4. Harvest each well into 15 ml conical tubes keeping each treatment group separated. Rinse each well with 1–2 ml PBS and collect with the harvested cells. Centrifuge for 5 min at 1,800 × *g*. Wash twice with 10 ml PBS. During the second wash, take a sample to count cells by trypan blue exclusion.
5. Re-plate cells into 12 well plates at 8×10^6 cells/well in 4 ml complete media plus 10 IU/ml of IL-2. Incubate at 37°C for 3 days.
6. Harvest cells from each well, wash and analyze each treatment group for CD4+CD25+FoxP3+ expression frequency as described in Subheading 3.1. Proceed with proliferation and/or expansion experiment as described in Subheadings 3.3 and 3.4.

3.3. Proliferation Assay

1. The proliferative capacity of PBMC depleted of Treg can be determined by assessing non-specific proliferation following treatment with denileukin diftitox or media alone and recovery in IL-2 as described in Subheading 3.2.

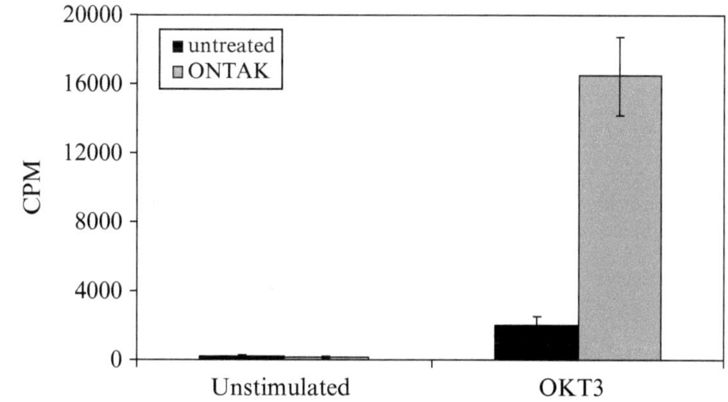

Fig. 1. Fresh PBMC from a normal donor were treated 18 h with 5 nM denileukin diftitox (ONTAK) or media alone (untreated). Cells were then harvested, washed and recultured in 10 U/ml IL-2 for 72 h. Cells were then stimulated in a proliferation assay with media alone (unstimulated) or 0.5 mg/ml soluble OKT3 for 72 h. Proliferation was measured by [^{3}H]Thymidine incorporation and the mean cpm +/– SD is presented.

2. Prepare a flat bottom 96 well plate on the day preceding initiation of proliferation assay by mixing 1 μg anti-CD3 (OKT3 monoclonal antibody) per ml of carbonate-bicarbonate buffer (pH 9.6). Coat plate using 200 μl/well of the diluted anti-CD3 and incubate at 4°C overnight (see Note 5).
3. Just prior to use, aspirate anti-CD3 coating mixture and wash twice with approximately 300 μl/well PBS. Make sure wells are not allowed to dry out prior to use.
4. Add 10^5 cells/well from each treatment group in a total volume of 200 μl/well of complete media to the anti-CD3 coated plate. Negative controls should be performed in parallel in uncoated plate/wells. Incubate at 37°C for 96 h.
5. Add 1 μCi/well of [^{3}H]Thymidine diluted in complete media and incubate at 37°C for 12–18 h.
6. Harvest cells from the plate using a cell harvester and determine total counts per minute (+/– standard deviation) on a scintillation counter. An example result is shown in Fig. 1.

3.4. Antigen Specific T Cell Expansion

3.4.1. Donor Cells

1. Human donors must be MHC genotyped prior to T cell stimulation with single peptide antigens. Donors must be class I HLA-A*0201 when using the CMV pp65(495-503) peptide or MART 1(27-35) peptide.
2. Once treated with media alone or denileukin diftitox as described in Subheading 3.2, culture PBMC from each treatment group in 24 well plates at 2×10^6 cells/well in complete media at 1 ml/well.

3.4.2. CMV Peptide T Cell Expansion

1. Prior to use for CMV T cell expansion, donors should be tested for CMV positive status. The baseline CMV pp65(495-503) peptide-MHC tetramer response for the donor should be determined prior to beginning this protocol.
2. Add 1 ml complete media containing 2 μg/ml CMV pp65(495-503) peptide to each well containing cells from Subheading 3.4.1, step 2 for 1 μg/ml final concentration (the total volume should be 2 ml/well cells plus peptide). Mix cells and peptide gently with a 3 ml plastic transfer pipette. Incubate at 37°C for 2 days.
3. Add 300 IU/ml IL-2 per well, with minimal adjustment to the total volume in each well. Incubate an additional 2–3 days.
4. If culture medium in cell-containing wells is yellow or orange in color, remove half the media (1 ml) with a 3 ml transfer pipette, and replace with 1 ml fresh complete media. If media has not changed in color, do not add anything to the plate. Incubate 1–2 days (see Note 6).
5. If wells are reddish-orange (original media color) or orange and cells are sparse or slightly confluent, add 300 IU/ml IL-2 to each (as in step 3). If medium is yellow or yellow-orange and cell population is confluent, split the culture by using the rubber tip of a sterile syringe plunger to gently scrape the adherent cells off the bottom of the well. Then resuspend the contents of the well and remove 1 ml to a new unused well. Add 1 ml fresh medium per well and add IL-2 to give a final concentration of 300 IU/ml. Incubate an additional 1–3 days.
6. If the medium if yellow or yellow orange and total cell population is confluent, split as in step 5.
7. After 10–14 total days in the expansion culture, depending on the T cells expansion rate, harvest $3–5 \times 10^6$ cells for peptide-MHC tetramer staining (see Subheading 3.4.3) to determine percent T cells that are CMV pp65(495-503) specific. Keep remaining cells in culture to continue to expand or cryopreserve for future use.

3.4.3. MART-1 Peptide T Cell Expansion

1. The baseline MART-1(27-35) peptide-MHC tetramer response for the donor should be determined prior to beginning this protocol.
2. Add 1 ml complete media containing 2 μg/ml MART-1(27-35) peptide to each well containing cells from Subheading 3.4.1, step 2 for 1 μg/ml final peptide concentration. Proceed with an initial 10–11 days stimulation as described in Subheading 3.4.2, steps 2–7 and determine the percentage of T cells that are MART-1(27-35) specific by peptide-MHC tetramer staining (see Subheading 3.4.3).
3. Plan to restimulate the primary MART-1(27-35) peptide stimulated culture following 10–11 days peptide stimulation.

Four days prior to the restimulation, prepare stimulator cells by collecting fresh or thawing cryopreserved *autologous* PBMC. Culture these PBMC with media alone or denileukin diftitox as described in Subheading 3.2.

4. Set up the restimulatation of the primary cultured cells by harvesting each well of the 24 well plate containing the day 10 or 11 primary MART-1(27-35) culture using a 3 ml transfer pipette. Pool all cells from each denileukin diftitox or media alone treatment group into a separate 50 ml conical tube. Wash each with 10 ml PBS, centrifuge and count using trypan blue exclusion.
5. Resuspend cells at 5×10^5 cells/ml and plate 1 ml/well of a 24 well plate in complete media. Based on number of wells, determine number of stimulators required to have 1:8 ratio of responders:stimulators.
6. Prepare stimulator cells by harvesting the autologous PBMC treated with denileukin diftitox and media from step 3. Wash, count, and suspend each treatment group at $\leq 6 \times 10^6$ cells/ml complete media and 1 μg/ml MART-1(27-35) peptide in a 50 ml conical tube.
7. Incubate stimulators at 37°C for 3–5 h with gentle mixing. Wash twice with PBS and resuspend in complete media at 4×10^6 cells/ml.
8. Irradiate peptide pulsed stimulator cells with 10,000 rads. Add 1 ml of stimulators pre-treated with denileukin diftitox to each well containing primary expanded cells that were also pre-treated with denileukin diftitox (from step 3) and gently mix with a 3 ml transfer pipette. Repeat with stimulators pre-treated with media alone and add to primary expanded culture of cells pretreated with media only. This will result in a 1:8 ratio responders to stimulators. Incubate at 37°C for 18–24 h.
9. Add 300 IU/ml IL-2 per well with minimal adjustment to the total volume in each well. Incubate at 37°C.
10. Observe cultures daily. Change media on cells as described in Subheading 3.4.2 and split as necessary based on media color and cell confluency. Add IL-2 every 3–4 days as needed (see Note 7).
11. Screen cultures for MART-1(27-35) reactivity by harvesting $3–5 \times 10^6$ cells for peptide-MHC tetramer staining (see Subheading 3.4.3) to determine percent T cells that are MART-1(27-35) specific. Keep remaining cells in culture to continue to expand or cryopreserve for future use (see Note 8).

3.4.4. Peptide-MHC Staining of Peptide Expanded T Cells

1. Prepare antibody cocktails for peptide-MHC tetramer staining in 0.5 ml microfuge screw cap tubes. The antibody cocktail for peptide-MHC tetramer staining of a single expanded T cell

sample should be prepared using the following: 10 µl anti-CD25-FITC, 5 µl anti-CD8-PerCP, 5 µl each anti-CD4-APC, anti-CD14-APC, and anti-CD19-APC. The peptide-MHC tetramer is supplied as tetramer-PE (see Note 9).

2. Add 1×10^6 expanded T cells in 100 µl 1% BSA/PBS to 5 ml snap cap culture tubes (or 96 well plate if large number of samples to stain) for positive and negative staining controls and for each peptide-MHC tetramer stain.
3. Add 25 µl of the antibody cocktail as described in step 1 plus 2 µl of peptide-MHC tetramer to appropriate tubes. Gently mix the cocktail with the cells thoroughly, so that the cell pellet is resuspended. Incubate the plate for 30 min at room temperature in the dark (see Note 10).
4. Following incubation, wash cells twice in 1 ml 1% BSA/PBS.
5. Resuspend cells with 200 µl 1%BSA/PBS. Cells can be kept at 4°C for up to 4 h prior to reading by FACS or they can be fixed by resuspending in 200 µl 1% paraformaldehyde/BSA/PBS for up to 4 days.
6. To analyze peptide-MHC tetramer positive cells by flow cytometry, gate by forward and side scatter on live lymphocytes (a viability dye should be used if possible to exclude dead cells. We recommend Invitrogen LIVE/DEAD Fixable Violet Dead Cell Stain Kit catalog number L34955), positively for CD3 cells, and negatively on CD4, CD14, and CD19 cells. A dot plot of CD8+ cells verses peptide-MHC Tetramer positive cells can be used show the CD8+CMVpp65(495-503) or MART-1(27-35) Tetramer positive population. An example result is shown in Fig. 2.

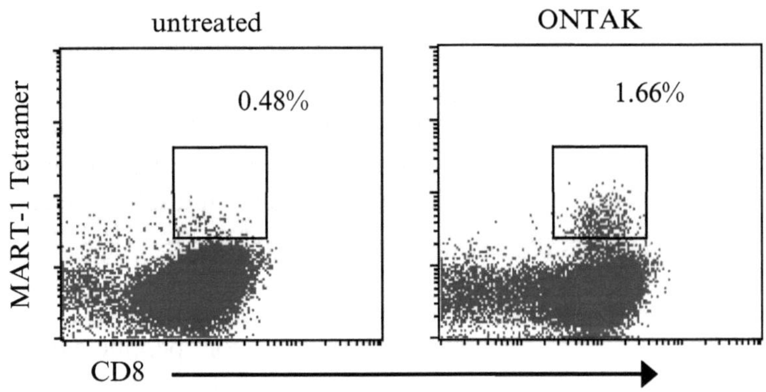

Fig. 2. Normal healthy volunteer PBMC pre-treated with denileukin diftitox (ONTAK) or media alone were cultured with MART-1(27-35) peptide and analyzed for antigen specific CD8+ cells by MART-1(27-35) peptide-MHC tetramer day 10 of culture. These results demonstrate an increase in the percent CD8+tetramer+ lymphocytes specific for the MART-1 tumor antigen in ONTAK treated PBMC over untreated PBMC following a single in vitro stimulation with MART-1 peptide.

4. Notes

1. Antibody cocktails for staining can typically be mixed several days ahead of time. The advantage to this is a consistent preparation across multiple samples and experiments. Positive and negative controls, as well as set up controls for FACS analysis, will depend upon the personal preference of the FACS operator and the type of flow cytometer used. Various antibody fluorochrome combinations can be used. Each should be optimized for a given set of experiments.
2. At this incubation stage, the tubes can be left up to 20 h in the Fix/Perm buffer at 4°C and get results comparable to the suggested 30 min incubation.
3. PBMC freshly isolated from whole blood are ideal for these experiments. However, PBMC that have been cryopreserved can also be used with good results. We recommend optimizing freezing and thawing procedures for PBMC (30).
4. We found 5 nM denileukin diftitox to be ideal for our purposes. The molecular weight for denileukin diftitox is 58 kD. Denileukin diftitox (ONTAK) is a recombinant DNA-derived cytotoxic protein composed of the amino acid sequences for diphtheria toxin fragments A and B (Met_1-Thr387)-His and the sequences for human interleukin-2 (IL-2; Ala_1-Thr_{133}).
5. Anti-CD3 coated plates can be left at 4°C for up to 1 week.
6. The expansion of T cells with peptide must be monitored carefully to avoid over growth and media depletion in the cultures, especially following depletion of Treg with denileukin diftitox. The expansion should ideally be watched daily to change media and split expanding cultures to avoid overgrowth. If media appears yellow and cells growth is rapid, changing of media and splitting of wells should happen more frequently than outlined here. For larger scale T cell expansions, vented T25 flasks can also be used and this protocol brought up to scale. An approximate concentration of 10^6 cells/ml should be maintained when using a flask to avoid overgrowth.
7. With a second stimulation, adherent cells should be in the minority, so cultures can be split without scraping the adherents cells from the bottom of the well.
8. We recommend testing cultures for peptide specificity by ELISpot, ELISA, Intracellular cytokine staining, or peptide-MHC Tetramer analysis depending upon reagents available and experience of the individual lab.
9. Do not dilute or add peptide-MHC Tetramers to antibody cocktails until just prior to use.

10. Optimal concentration of the peptide-MHC tetramer should be determined for each tetramer prior to use. When available, T cell clones that are specific for the MHC restricted epitope recognized by peptide-MHC tetramer are ideal for determining optimal tetramer concentration.

References

1. Shevach, E. M., DiPaolo, R. A., Andersson, J., Zhao, D. M., Stephens, G. L., and Thornton, A. M. (2006) The lifestyle of naturally occurring CD4+ CD25+ Foxp3+ regulatory T cells. *Immunol. Rev.* **212,** 60–73.
2. Ahmad, M., Rees, R. C., and Ali, S. A. (2004) Escape from immunotherapy: possible mechanisms that influence tumor regression/progression. *Cancer Immunol. Immunother.* **53,** 844–854.
3. Terabe, M., and Berzofsky, J. A. (2004) Immunoregulatory T cells in tumor immunity. *Curr. Opin. Immunol.* **16,** 157–162.
4. Onizuka, S., Tawara, I., Shimizu, J., Sakaguchi, S., Fujita, T., and Nakayama, E. (1999) Tumor rejection by in vivo administration of anti-CD25 (interleukin-2 receptor α) monoclonal antibody. *Cancer Res.* **59,** 3128–3133.
5. Sakaguchi, S., Sakaguchi, N., Asano, M., Itoh, M., and Toda, M. (1995) Immunologic self-tolerance maintained by activated T cells expressing IL-2 receptor a-chains (CD25). Breakdown of a single mechanism of self-tolerance causes various autoimmune diseases. *J. Immunol.* **155,** 1151–1164.
6. Sakaguchi, S., Sakaguchi, N., Shimizu, J., Yamazaki, S., Sakihama, T., Itoh, M., et al. (2001) Immunologic tolerance maintained by CD25+CD4+ regulatory T cells: their common role in controlling autoimmunity, tumor immunity, and transplantation tolerance. *Immunol. Rev.* **182,** 18–32.
7. Shimizu, J., Yamazaki, S., and Sakaguchi, S. (1999) Induction of tumor immunity by removing CD25+CD4+ T cells: a common basis between tumor immunity and autoimmunity. *J. Immunol.* **163,** 5211–5218.
8. Asseman, C., Mauze, S., Leach, M. W., Coffman, R. L., and Powrie, F. (1999) An essential role for interleukin 10 in the function of regulatory T cells that inhibit intestinal inflammation. *J. Exp. Med.* **190,** 995–1004.
9. Lombardi, G., Sidhu, S., Batchelor, R., and Lechler, R. (1994) Anergic T cells as suppressor cells in vitro. *Science* **264,** 1587–1589.
10. Seddon, B., and Mason, D. (1999) Regulatory T cells in the control of autoimmunity: the essential role of transforming growth factor beta and interleukin 4 in the prevention of autoimmune thyroiditis in rats by peripheral CD4(+)CD45RC- cells and CD4(+)CD8(-) thymocytes. *J. Exp. Med.* **189,** 279–288.
11. Thornton, A. M., and Shevach, E. M. (1998) CD4+CD25+ immunoregulatory T cells suppress polyclonal T cell activation in vitro by inhibiting interleukin 2 production. *J. Exp. Med.* **188,** 287–296.
12. Morse, M. A., Clay, T. M., Hobeika, A. C, Osada, T., Khan, S., Chui, S., et al. (2005) Phase I study of immunization with dendritic cells modified with fowlpox encoding carcinoembryonic antigen and costimulatory molecules. *Clin. Cancer Res.* **15,** 3017–3024.
13. Curiel, T. J., Coukos, G., Zou, L., Alvarez, X., Cheng, P., Mottram, P., et al. (2004) Specific recruitment of regulatory T cells in ovarian carcinoma fosters immune privilege and predicts reduced survival. *Nat. Med.* **10,** 942–949.
14. Liyanage, U. K., Moore, T. T., Joo, H. G., Tanaka Y., Herrmann, V., Doherty, G., et al. (2002) Prevalence of regulatory T cells is increased in peripheral blood and tumor microenvironment of patients with pancreas or breast adenocarcinoma. *J. Immunol.* **169,** 2756–2761.
15. Small, E. J., Schellhammer, P. F., Higano, C. S., Redfern, C. H., Nemunaitis, J. J., Valone, F. H., et al. (2006) Placebo-controlled phase III trial of immunologic therapy with sipuleucel-T (APC8015) in patients with metastatic, asymptomatic hormone refractory prostate cancer. *J. Clin. Oncol.* **24,** 3089–3094.
16. Wolf, A. M., Wolf, D., Steurer, M., Gastl, G., Gunsilius, E., and Grubeck-Loebenstein, B. (2003) Increase of regulatory T cells in the peripheral blood of cancer patients. *Clin. Cancer Res.* **9,** 606–612.
17. Viguier, M., Lemaitre, F., Verola, O., Cho, M. S., Gorochov, G., Dubertret, L., et al. (2004) Foxp3 expressing CD4+CD25(high) regulatory T cells are overrepresented in human metastatic melanoma lymph nodes and inhibit the function of infiltrating T cells. *J. Immunol.* **173,** 1444–1453.
18. Woo, E. Y., Yeh, H., Chu, C. S., Schlienger, K., Carroll, R. G, Riley, J. L., et al. (2002) Cutting

edge: regulatory T cells from lung cancer patients directly inhibit autologous T cell proliferation. *J. Immunol.* **168,** 4272–4276.

19. Hussain, S. F., and Paterson, Y. (2004) CD4+CD25+ regulatory T cells that secrete TGFbeta and IL-10 are preferentially induced by a vaccine vector. *J. Immunother.* **27,** 339–346.
20. Casares, N., Arribillaga, L., Sarobe, P., Dotor, J., Lopez-Diaz de Cerio, A., Melero, I., et al. (2003) CD4+/CD25+ regulatory cells inhibit activation of tumor-primed CD4+ T cells with IFN-gamma-dependent antiangiogenic activity, as well as long-lasting tumor immunity elicited by peptide vaccination. *J. Immunol.* **171,** 5931–5939.
21. Nagai, H., Horikawa, T., Hara, I., Fukunaga, A., Oniki, S., Oka, M., et al. (2004) In vivo elimination of CD25+ regulatory T cells leads to tumor rejection of B16F10 melanoma, when combined with interleukin-12 gene transfer. *Exp. Dermatol.* **13,** 613–620.
22. Turk, M. J., Guevara-Patino, J. A., Rizzuto, G. A., Engelhorn, M. E., and Houghton, A. N. (2004) Concomitant tumor immunity to a poorly immunogenic melanoma is prevented by regulatory T cells. *J. Exp. Med.* **200,** 771–782.
23. Vieweg, J., Su, Z., and Dannull, J. (2004) Enhancement of antitumor immunity following depletion of $CD4^{+}CD25^{+}$ regulatory T cells [Abstract]. *J. Clin. Oncol. 2004 ASCO Annual Meeting Proceedings.* **22,** 2506.
24. Morse, M. A., Hobeika, A. C., Osada, T., Serra, D., Niedzwiecki, D., Lyerly, H. K., et al. (2008) Depletion of human regulatory T cells specifically enhances antigen-specific immune responses to cancer vaccines. *Blood* **112,** 610–618.
25. Almeida, A. R., Legrand, N., Papiernik, M., and Freitas, A. A. (2002) Homeostasis of peripheral CD4+ T cells: IL-2α and IL-2 shape a population of regulatory cells that controls CD4+ T cell numbers. *J. Immunol.* **169,** 4850–4860.
26. Taniguchi, T., and Minami, Y. (1993) The IL-2/IL-2 receptor system: a current overview. *Cell* **73,** 5–8.
27. Duvic, M., Cather, J., Maize, J., and Frankel, A. E. (1998) $DAB_{389}IL2$ diphtheria fusion toxin produces clinical responses in tumor stage cutaneous T cell lymphoma. *Am. J. Hematol.* **58,** 87–90.
28. Olsen, E., Duvic, M., and Frankel, A. (2001) Pivotal phase III trial of two dose levels of denileukin diftitox for the treatment of cutaneous T-cell lymphoma. *J. Clin. Oncol.* **19,** 376–388.
29. Waters, C. A., Schimke, P. A., Snider, C. E., Itoh, K., Smith, K. A., Nichols, J. C., et al. (1990) Interleukin 2 receptor-targeted cytotoxicity. Receptor binding requirements for entry of a diphtheria toxin-related interleukin 2 fusion protein into cells. *Eur. J. Immunol.* **20,** 785–791.
30. Disis, M. L., dela Rosa, C., Goodell, V., Kuan, L. Y., Chang, J. C., Kuus-Reichel, K., et al. (2006) Maximizing the retention of antigen specific lymphocyte function after cryopreservation. *J. Immunol. Methods* **308,** 13–18.

Chapter 15

Assessment of Suppressive Capacity by Human Regulatory T Cells Using a Reproducible, Bi-Directional CFSE-Based In Vitro Assay

Anya Schneider and Jane H. Buckner

Abstract

Regulatory T cells are involved in the maintenance of tolerance. Alterations in their functional capacity are implicated in the development of autoimmunity. In the case of common autoimmune disorders the defects in suppression may be partial, and may be due to a loss of Treg function, or a resistance to suppression by responder T cells. Thus in order to assess Treg function, an in vitro assay that is sensitive enough to demonstrate modest alterations in suppression, and which can differentiate between impaired suppression due to Treg dysfunction, and responder cell resistance is ideal. In this chapter we describe a CFSE based proliferation assay that utilizes a bead based activation system, which is reproducible, consistent and able to distinguish between defects in Treg function and the resistance of responder T cells.

Key words: Adaptive Treg, T effector cells, CFSE-assay, Regulatory function, CD4+CD25+FOXP3+ Treg, Suppression, Immunoregulation

1. Introduction

Regulatory T cells suppress both proliferation and cytokine production of effector T cells (1). A lack of Treg results in overwhelming autoimmunity, in both mouse and man (2, 3). Thus it is thought that defects in suppression by Treg may contribute to autoimmunity (4–6), either due to impaired function of the Treg themselves, or due to a resistance to the suppressive function of Treg by pathogenic effector T cells. In order to determine whether either of these potential defects in regulation is present in human disease, an in vitro assay that is reproducible and sensitive is needed. Several types of suppression assays have been utilized for the purpose of measuring the impact of Treg on

George Kassiotis and Adrian Liston (eds.), *Regulatory T Cells: Methods and Protocols*, Methods in Molecular Biology, vol. 707, DOI 10.1007/978-1-61737-979-6_15, © Springer Science+Business Media, LLC 2011

responder T cell proliferation: [3H] thymidine incorporation and 5.6-carboxyfluorescein diacetate succinimidyl ester (CFSE) dye dilution (7). [3H] thymidine been used most frequently due to the ease of these studies, and the ability to utilize very low cell numbers to perform these assays. However, a shortcoming of these assays is their inability to distinguish which cells in the coculture have incorporated [3H] thymidine, which can result in an underestimated Treg mediated suppression (8). In addition, [3H] based assays can only give a snapshot of proliferation during the period of time that the H3T is present in the culture. The limitation of CFSE dye dilution assays is that they require a larger cell number than H3T assays, however, the advantages of this approach include the ability to specifically evaluate the proliferation of the responder T cell population, and to examine the number of cell divisions throughout the culture period (see Fig. 1).

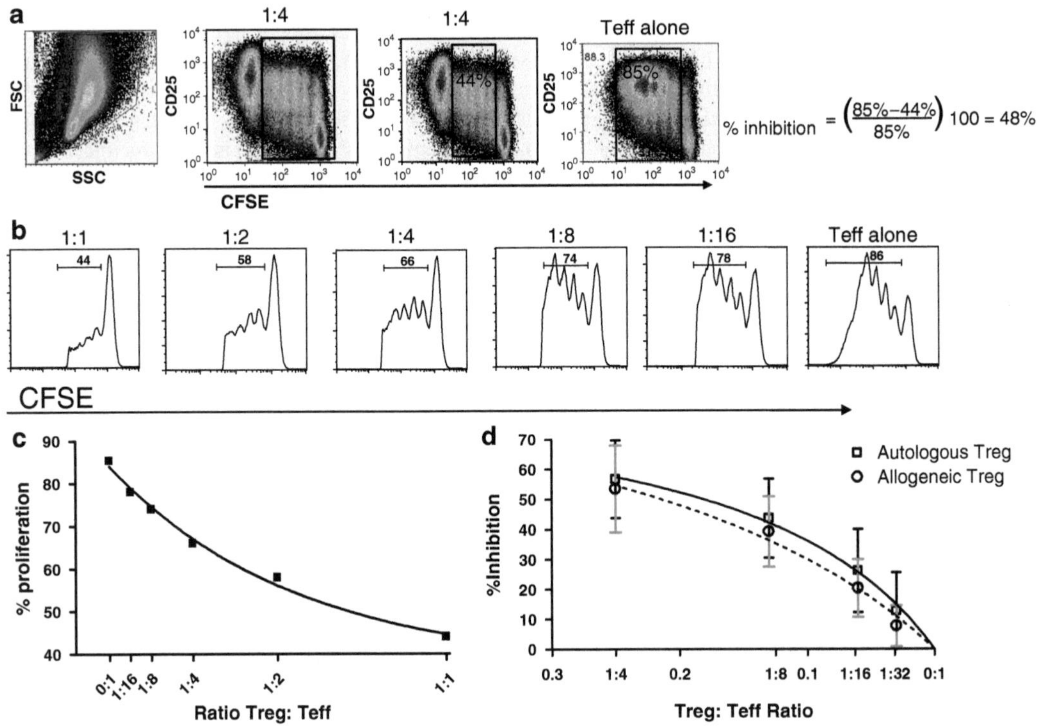

Fig. 1. (**a**) Gating strategy and calculation of percent inhibition for suppression assay: FACS analysis was performed on day 4 by gating on live cells and excluding the CFSE low population (unstained Treg) to determine the percentage of the responder cells that have diluted CFSE. Percentage inhibition was determined by comparing the percentage of proliferating Teff cells cultured alone to the percentage of proliferating Teff cells in coculture with in vitro generated Treg at a ratio 1:4 (Treg:responder cells). (**b**) These histograms illustrate the correlation between the percent inhibition and the Treg number in the coculture at a range of Treg: responder ratios as indicated. The histogram on the far right shows CFSE-labeled Teff cells when they were cultured alone with Dynabeads. (**c**) Percentage proliferation of CFSE-labeled responder cells is plotted against Treg: responder ratio. (**d**) Comparison of percentage inhibition between suppression assays performed with autologous Treg/Teff cocultures (□, $n=6$) and those that were performed with allogeneic cells (○, $n=6$).

Second variable to be considered in suppression assays is the type of T cell activation to be used, and the role of antigen presenting cells in the cultures, both of which can impact the outcome of suppression assays (9). The assay that we describe in this chapter was developed first in the absence of APC with a bead based activation system (10). This has the advantage of calibrating the strength of signal, and keeping the level of activation consistent between samples. In addition, we have found that this approach allows us to perform these studies with cocultures containing allogeneic Treg and T responder, as well as autologous Treg and T responder cells. This has the advantage of allowing investigators to assess the function of Treg, and the responsiveness of Teff to Treg of an individual independently.

The technique described in this chapter, can be extended to other coculture and activation conditions. We describe the use of both in vitro generated Treg (aTreg) (11, 12) and use of Treg directly isolated from the peripheral blood (nTreg). We have verified that suppression is no different between these two types of Treg (10). Further one can alter the protocol with respect to the responder cell used, such as CD8 T cells, or the form of activation, including use of autologous irradiated APC and peptide antigen or soluble anti-CD3. However, for an analysis of any subject population, a well matched set of samples from healthy controls must be used as the comparison group. This will establish the "normal" level of suppression for this assay system.

2. Materials

1. Complete medium: RPMI 1640 HEPES medium (Thermo Scientific), 10% PHS (human serum off-clot, sterile filtered untransfused male donors, MP Biomedicals), 1 mM penicillin/streptomycin (Thermo Scientific), 1 mM Na-pyruvate (Thermo Scientific) and l-Glutamine 2 mM (Gibco Scientific), store at 37°C.
2. PBS: 1× Gibco PBS (calcium chloride, magnesium chloride), store at room temperature (RT) and at 4°C.
3. FACS buffer: PBS, 1% Fetal bovine serum (Atlas biologicals company), 0.1% Na N_3 (Sigma-Aldrich), store at 4°C.
4. MACS buffer: PBS, 2 mM EDTA (Sigma-Aldrich), 0.5% BSA, store at 4°C.
5. IL-2 (Chiron), final concentration: 200 IU/ml in complete medium, store at 4°C.
6. CFSE-staining solution (Invitrogen): store at −20°C, make up 2× staining solution (1.6 μM) in 1× PBS at 1 ml per sample to be stained, light sensitive, make fresh as required.

7. Freezing medium: 10% FBS in RPMI 1640 HEPES complete medium, make fresh as required.
8. FOXP3-intracellular staining reagents: 1× Biolegend FOXP3 Fix/Perm solution in 1× PBS, make fresh as required; 1× FOXP3 Perm buffer in 1× PBS, can be stored at 4°C for a couple of months.
9. Antibodies: anti-CD4 and anti-CD25 (available through multiple vendors), isotype controls, anti-Alexa Fluor® 647 antihuman FOXP3 (clone: 206D, Biolegend), FOXP3 IgG1 control (clone: MOPC-21, Biolegend), anti-CD3 (UCHT1, BD Pharmingen), anti-CD28 (CD28.2, BD Pharmingen), store at 4°C.
10. MACS separation: CD4+ T cell isolation kit II, CD25 Microbeads II (Miltenyi Biotec), store at 4°C.
11. Dynabeads® M-280 Tosylactivated store at 4°C, Buffer B: 0.1 M borate buffer pH 9.5 dissolve in distilled water; Buffer C: PBS pH 7.4 with 0.1 % (w/v) BSA, Buffer D: 0.2 M Tris pH 8.5 with 0.1 (w/v) BSA dissolve in distilled water. Keep sterile filtered at 4°C.

3. Methods

3.1. Dynabeads Coating Procedure

1. Pipette the volume of beads to be used into tube and place on magnet. Pipette off supernatants and wash beads twice in Buffer B. Release from magnet. Dilute Dynabeads to a concentration of $1\text{–}2 \times 10^9$ beads per ml in Buffer B. Then add equal amounts of anti-CD3 and anti-CD28 (5 μg/ml). Incubate at room temperature for 24 h with slow tilt rotation.
2. After incubation, place tube on magnet and remove the supernatants with a pipette. Wash beads four times in the following order: twice in Buffer C at 4°C for 5 min, once in Buffer D for 4 h at 37°C (Tris will block free tosyl-groups), once in Buffer C at 4°C for 5 min. Coated Beads can be stored in Buffer C at 4°C (see Note 1).

3.2. Isolation of Treg Populations

In these studies, one can use Treg that are isolated directly from the peripheral blood or which are generated in vitro. We have found that both types of CD4+CD25+FOXP3+ Treg demonstrate similar levels of suppression in this assay system (10). Isolation of the Treg population to be studied must be performed on the day of the assay, making it the most time sensitive aspect of these studies. We outline below in Subheading 3.2.1 an approach to isolation of the Treg directly from PBMC. In Subheading 3.2.2 we describe how to generate Treg using an in vitro generation culture system. Once Treg are isolated, we perform flow cytometry on a portion of the cells to determine the purity of the Treg population. This includes characterizing both their cell surface markers and expression of FOXP3.

3.2.1. Isolation of nTreg from the Peripheral Blood

1. PBMC are isolated from human peripheral blood by centrifugation over Ficoll Hypaque gradients.
2. CD4+ T cells are isolated via negative selection using CD4 T cell isolation kit (Miltenyi Biotec) (see Note 2). Resuspend cell pellet in 40 μl of cold MACS buffer per 10^7 total cells and add 10 μl of Biotin-antibody cocktail per 10^7 total cells. Incubate for 10 min at 4–8°C. Add 30 μl of cold MACS buffer per 10^7 total cells and 20 μl of antibiotin microbeads per 10^7 total cells. Incubate for 15 min at 4–8°C. Wash cells in cold MACS buffer for 5 min at 1,000 rpm (228 RCF), utilizing brake on low setting and resuspend up to 10^8 cells in 500 μl of cold MACS buffer and proceed to magnetic separation. Run the "deplete program" on the autoMACS.
3. Stain CD4+ T cells with anti-CD4 and anti-CD25 in complete medium on ice for 30 min. Include samples stained with isotype controls as well.
4. Wash fluorophore labeled cells with complete medium, centrifuge for 5 min at 1,000 rpm, utilize brake on low setting. Resuspend cells in about 1–2 ml complete medium to isolate nTreg. Select the 5% of T cells with the highest CD25 expression and sort via cell sorter. Store isolated Treg in complete medium, on ice until their addition to coculture assay.

3.2.2. In Vitro Generation of Adaptive Treg

1. Isolate PBMC from human peripheral blood by centrifugation over Ficoll Hypaque gradients.
2. Isolate CD4+ T cells from the PBMC via negative selection using CD4 T cell isolation kit (Miltenyi Biotec). As described in 3.2.1 (Subheading 2).
3. Save positive fraction from cell separation to use as APC. Store all cells in complete medium at 37°C.
4. Isolate CD4+CD25− T cells from the CD4+ T cells obtained above by removing the CD25+ cells. We use the CD25 Microbeads II Kit (Miltenyi Biotec) for this purpose. Resuspend cell pellet in 90 μl of cold MACS buffer per 10^7 total cells and add 10 μl of CD25 Microbeads II per 10^7 total cells. Incubate for 15 min at 4–8°C. Wash cells in cold MACS buffer for 5 min at 1,000 rpm, low brake and resuspend up to 10^8 cells in 500 μl of cold MACS buffer and proceed to magnetic autoMACS separation and run the "deplete program."
5. Irradiate autologous APC obtained in step 3 with 5,000 rad.

6. Coculture CD4+CD25− T cells in 24-well plates with irradiated autologous APC (5,000 rad) at a ratio of 2:1 (APC: responder cells) and add soluble anti-CD3 (5 μg/ml) to activate. Maintain for 10 days in 2 ml of complete medium/ well adding IL-2 at a concentration of 200 IU/ml on day 6 (see Note 3).
7. Isolate aTregs by FACS staining and sorting as described for nTreg in Subheading 3.2.1.

3.3. Determination of the Purity of Treg Population by Intracellular Anti-Human FOXP3 Staining

1. Save an aliquot of Treg population for FOXP3 analysis (about $1–2\times10^6$ cells). Perform surface staining in FACS buffer for 30 min on ice with anti-CD4, anti-CD25 and additional markers if desired to determine the % of isolated CD25+ cells that express FOXP3. Include samples stained with isotype controls.
2. Wash cells in FACS buffer for 5 min at 1,000 rpm, utilizing brake on low settings. Add 250–300 μl 1× FOXP3 Fix/Perm solution to each tube, vortex gently and incubate at RT for 20 min.
3. Wash cells with FACS buffer, then centrifuge for 5 min at 1,000 rpm, low brake. Add 1 ml 1× FOXP3 Perm buffer and centrifuge for 5 min at 1,000 rpm, low brake. Resuspend in 1 ml 1× FOXP3 Perm buffer, vortex gently and incubate at RT for 15 min and then centrifuge for 5 min at 1,000 rpm, low brake.
4. Discard the supernatant, resuspend in 100 μl FACS buffer and add 2 μl anti-human fluorochrome conjugated FOXP3 and 2 μl FOXP3 IgG1 control and incubate at RT for 30 min, respectively.
5. Wash cells once in FACS buffer, centrifuge for 5 min at 1,000 rpm, low brake. Discard the supernatant and resuspend the cells in 200 μl of FACS buffer. Analyze by FACS (FACS Calibur).

3.4. Isolation of Responder Cells

1. Responder cells can be obtained in several ways (see Note 4). We typically use CD4+CD25− T cells isolated from PBMC using the no-touch approach described in Chapter 3.2.2. Once isolated $10–20\times10^6$ responder cells are frozen in 10% freezing medium per 1.8 ml cryovials then stored in liquid nitrogen.

3.5. CFSE-Based Polyclonal Suppression Assay

1. Thaw autologous or allogeneic CD4+CD25− responder cells in 100% FBS in a 15-ml conical tube. Add cold PBS and wash pellet immediately, centrifuge for 5 min at 1,000 rpm at 4°C, low brake. Discard the supernatant, add cold PBS and wash again followed by centrifugation for 5 min at 1,000 rpm at 4°C, low brake. Discard the supernatant and resuspend the pellet in 1–2 ml of complete medium, rest cells for about 10–15 min at 37°C.

2. Wash cells with PBS for 5 min at 1,000 rpm, low brake, discard the supernatant and resuspend pellet in 1 ml PBS. Add 1 ml of 2× CFSE staining solution (1.6 μM) in 1× PBS, so that the final CFSE staining concentration is 0.8 μM. Perform CFSE staining in the dark and incubate for 6 min at 37°C at a total volume of 2 ml 1× PBS.
3. Add 4 ml of 100% FBS and incubate at RT for 2 min. Add 6 ml of complete medium, wash cells for 5 min at 1,000 rpm, low brake. Discard the supernatant, resuspend CFSE-labeled responder cells in complete medium and count cells (see Note 5).
4. Place 1.5×10^5 CFSE-labeled responder cells in a 96-round bottom well at a final volume of 200 μl/well in complete medium. One condition should be responders alone, additional condition should include Treg at a range of ratios (we have had success with ratios of 1:1 to as low as 1:64). When possible perform each condition in duplicate. Add anti-CD3/anti-CD28 coated Dynabeads at a ratio of 1:2 (T cells: Dynabeads) to each condition (see Note 6).
5. To measure proliferation perform Flow Cytometry on day 4. Pool duplicate wells, surface stain as above for cell surface markers of interest, we typically use CD4 and CD25. We acquired data using CellQuest Pro software (BD Bioscience) and analyzed by FlowJo (Tree Star).
6. Calculation of percent inhibition (see Fig. 1): determine the percentage of dividing CFSE-labeled CD4+CD25− T cells in each coculture as compared with the percentage of dividing CFSE-labeled CD4+CD25− T cells when cultured alone in the presence of coated Dynabeads. If the proliferation of the responder plus dynabeads condition is less than 20%, we consider the experiment to be a failure and do no further analysis. In our experience, suppression is inconsistent if the responder proliferation is below 20% (see Note 7).
7. Statistical analysis: These analyses were performed using PRISM. Statistical relevance was determined by student's *t*-test with Welch's correction, and an ANOVA using Tukey's multiple comparison test and linear regression model. All curves were based on a nonlinear regression analysis performed with one site binding hyperbole with 95% confidence interval.

4. Notes

1. Setting up the functional CFSE-Assay required a consistent proliferation of CFSE-labeled responder cells in the presence of coated Dynabeads. We tested different stimuli at different

concentrations to get a reproducible percentage of proliferating responder cells. Among these were Dynabeads M-450 CD3/CD28 T (Xcyte beads, already coupled to anti-CD3/anti-CD28), plate-bound anti-CD3 and soluble anti-CD28, and irradiated APCs and anti-CD3. Results were consistent when either Dynabeads® M-280 Tosylactivated or irradiated APC and anti-CD3 were used. For the methods described, we utilized Dynabeads to take advantage of the APC-free aspect of the system.

2. CD4+ T cells are isolated from PBMC using a no touch approach. This allows for more rapid acquisition of CD4+CD25^{bright} cells from the sorter. Several commercial products are available for this purpose. We describe isolation using Miltenyi MACS separation kits as this is our current approach.
3. When aTregs differentiate and proliferate well, the culture medium may turn yellow before the sort on day 10. In this case, do not split the cells but instead replace 50% of the medium while maintaining the total volume and leaving the cells undisturbed.
4. Instead of CFSE labeling previously frozen CD4+CD25− T cells (see Chapter 3.4), the functional Assay can be set up with freshly isolated responder cells. This procedure is described in Chapter 3.2.1 and the FACS sorted CD4+CD25− fraction will be obtained and CFSE labeled. Alternatively, CD4+CD25− T cells can be isolated from aliquoted total PBMC via magnetic separation with MACS Columns (Miltenyi Biotec). For this procedure thaw PBMC and magnetically label the cells (described in Chapter 3.2.2). According to the total number of cells, choose an appropriate Column (MS, LS or XS Column), rinse it with MACS buffer and then apply the cell suspension (up to 10^8 cells in 500 μl of MACS buffer). Collect the enriched CD4+ fraction which will pass through the Column. Incubate the obtained negative fraction with CD25 Microbeads II Kit to isolate the CD4+CD25− T cells by applying them onto the Column.
5. We recommend to set up the functional CFSE-Assay on one day. Do not rest sorted aTreg or CFSE-labeled responder cells in complete medium over night.
6. Since the number of Treg is typically the limiting factor in these assays, we usually start at a ratio of 1:4 (Treg: Teff cells) for a dose-titration. If responder T cells are limiting, the number of CFSE-labeled responder cells per well can be further reduced and the FACS analysis would still be interpretable.
7. The cut-off for the FACS analysis was based on at least 20% proliferation of CFSE-labeled responder cells in the presence

of coated Dynabeads. Less than 20% proliferation resulted in an overestimated Treg mediated inhibition that varied along with a lack of reproducibility for this CFSE-based assay. Above the 20 % cut-off, the results were consistent over a range of Treg: Teff ratios.

References

1. Sakaguchi S. (2000) Regulatory T cells: key controllers of immunologic self-tolerance. Cell; 101(5):455–458.
2. Wildin RS, Smyk-Pearson S, Filipovich AH. (2002) Clinical and molecular features of the immunodysregulation, polyendocrinopathy, enteropathy, X linked (IPEX) syndrome. J Med Genet; 39(8):537–545.
3. Khattri R, Kasprowicz DJ, Cox T, Yasayko S-A, Ziegler SF, Ramsdell F. (2001) The amount of scurfin protein determines peripheral T cell number and responsiveness. J Immunol; 167:6312–6320.
4. Viglietta V, Baecher-Allan C, Weiner HL, Hafler DA. (2004) Loss of functional suppression by CD4+CD25+ regulatory t cells in patients with Multiple sclerosis. J Exp Med; 199(7):971–979.
5. Nosaka Y, Nishio J, Nanki T, Koike R, Kubota T, Miyasaka N. [A refractory case of relapsing polychondritis (published erratum appears in Nihon Rinsho Meneki Gakkai Kaishi (1998) Jun;21(3):following 144)]. Nihon Rinsho Meneki Gakkai Kaishi; 21(2):80–86.
6. Shevach EM. (2002) CD4+ CD25+ suppressor T cells: more questions than answers. Nat Rev Immunol; 2(6):389–400.
7. Lyons AB. (2000) Analysing cell division in vivo and in vitro using flow cytometric measurement of CFSE dye dilution. J Immunol Methods; 243(1–2):147–154.
8. Venken K, Thewissen M, Hellings N et al. (2007) A CFSE based assay for measuring CD4(+)CD25(+) regulatory T cell mediated suppression of auto-antigen specific and polyclonal T cell responses. J Immunol Methods; 322(1–2):1–11.
9. Tree TI, Roep BO, Peakman M. (2006) A mini meta-analysis of studies on CD4+CD25+ T cells in human type 1 diabetes: report of the Immunology of Diabetes Society T Cell Workshop. Ann N Y Acad Sci; 1079:9–18.
10. Schneider A, Rieck M, Sanda S, Pihoker C, Greenbaum C, Buckner JH. (2008) The effector T cells of diabetic subjects are resistant to regulation via CD4+ FOXP3+ regulatory T cells. J Immunol; 181(10): 7350–7355.
11. Long SA, Buckner JH. (2008) Combination of rapamycin and IL-2 increases de novo induction of human CD4(+)CD25(+) FOXP3(+) T cells. J Autoimmun; 30(4): 293–302.
12. Walker MR, Carson BD, Nepom GT, Ziegler SF, Buckner JH. (2005) De novo generation of antigen-specific CD4+CD25+ regulatory T cells from human CD4+CD25- T cells. Proc Natl Acad Sci U S A; 102(11):4103–4108.

Chapter 16

Measurement of Proliferation and Disappearance of Regulatory T Cells in Human Studies Using Deuterium-Labeled Glucose

Milica Vukmanovic-Stejic, Yan Zhang, Arne N. Akbar, and Derek C. Macallan

Abstract

The in vivo proliferation and disappearance kinetics of lymphocytes may be estimated in humans from rates of deuterium-labeled glucose ($^{2}H_{2}$-glucose) incorporation into DNA. This protocol describes its application to regulatory T cells (Treg). Because Treg divide frequently, $^{2}H_{2}$-glucose is a suitable precursor, achieving high levels of enrichment over a short period. Being nonradioactive and readily administered, it is appropriate for human studies.

There are four phases to the method: labeling, sampling, analysis and modeling. Labeling consists of administration of $^{2}H_{2}$-glucose, either intravenously or orally; during this phase, small blood samples are taken to monitor plasma glucose enrichment. Sampling occurs over the ensuing ~3 weeks; PBMC are collected and sorted according to surface marker expression. Cell separation can be achieved by fluorescence-activated cell sorting (FACS) using CD4, CD45RA and CD25 to define memory Treg ($CD4^{+}CD25^{hi}$), or by a combination of magnetic bead separation and FACS. Analysis consists of DNA extraction, hydrolysis, derivatization to the pentafluoro tri-acetate (PFTA) derivative, and quantitation of deuterium content by gas-chromatography mass-spectrometry (GC/MS). The ratio of deuterium enrichment in cellular DNA relative to plasma glucose is used to derive the fraction of new cells in the sorted population, and this is modeled as a function of time to derive proliferation and disappearance kinetics.

Key words: Lymphocyte, Regulatory T cell, Treg, Kinetics, Isotope, Tracer, Proliferation, Death

1. Introduction

1.1. Regulatory T Cells and Kinetics

Regulatory T cells (Treg) are $CD4^{+}CD25^{+}$ cells, which specifically express the transcription factor Foxp3. They play a significant role in both normal immune homeostasis, and in many immunopathological processes, where they down-regulate responses to

George Kassiotis and Adrian Liston (eds.), *Regulatory T Cells: Methods and Protocols*, Methods in Molecular Biology, vol. 707, DOI 10.1007/978-1-61737-979-6_16, © Springer Science+Business Media, LLC 2011

both self and foreign antigens (1–3). Treg behavior can be defined in a number of ways. Changes in phenotype may be assessed by flow cytometry. Functional pathways may be investigated by microarray analysis and confirmation of gene product activities may be explored using knockdown experiments, e.g., (4). The most widely-used approach is to take the abundance of $CD25^{hi}$/$FoxP3^{+}$ cells in circulating blood as a surrogate for Treg activity, and there are many examples of this approach in the literature, for example in infections such as hepatitis (5, 6) and malignancies such as renal cell cancer (7). Cell numbers, however, only give limited information on the behavior of subpopulations of T cells such as Treg. More in-depth information on rates of cellular production and disappearance can only be obtained by use of an index of division and death such as in vivo labeling of cell proliferation using stable isotopes.

Stable isotopes have the great advantage of being suitable for human clinical investigation, since they are nonradioactive and have no inherent toxicity (unless administered in massive doses). In this paper, we describe the use of deuterium-labeled glucose to quantify proliferation and death of regulatory T cells in vivo, and demonstrate how mathematical modeling may be used to enable interpretation of experimental data (8, 9). Although this approach has many potential applications (10), this protocol is limited to its application to the study of human Treg kinetics. Modification for animal studies is also possible. Using this approach, we were able to demonstrate that $CD4^{+}CD45R0^{+}Foxp3^{+}CD25^{hi}$ T lymphocytes are highly proliferative, with a doubling time of 8 days, compared with memory $CD4^{+}CD45R0^{+}Foxp3^{-}CD25^{-}$ (24 days) or naive $CD4^{+}CD45RA^{+}Foxp3^{-}CD25^{-}$ populations (199 days) (11). We also argued, on the basis of extremely close TCR clonal homology between regulatory and memory CD4+ T cells, that at least some human $CD4^{+}CD25^{+}Foxp3^{+}$ Tregs are likely to be generated from rapidly dividing, highly differentiated memory $CD4^{+}$ T cells (11).

1.2. Principles of Isotopic Labeling

Stable isotopic approaches depend upon the generic principle that monitoring the incorporation of a label from a precursor into a product gives an index of the production rate of the product (Fig. 1). In this case the precursor is glucose and the product DNA. Glucose is converted within cells to pentose moieties which form the building blocks for nucleotide synthesis. Hence, cells that replicate in the presence of deuterium-labeled glucose will incorporate deuterium into sugar-phosphate backbone of the DNA of their progeny, whilst nondividing cells remain unlabeled. Because of its rapid "on" and "off" kinetics, glucose is an excellent candidate precursor for analysis of proliferation rates in cells with rapid proliferation. (A similar approach may be taken with deuterated water which is incorporated at multiple sites within newly synthesized

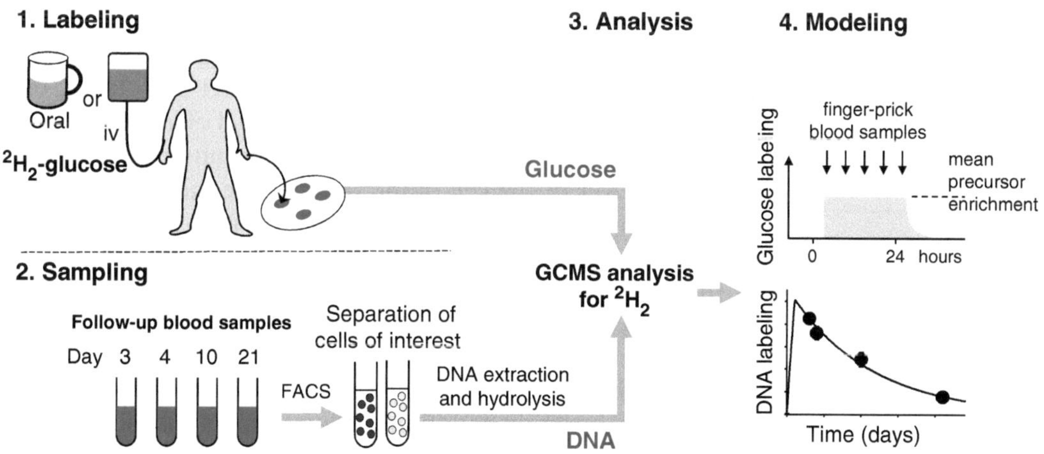

Fig. 1. General schematic of protocol for analysis of lymphocyte kinetics. The example shown illustrates how either oral or intravenous administration of [6,6-$^2H_2$]-glucose may be used to label dividing lymphocytes in vivo. The deuterium content of DNA, analyzed by gas chromatography-mass spectrometry (GCMS) is compared to the average glucose enrichment in plasma, E_{Glu}, to derive fractional labeling curves over time. *FACS* fluorescence activated cell sorting. Modified from Macallan et al. (31).

DNA (12), but this is less well suited to the study of rapidly dividing cells such as Treg.) The label is retained within the progeny, even if the cell divides again (13), until the cell dies or leaves the pool of cells under investigation, either by localization in tissues or by phenotype transition.

Quantitation is achieved by analyzing the fraction of deuterium-labeled molecules within the DNA of sampled cell populations at follow-up points over several weeks, using gas-chromatography mass-spectrometry (GCMS). In this respect, the method resembles more familiar "pulse-chase" experiments. Analysis only allows conclusions to be drawn about populations of cells. A population of cells with a high rate of turnover will incorporate large amounts of the isotope deuterium, whereas one with few mitoses will incorporate little. Conclusions about individual cells cannot be drawn using this approach, by contrast with ex vivo approaches such as BrdU (14, 15), ^{3}H-thymidine or Ki67 (16), labeling or dilution of cytoplasmic stains such as CFSE (17–19), which enable cell-by-cell analysis and which should be seen as complementary. However, significant toxicities limit their application to human studies. It is assumed that labeling due to nonreplicative DNA repair, or nucleoside substitutions due to processes such as RNA–DNA interactions during message transcription or DNA unfolding are quantitatively insignificant compared to the DNA synthesis that accompanies S-phase transition (12, 20).

1.3. Selection of Protocols

1.3.1. Labeling

Two alternative labeling strategies are described, one using oral and the other intravenous administration of labeled glucose. Each has its respective advantages. In order to select a protocol, one must estimate the desired target minimum level of glucose enrichment x duration of labeling; this will depend upon the expected rate of turnover of the cell of interest and the minimum level of DNA enrichment that can be reliably measured. Oral labeling is less invasive, but requires at least half-hourly administration, and for this reason is difficult to administer overnight. Labeling orally for 10 h gave sufficient signal to measure the kinetics of regulatory T cells (11), but in these experiments signals in slowly-dividing $CD4^{+}CD45RA^{+}$ T cells were only just detectable and not adequately quantifiable. Intravenous infusion allows longer labeling phases, but is more invasive and requires more intensive attention to issues such as the sterility and pyrogenicity of the infusate. Labeling for 24 h enables one to capture the kinetics of memory and naïve T cells (9, 21–23), B-cells (24), and leukaemic cells in chronic lymphocytic leukemia (25). Longer infusions have been described; some studies have extended to 48 h (26–28) or 5 days (29, 30). The higher level of enrichment x duration achieved allows analysis of less-rapidly dividing cell populations (31), but for very slowly dividing populations, the heavy water approach should be preferred (12).

1.3.2. Sampling

We assume that maximal intracellular labeling occurs at or very shortly after the end of the glucose infusion/oral administration on the basis that blood glucose and intracellular dNTP pools are likely to be small and short-lived ($t½$ for blood glucose is <2 h). However, maximal labeling of circulating cells in blood does not occur until later because most cell division occurs in the lymphoid compartment or in tissues. The length of this “lag” phase has not been well-defined for Treg, but appears to be similar to other activated or memory T cells, in that labeling at day 3 postinfusion may be lower than day 4 (11). (For a fuller discussion see Macallan et al. (31).) Finding the precise timing of the peak is not critical if modeling is used to estimate maximum enrichment at the end of the labeling period. For Treg, we therefore suggest delaying initial sampling until ≥ day 3, assuming that by this time recirculation and mixing has occurred.

2. Materials

For human studies, prior institutional and ethical approval should be obtained in accordance with local and national guidelines and regulations; informed consent must be obtained from all subjects before any interventions. If applied to animal models, experiments must be performed in accordance with relevant guidelines and regulations. To avoid contamination, use reagents of the highest quality commercially available.

2.1. Clinical Studies

1. [6,6-$^2H_2$]-glucose (Cambridge Isotopes Inc, MA, USA, or Isotec (Sigma-Aldrich), St Louis, MI, USA); should be sterile and certified pyrogen-free if following Option B (see Note 1).
2. Infusion fluids: water for injections or 0.45% saline for injection (Baxter Healthcare UK).

Option A – Oral labeling

1. Weigh out 0.6–1.0 g/kg body weight deuterated glucose.
2. Make up to 240 mL with water. Once reconstituted, use fresh or store at 4°C. Glucose is chemically stable; the shelf-life is determined by local pharmacy guidelines and relates primarily to possible microbial contamination.

Option B – Intravenous labeling

1. Prepare [6,6-$^2H_2$]-glucose infusate by taking 1 g/kg body weight [6,6-$^2H_2$]-glucose and reconstituting into a 1,000 mL (or two 500 mL) bag of 0.45% saline or water for injection (see Note 2). This is most readily achieved by withdrawing approximately 200 mL (or 100 mL from each of two bags), dissolving the glucose powder and reinjecting into infusate bags through a 0.2-μm filter. (If using two bags, ensure equal volumes are replaced; note that the volume returned will be greater than the volume aspirated and that the final volume will be >1,000 mL due to volume expansion on glucose dissolution.) Prepare shortly before use and store at 4°C (see Note 3).

   Observe sterility precautions throughout to avoid contamination.

2.2. Cell Separation Reagents

1. Reagents for magnetic separation for example CD4 Isolation kit (Miltenyi biotec), columns (MS or LS) and a Vario MACS magnet (or similar).
2. Monoclonal antibodies, such as CD4 PerCP, CD25PE and CD45RAFITC; see for examples (9, 11, 21–25).

2.3. DNA Extraction Reagents

1. Qiagen QiaAmp Micro DNA extraction kit (Qiagen), for low cell number samples, or Qiagen DNeasy kit (Qiagen), if cellular material/DNA abundant.
2. Qiagen Flexigene kit (Qiagen), to extract DNA from baseline whole blood.

2.4. DNA Hydrolysis Reagents (Following Published Protocols (12))

1. Water, molecular biology grade.
2. Sodium acetate (Sigma).
3. Acetic acid (Sigma-Aldrich).
4. Zinc sulphate (Sigma).

5. Acid phosphatase (potato, 1 kU; Calbiochem).
6. S1 nuclease (Sigma).

2.5. GC/MS Derivitization Reagents for DNA Analysis

Reagents marked with asterix have significant toxicities – use protective equipment and follow local and national guidelines.

1. Pentafluorobenzyl hydroxylamine* (PFBHA, 1 mg/mL aqueous solution; Sigma-Aldrich, cat no 194484). Prepare solution fresh; store at 4°C for <1 week.
2. Acetic anhydride* (Sigma-Aldrich).
3. *N*-methylimidazole* (Sigma-Aldrich). Store dry at 4°C.
4. Sodium sulphate, granular, anhydrous (Sigma).
5. Dichloromethane* (Sigma-Aldrich).
6. Ethyl acetate (VWR International).

2.6. GC/MS Derivitization Reagents for Glucose Analysis

For glucose analysis:

1. Hydroxylamine hydrochloride* (Sigma-Aldrich).
2. Pyridine* (Sigma-Aldrich).

2.7. Standard Solutions

1. DNA (Source not critical but use molecular biology grade, eg. Calf thymus, Sigma-Aldrich).
2. [5,5-$^2H_2$]-2-deoxyribose (Cambridge Isotopes Inc, MA, USA) and unlabeled deoxyribose (Sigma) for GC/MS standards for deoxyadenosine. (Deoxyribose produces the same compound on derivitization.) Prepare solutions of known concentrations of labeled and unlabeled material and combine to prepare a series of standard solutions containing between 0 and 1% molar-enriched deoxyribose.
3. Glucose (Sigma). Combine with [6,6-$^2H_2$]-glucose from Subheading 2.1.1 to prepare a standard curve of 0–50% enriched glucose.

2.8. Equipment

1. 0.2-μm filters (Sartorius Stedim Biotech GmbH).
2. Infusion pump (IVAC 590 volumetric pump, or equivalent), calibrated gravimetrically (see Note 4).
3. Canula and intravenous administration sets (Venflon, BD Medical or equivalent).
4. Lancets.
5. Filter paper.
6. Heat-block/sample concentrator to dry under nitrogen gas.
7. Cell sorting equipment (MoFlo high speed sorter, Dako-Cytomation, or equivalent) and Vario MACS magnet for cell separation (or similar).

8. Gas Chromatograph Mass Spectrometer (GC/MS; Agilent 5973/6890 with DB-225MS or DB-17 column, Agilent Technologies, or equivalent).

3. Methods

3.1. $^2H_2$-Glucose Labeling

Glucose can be administered orally using option A, or intravenously using option B.

(A) Oral deuterium-labeled glucose administration

1. Take baseline blood sample by venepuncture, ~12 mL; aliquot as follows:
   i. 1 mL Heparinized whole blood for baseline DNA enrichment, from which
   ii. Put 3–4 blood spots on filter paper for baseline glucose enrichment
   iii. Complete blood count/differential lymphocyte count (5 mL) to aid interpretation of results
   iv. Other clinical samples, e.g., biochemistry (~5 mL) required for clinical interpretation

   The baseline DNA sample may be processed straightaway or frozen at −20°C and DNA extracted later as described in Subheading 3.3.2; the filter paper blood spots should be air-dried then stored at 4°C until analysis.
2. Take oral glucose solution; concentration should be about 200 g/L and volume 240 mL. Administer 36 mL oral glucose solution at time zero (*T*0). Aliquot the glucose solution into a disposable cup from which the subject drinks. Follow with >2 rinses of the cup with water, each of which is drunk. Do this for all doses.
3. Administer further 10 mL doses every half-hour thereafter until 10 h later (*T*10). Meals should be restricted to ≤200 kcal (low glycaemic index foods preferred) to avoid large changes in glucose enrichment. Commercially available "diet" meals typically comprising about 200 kcal each, (~15 g carbohydrate, ~8 g fat) are suitable. Meals may be given every 2–3 h. Discourage physical activity.
4. Monitor glucose enrichment with finger-prick blood samples, as below, step 1 in Subheading 3.2.
5. After administering last dose of labeled glucose, check remaining volume of glucose – should be approximately 4 mL.

(B) Intravenous deuterium-labeled glucose infusion

1. Insert intravenous canula – take baseline blood sample by venepuncture (as detailed in step 3.1A); flush canula with 0.9% saline.
2. Take intravenous glucose solution from Subheading 2.1, Option B; concentration should be about 60 g/L and volume about 1,100 mL (see Note 2).
3. Set up infusion equipment. Start infusion at 300 mL/h, recording exact start time.
4. Run at 300 mL/h for 15 min (75 mL), as priming dose (see Note 5), then at ~43 mL/h (see Note 6).
5. Administer small regularly-spaced meals as in step A.3 in Subheading 3.1. We typically give four meals (each ≤200 kcal) at ~3, 6, 9, and 12 h with a snack at 15 h and a smaller meal at 23 h. Discourage excessive physical activity.
6. Monitor subject at least four-hourly for temperature, pulse and blood pressure (see Note 7).
7. Monitor glucose enrichment with finger-prick blood samples, as below, Subheading 3.2.1.
8. At end of infusion, record exact time. This is needed to calculate the area under curve of glucose enrichment vs. time, against which DNA labeling is compared (see step 16, Subheading 3.3.1).

3.2. Sampling

3.2.1. Blood Sampling to Monitor Blood Glucose Enrichment

1. Pinprick carefully cleaned finger/thumb at time points:
   1, 4, 7, and 10 h for Option A (oral protocol).
   1, 4, 8, 12, 20 and 23 for Option B (intravenous protocol).
   Blot 3 or 4 spots on to filter paper; mark the paper with the time-point. Note that baseline ($t = 0$) should already have been taken (see Subheading 3.1, section A1.ii) (see Note 8).
2. Leave to air-dry (≥10 min).
3. Store dried blood spots on filter paper at 4°C until glucose extraction.

3.2.2. Blood Sampling and Cell Sorting to Monitor DNA Deuterium Enrichment

1. Take follow-up blood samples (50 mL) into preservative-free heparin (≥20 U/mL blood) at 3, 4, 10 and 21 days postlabeling. (see Note 9)
2. Isolate PBMC by density centrifugation on Ficoll-Paque (Amersham Biosciences).
3. Isolate $CD4^+$ T cells by negative selection using the CD4 isolation kit for magnetic separation (Miltenyi Biotech) (see Note 10). Add CD4 isolation kit antibody reagent to peripheral blood lymphocytes (10 μL per 10^7 cells) incubate, 4°C,

10 min. Add CD4 isolation kit beads to the mixture (20 μL per 10^7 cells) and incubate for a further 15 min at 4°C.

4. Wash cells with PBS/BSA/EDTA and pass through a *MACS* magnet using LS or MS column (depending on cell numbers): Collect the $CD4^+$ cells which pass through the column.
5. Stain purified $CD4^+$ T cells with a mixture of antibodies:
   (a) Anti-CD4-PerCP (Becton Dickinson)
   (b) Anti-CD45RA-FITC (Pharmingen)
   (c) Anti-CD25PE (Dako)

   Incubate for 30 min on ice. Wash in PBS/2%BSA.
6. Filter cells and sort on a MoFlo flow cytometer (Cytomation) (see Note 10), with gates set so that purified $CD4^+CD25^{hi}$ cells represent approximately the brightest 2% of total CD4 population. Collect both $CD4^+CD25^{hi}$ and $CD4^+CD25^-$ population from the $CD45RA^-$ fraction (designated as $CD45R0^+$ for clarity) (Fig. 2). Other subsets of interest can be collected

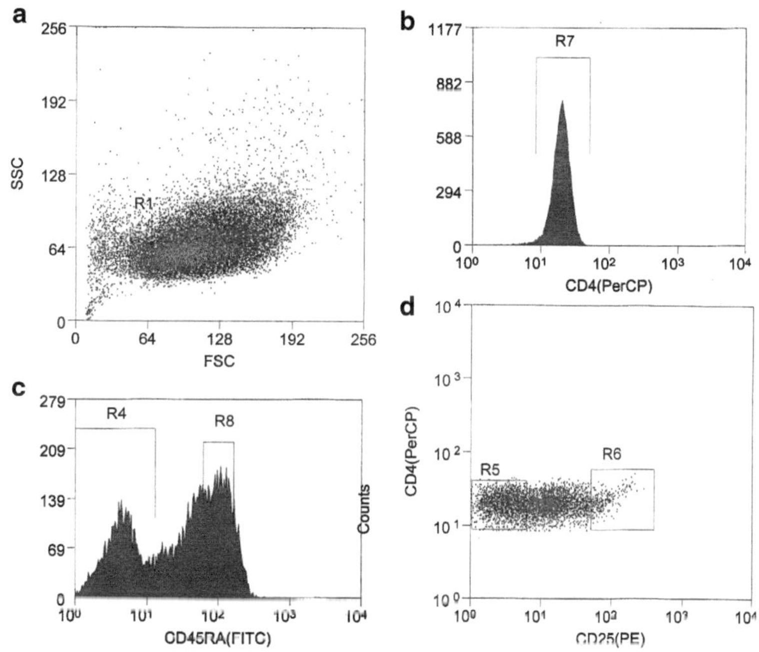

Fig. 2. Example of Treg flow cytometry sorting plots. Typical flow cytometry data plots showing the gating regions used to cell sort CD4 Tcells that had been purified from freshly isolated PMBC using magnetic bead separation (Miltenyi Biotec). Cells (10^7/mL in PBS + 0.2% BSA) were labeled with CD45RA-FITC, CD4-PerCP and CD25-PE for 30 min on ice, then sorted into $CD4^+CD45RA^+$ (using sort regions R1 + R2 + R7 + R8), $CD4^+CD45RA^-CD25^{hi}$ (sort regions: R1+ R2 + R4 + R7+ R6) and $CD4^+CD45RA^-CD25^-$ (sort regions: R1+ R2 + R4 + R7+ R5) using MoFlo cytometer (Beckman Coulter). (**a**) Forward and side-scatter plot; R1 represents gate for lymphocytes, R2 excludes doublets (not shown). (**b**) Histogram plots showing CD4 staining and gate R7 (**c**) histogram plot showing CD45RA staining and gates R4 and R8 (**d**) CD25 (*x* axis) vs. CD4 staining (*y* axis) showing gates R5 and R6 ($CD25^-$ and $CD25^{hi}$ populations respectively).

by the addition of appropriate phenotypic markers, eg. naive $CD4^{+}CD25^{-}$ and naïve $CD4^{+}CD25^{hi}$ populations may be sorted from the CD45RA+ fraction (see Note 11).

7. Proceed to DNA extraction (Subheading 3.3.2) or store separated cells frozen at −70°C until DNA extraction.

3.3. Analysis

3.3.1. Blood Glucose Enrichment Analysis

This is achieved by first extracting glucose from filter paper blood spots, then derivatizing to the aldonitrile triacetate (ATA) derivative and finally GC/MS analysis. (see Note 12).

1. Cut at least two blood drops on filter paper from Subheading 3.2 into a 1.5-mL centrifuge tube and add 1 mL 50% ethanol. Ensure that there is no contamination in any of these steps, wear gloves and use scissors designated for this purpose only. Clean blades between samples (see Note 8).
2. Leave at ambient temperature (approximately 20°C) for ~30 min.
3. Vortex and transfer supernatant to another microcentrifuge tube.
4. Centrifuge (>15,000 × *g*, 10 min, ambient temperature, in a 1.5-mL centrifuge tube) to remove any precipitate; transfer to a clean tube and dry under nitrogen gas at 50°C. Ethanol extract can be stored at −20°C until derivitization.
5. Make a fresh solution of 1% w/v hydroxylamine.HCl in pyridine (1 mg/100 μL).
6. To the dried samples from step 4, Subheading 3.3.1 and to standards of glucose of known enrichment (from Subheading 2.7.3) add 25 μL of hydroxylamine/pyridine reagent, seal and mix gently.
7. Heat the samples at 100°C for 60 min in a dri-block.
8. Cool and pulse microfuge the samples.
9. At ambient temperature, add 25 μL of acetic anhydride and seal the tube. Mix gently.
10. Incubate at room temperature for 30 min. To ensure completion of the derivatization reaction, heat for the last 10 min at 70°C in a heat block.
11. Pulse microfuge the samples, to ensure all the solution is at the bottom of the tube, then dry at 50°C under nitrogen.
12. Resuspend the derivatised samples and standards in 400 μL of ethylacetate.
13. Vortex briefly, then pulse microfuge to remove any particulate matter or precipitate, transferring the supernatant to vials ready for analysis by GC/MS.

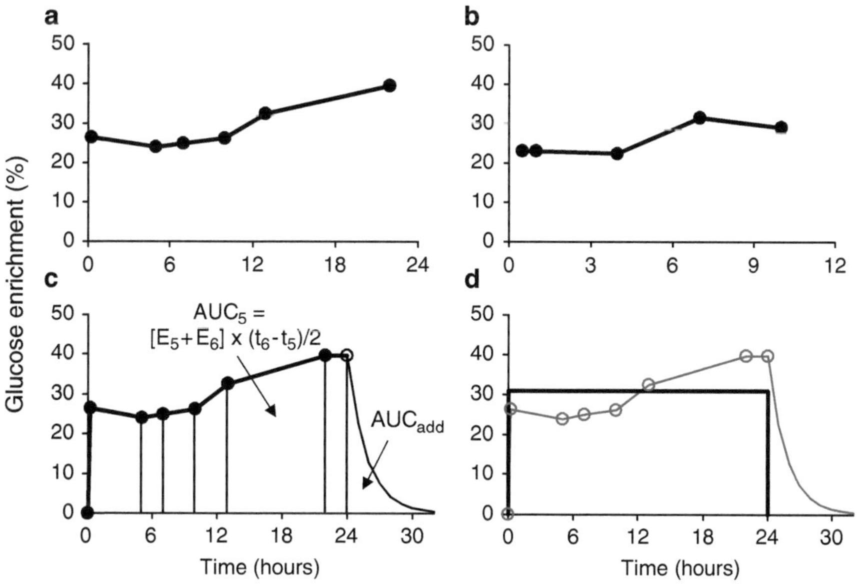

Fig. 3. Glucose enrichment curves. Typical glucose enrichment curves during (**a**) primed intravenous infusion for 24 h, and (**b**) primed half-hourly oral administration for 10 h, of [6,6-$^2H_2$]-glucose. (**c**) Detail of (**a**) showing estimation of area-under-curve (AUC) by trapezoid measurement with AUC_5, the area of the fifth trapezoid, formula illustrated (where *E* is the enrichment, *t* is time and *5* and *6* refer to the fifth and sixth time points). Estimated additional postlabeling AUC is shown as AUC_{add} and is included to derive the mean AUC for the whole infusion period shown in (**d**) as a bold line indicating a constant enrichment for 24 h which gives the same AUC as measured labeling; this level is used in subsequent calculations.

14. Analyse by GC/MS, monitoring in SIM mode for ions *m/z* 328 and 330. We use an Agilent 5973/6890 with DB-225MS column (Agilent Technologies). (see Note 13).
15. Determine enrichment from the ratio of ions $M_{+2}/[M_{+0}+M_{+2}]$, calibrating against standard glucose samples of known enrichment from Subheading 2.7.3.
16. Calculate the area under curve (AUC) for glucose enrichment vs, time by the trapezoid method (see Fig. 3c, d, and Note 14). The corrected value gives the mean glucose × time enrichment value, *b*.

3.3.2. DNA Enrichment Analysis

This section of the protocol follows the method described by Busch et al. (12), digesting DNA to deoxyadenosine, then derivatizing to the pentafluoro tri-acetate (PFTA) derivative before GC/MS.

1. Take baseline heparinised whole blood from step 3.1.A.1 (before labeling) and extract DNA using the Qiagen Flexigene kit (see Note 15); follow manufacturer's instructions except for final resuspension, at which point dissolve DNA in 1 mL water, not buffer, and divide into five 200 μL aliquots. Store DNA at −20°C or proceed directly to hydrolysis, step 3, Subheading 3.3.2.

2. Take sorted cells from Subheading 3.2.2 and extract DNA using Qiagen QiaAmp Micro DNA extraction kit (see Note 15), following the protocol modifications described by Busch et al. (12); specifically, note that, at the end of extraction, DNA should be suspended in 200 µL water, not buffer. Store DNA at −20°C or proceed directly to hydrolysis, step 3, Subheading 3.3.2.
3. To digest the extracted DNA sample to deoxyadenosine, as described by Busch et al. (12), add 50 ul of hydrolysis cocktail and incubate at 37°C overnight with shaking.
4. Derivatize the deoxyadenosine to the pentafluoro tri-acetate (PFTA) derivative as follows, alongside [5,5-$^2H_2$]-ribose standard solutions of known enrichment from item 2, Subheading 2.7 to calibrate isotope ratios.
5. Transfer the digested samples into 16 × 100 mm screw-capped glass tubes. To each sample/standard add:
   i. 100 µL of freshly made aqueous pentafluorobenzyl hydroxylamine solution (1 mg/mL).
   ii. 75 µL of glacial acetic acid.
6. Cap the tubes and incubate on heating block for 30 min at 100°C.
7. Remove samples and allow to cool to room temperature.
8. Add:
   i. 1 mL of acetic anhydride.
   ii. 100 µL of *N*-methylimidazole.
   iii. Mix immediately. Perform this step in a fume hood, wearing protective goggles and pointing the opening of the tube away from you. Samples may splash as the exothermic acetylation reaction proceeds due to sudden overheating. Allow the reaction to proceed at ambient temperature for 15–20 min during which the samples will cool down.
9. Add 2 mL of water to the reactions, vortex for 10 s.
10. Add 750 µL of dichloromethane to the tubes and vortex for 5 s. Allow phases to separate (~1 min).
11. Set up and label a series of tubes, each containing sufficient sodium sulphate (as a dessicant) to cover the bottom of the tube. (We use 13 × 100 mm disposable polypropylene culture tubes.)
12. Transfer 500 µL of the bottom (organic) layer (from step 10, Subheading 3.3.2) into the tubes containing sodium sulphate. Avoid transferring any of the aqueous phase, which may introduce contaminants. (Wetting the pipette tip with dichloromethane before transfer helps reduce inadvertent mixing.)

13. Add a further 750 μL of dichloromethane to the reaction tubes and repeat the dichloromethane extraction (vortex for 5 s, then allow phases to separate, ~1 min), adding the organic layer to that already extracted. Vortex gently, then allow sodium sulphate crystals to settle.
14. Transfer the supernatant to a clean, labeled microcentrifuge tube, avoiding transfer of any sodium sulphate crystals.
15. Dry the microcentrifuge tubes. (We use a SpeedVac at ambient temperature for ≥4 h, ≤ overnight.) Note: Avoid heating the sample as this may cause evaporative loss of the derivative. Verify complete drying visually; avoid residual moisture or acid as this may damage the GC column. Drying in a stream of nitrogen is not sufficient to remove residual acetic acid.
16. Resuspend each sample in 250 μL of ethyl acetate, vortex and pulse centrifuge to remove any precipitate or solid material. Transfer the supernatant to a GC glass insert. Take care not to transfer any precipitate or solid material. Evaporate ethyl acetate (SpeedVac, ambient temperature, ~1 h).
17. Resuspend each sample in 50 μL of ethyl acetate. Place the glass insert and into a labeled GC vial and cap immediately.
18. Analyze by GC/MS using selective ion monitoring (SIM) quantifying ions 435 and 437, the M+0 and M+2 ions respectively (see Note 16). Use the ratio $M_{+2}/[M_{+0}+M_{+2}]$ of abundance-matched samples to calculate the enrichment of deuterated deoxyadenosine, calibrating against standard curves of known enrichment from item 2, Subheading 2.7 (see Note 17).

3.4. Data Analysis and Modeling of DNA Production and Disappearance

1. Calculate the fraction new DNA, *F*, by dividing DNA enrichment levels at time points postlabeling (from step 4; Subheading 3.5) by *b*, mean precursor enrichment from, step 16, Subheading 3.3.1 ($F = E/b$, as in, where $E = A^*/A$, eq. 1).
2. Plot the DNA/time profile as a graph and assess profile. Enrichment should start at zero at baseline, rise to a peak (usually day 3 or 4 postlabeling) then fall thereafter.
3. Fit data to equations (8, 9):

   $F(t) - p / d^*(1 - e^{-d^*t})$, where $t \leq \tau$, during the labeling period (eq. 1), and

   $F(t) = p / d^*(1 - e^{-d^*\tau}) \cdot e^{-d^*(t-\tau)}$, where $t > \tau$, after the labeling period (eq. 2), where t is time and τ is the length of the labeling period (see Note 18).

   τ is taken as 12 h for the oral protocol (Option A) and 24 h for the intravenous protocol (Option B).

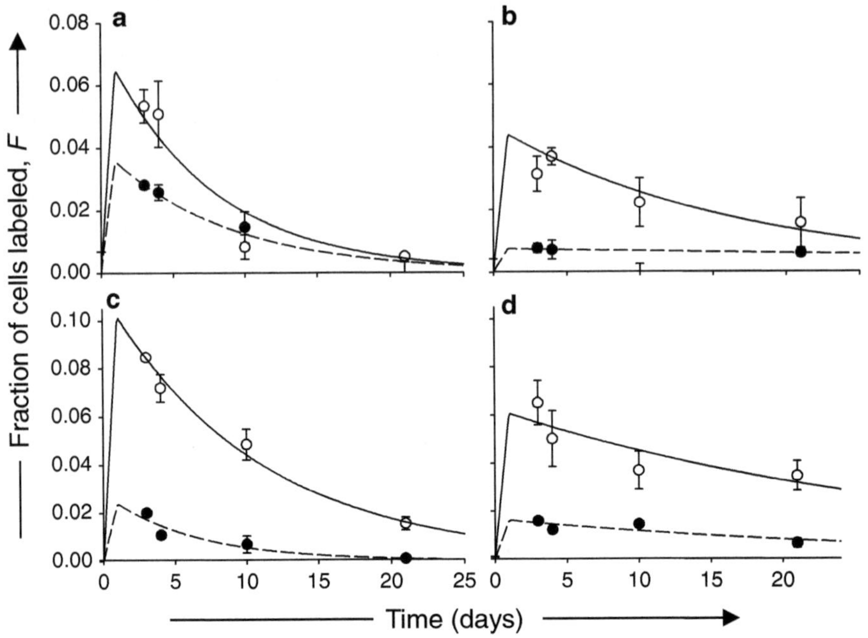

Fig. 4. Treg labeling curves. Fraction of labeled cells (*F*, per day) from DNA labeling of Treg lymphocytes subsets in two young (**a**, **b**) and two elderly (**c**, **d**) subjects following oral labeling with $^2H_2$-glucose. $CD4^+CD25^{bright}$ Treg cells (*open circles*) have high turnover compared to $CD4^+CD25^-$ cells (*filled circles*, *dashed line*). Error bars are the standard deviation of repeated GC/MS enrichment measurements.

Curve fitting using nonlinear least squares regression allows estimation of the parameters of the model namely p, the average rate of proliferation of the lymphocyte population and d^*, the average rate of disappearance of labeled lymphocytes. This can be done with a number of software packages that allow user-defined equations, such as *Sigmaplot* (Systat Software Inc, San Jose, CA). Typical plots and curve-fitting are shown in Fig. 4.

Proliferation and disappearance rate constants (the "doubling-time," T_2, and "half-life," $T_{½}$) can then be calculated as ln $2/p$ and ln $2/d^*$, respectively.

4. Notes

1. [6,6-$^2H_2$]-glucose should be sterility and pyrogenicity tested and certified as such by the manufacturer. Microbial contamination of the infusate must be avoided. Reconstitution as an infusate must be performed under sterile conditions. Passage of reconstituted glucose through a 0.2-μm filter is advisable to ensure sterility.

2. Hyperosmolar solutions may cause phlebitis. The final osmolarity of the solution will be about 450 mOsmol/L; this should not cause phlebitis. Avoid using infusion volumes of less than 1,000 mL.
3. Once reconstituted, use fresh or store at 4°C. Glucose is chemically stable; the shelf-life is determined by local pharmacy guidelines and relates primarily to possible microbial contamination.
4. Infusion pumps may not always run at the stated rate and should be calibrated gravimetrically prior to research use.
5. A priming dose of about 1.8 times the hourly dose was found empirically to reach plateau levels in most subjects; under-priming or over-priming result in rising or falling glucose enrichments in plasma (respectively) and should be avoided. Glucose enrichment tends to rise overnight as feeding is not continued during this time. Such curves can be converted to a "square wave" by calculating AUC and expressing this as the mean enrichment achieved throughout the duration of label administration. However, results may be slightly biased either towards the earlier or later part of the day; this is not considered a significant bias.
6. The infusion rate may need to be modified empirically as it will depend upon the infusion pump characteristics (which should be calibrated) and because the volume of the infusate is difficult to predict; bags often contain slightly more that the stated volume to allow for line priming and adding glucose expands the volume, depending upon the amount added. Aim to give the whole dose over 24 h.
7. In order to monitor for the possibility of a pyrogen or infusion reaction, subjects should be closely monitored by a clinically trained staff member and pulse, temperature and blood pressure recorded at least four-hourly.
8. Aberrent or unexpectedly high glucose enrichment levels in blood spots may indicate contamination. Always plot the glucose time-profile. Avoid any possible contamination from the infusate or oral solution onto filter paper blood spots. This is a particular risk with the oral solution which is easy to transfer from mouth or cup via hand to filter paper and is highly concentrated (even uL contamination will give aberrant results). Use disposable cups; ensure both subject and operator wear gloves when handling oral glucose solution even though it is nonhazardous. Clean subjects finger carefully prior to pinprick blood testing.
9. We recommend sorting cells fresh. It may be possible to freeze the cells at −70°C prior to sorting. If cells are frozen, it needs to be established that freezing does not preferentially affect the recovery of the labeled cells of interest.

10. Alternative sorting strategies, products and equipment are available.
11. The use of Foxp3 (and other intracellular markers) for cell sorting has a negative impact on the DNA yield. This needs to be taken into consideration when sorting small cell populations.
12. There are several alternative derivitization and GC/MS protocols for analysis of isotopic enrichment of glucose.
13. Detailed GC/MS protocols are beyond the scope of this article, but note that optimization of chromatography is crucial for reliable quantitation.
14. Calculation of the area under curve (AUC) for glucose enrichment vs. time by the trapezoid method applies the equation:

$$\text{AUC} = [E_0 + E_1]*t_1/2 + [E_1 + E_2]*(t_2 - t_1)/2 + [E_2 + E_3]*(t_3 - t_2)/2 + \ldots t_n,$$

where E is the glucose enrichment at times, 0, 1, 2, 3 …*n*. Two assumptions are made: (1) that the first measured enrichment (E_1) is reached immediately at the end of the prime, at 0.25 h (hence t1 is taken as 0.25), and (2) that enrichment remains constant from the last measured enrichment (E_{final}) until the end time for the infusion (from step B.8, Subheading 3.1). For the oral administration administration, this "end-time" is taken as the time of the last oral dose plus 0.5 h (for absorption).

We also estimate an additional postinfusion area (AUC_{add}) in order to allow for additional labeling that occurs during the die-away of labeled glucose. AUC_{add} can be either (i) measured directly from plasma glucose enrichments taken postinfusion, or (ii) estimated assuming exponential disappearance. Rather than taking multiple extra blood samples, we use the latter estimation approach based on the standard formula for the area under an exponentially disappearing curve. Thus AUC_{add} is given by $\text{AUC}_{add} = E_{final}/k$, where E_{final} is the value of enrichment from which disappearance occurs, and k is the disappearance rate constant. E_{final} is measured but the rate constant for glucose disappearance, k, must be estimated. This can be done from the rate of glucose flux, Rd glucose, and the pool size, Q, since $k = \text{Rd}/Q$, where,$\text{Rd}_{glucose} \approx$ rate of administration of labeled glucose/mean enrichment, and pool size, $Q \approx$ amount of glucose given as prime/E_1, where E_1 is the first measured enrichment (Q should be about 20 g and k should be ~0.007/min).

To calculate the total AUC, add AUC_{add} to the AUC up to the end-time of labeling. Dividing this sum by the infusion/administration time gives the mean glucose enrichment.

Multiply this by the correction factor derived for this cell type (0.65 in our case (31); although Kovacs et al. have used a slightly lower value of 0.60 in their studies (29, 30, 32)) to obtain the mean precursor enrichment, *b*, represented by the line in Fig. 3d.

15. Alternative DNA extraction protocols and kits are available and may be used.
16. Detailed Gas chromatography mass spectrometry (GC/MS) analysis protocols for deoxyadenosine from heavy water studies have been described in Busch et al (12). For analysis of DNA from deuterated glucose studies, measurement of adenosine containing two deuterium atoms is required. The ratio M_{+2}/M_{+0} gives the enrichment; for the pentafluoro triacetate (PFTA) derivative, this represents ions *m/z* 437.0 and 435.0 respectively. An alternative microcapillary liquid chromatography-electrospray ionization (LC-ESI)/MS has been described (33), an advantage of which is that it does not require a derivatization step. However, it may require larger sample amounts or quantitation.
17. Isotopc ratios are susceptible to abundance artifacts. Run samples as single injections first, to check abundance; then dilute or adjust injection volume to ensure equal abundances are achieved for all samples and standards; then repeat analysis of ratios in triplicate.
18. The rate of change of labeled deoxyadenosine is described by eq. (1) during the labeling phase and eq. (2) during the postlabeling follow-up, as labeled cells are replaced by unlabeled ones Fitting this solution to the experimentally obtained data allows estimation of the parameters *p*, the average rate of proliferation of the lymphocyte population and d^*, the average rate of disappearance of labeled lymphocytes. In this model, constancy of pool size has been assumed, but no assumption of equality between *p* and d^* has been made. The proliferation rate measured will incorporate both proliferation of the population of interest as well as any immediate precursors which divided during the labeling period and subsequently matured or trafficked to the pool of interest. The disappearance rate measured will incorporate death of cells, either within the circulation or by migration of cells to lymph nodes prior to death, net trafficking of cells out of the peripheral blood and disappearance due to phenotype switching.

This model has been widely applied (22, 34, 35), but it should be noted that this is not the only model available. For a review of alternative models see Borghans et al. (36).

Acknowledgments

We acknowledge financial support from the Medical Research Council (UK), BBSRC (UK), the Wellcome Trust, Merck Serono and the Charitable Trustees of St George's Hospital, London during the execution of studies included in this report.

References

1. Sakaguchi S, Yamaguchi T, Nomura T, Ono M. (2008) Regulatory T cells and immune tolerance. *Cell* **133**, 775–787.
2. Maloy KJ, Powrie F. (2001) Regulatory T cells in the control of immune pathology. *Nat. Immunol.* **2**, 816–822.
3. Shevach EM. (2000) Regulatory T cells in autoimmmunity. *Annu. Rev. Immunol.* **18**, 423–449.
4. Pan F, Yu H, Dang EV et al. (2009) Eos mediates Foxp3-dependent gene silencing in CD4+ regulatory T cells. *Science* **325**,1142–1146.
5. Peng G, Li S, Wu W, Sun Z, Chen Y, Chen Z. (2008) Circulating CD4+ CD25+ regulatory T cells correlate with chronic hepatitis B infection. *Immunology* **123**, 57–65.
6. Rushbrook SM, Ward SM, Unitt E et al. (2005) Regulatory T cells suppress in vitro proliferation of virus-specific CD8+ T cells during persistent hepatitis C virus infection. *J. Virol.* **79**, 7852–7859.
7. Jeron A, Pfoertner S, Bruder D et al. (2009) Frequency and gene expression profile of regulatory T cells in renal cell carcinoma. *Tumour Biol.* **30**, 160–170.
8. Asquith B, Debacq C, Macallan DC, Willems L, Bangham C. (2002) Lymphocyte kinetics: the interpretation of labelling data. *Trends Immunol.* **23**, 596–601.
9. Macallan DC, Asquith B, Irvine A et al. (2003) Measurement and modeling of human T cell kinetics. *Eur. J. Immunol.* **33**, 2316–2326.
10. Asquith B, Borghans JA, Ganusov VV, Macallan DC. (2009) Lymphocyte kinetics in health and disease. *Trends Immunol.* **30**, 182–189.
11. Vukmanovic-Stejic M, Zhang Y, Cook JE et al. (2006) Human CD4+ CD25hi Foxp3+ regulatory T cells are derived by rapid turnover of memory populations in vivo. *J. Clin. Invest.* **116**, 2423–2433.
12. Busch R, Neese RA, Awada M, Hayes GM, Hellerstein MK. (2007) Measurement of cell proliferation by heavy water labeling. *Nat. Protoc.* **2**, 3045–3057.
13. Hellerstein MK. (1999) Measurement of T-cell kinetics: recent methodologic advances. Immunol. Today **20**, 438–441.
14. Tough DF, Sprent J. (1994) Turnover of naive- and memory-phenotype T cells. *J. Exp. Med.* **179**, 1127–1135.
15. Kovacs JA, Lempicki RA, Sidorov IA et al. (2001) Identification of dynamically distinct subpopulations of T lymphocytes that are differentially affected by HIV. *J. Exp. Med.* **194**, 1731–1741.
16. Schwarting R, Gerdes J, Niehus J, Jaeschke L, Stein H. (1986) Determination of the growth fraction in cell suspensions by flow cytometry using the monoclonal antibody Ki-67. *J. Immunol. Methods* **90**, 65–70.
17. Quah BJ, Warren HS, Parish CR. (2007) Monitoring lymphocyte proliferation in vitro and in vivo with the intracellular fluorescent dye carboxyfluorescein diacetate succinimidyl ester. *Nat. Protoc.* **2**, 2049–2056.
18. Hawkins ED, Hommel M, Turner ML, Battye FL, Markham JF, Hodgkin PD. (2007) Measuring lymphocyte proliferation, survival and differentiation using CFSE time-series data. *Nat. Protoc.* **2**, 2057–2067.
19. Lyons AB. (2000) Analysing cell division in vivo and in vitro using flow cytometric measurement of CFSE dye dilution. *J. Immunol. Methods* **243**, 147–154.
20. Macallan DC, Fullerton CA, Neese RA, Haddock K, Park S, Hellerstein MK. (1998) Measurement of cell proliferation by labeling of DNA with stable isotope-labeled glucose: Studies in vitro, in animals and in humans. *Proc. Natl. Acad. Sci. USA* **95**, 708–713.
21. Macallan DC, Wallace DL, Irvine A et al. (2003) Rapid turnover of T cells in acute infectious mononucleosis. *Eur. J. Immunol.* **33**, 2655–2665.
22. Macallan DC, Wallace D, Zhang Y et al. (2004) Rapid turnover of effector-memory

CD4(+) T cells in healthy humans. *J. Exp. Med.* **200**, 255–260.
23. Wallace DL, Zhang Y, Ghattas H et al. (2004) Direct measurement of T cell subset kinetics in vivo in elderly men and women. *J. Immunol.* **173**, 1787–1794.
24. Macallan DC, Wallace DL, Zhang Y et al. (2005) B cell kinetics in humans: rapid turnover of peripheral blood memory cells. *Blood* **105**, 3633–3640.
25. Defoiche J, Debacq C, Asquith B et al. (2008) Reduction of B cell turnover in chronic lymphocytic leukaemia. *Br. J. Haematol.* **143**, 240–247.
26. Hellerstein M, Hanley MB, Cesar D et al. (1999) Directly measured kinetics of circulating T lymphocytes in normal and HIV-1-infected humans. *Nat. Med.* **5**, 83–89.
27. Hellerstein MK, Hoh RA, Hanley MB et al. (2003) Subpopulations of long-lived and short-lived T cells in advanced HIV-1 infection. *J. Clin. Invest.* **112**, 956–966.
28. McCune JM, Hanley MB, Cesar D et al. (2000) Factors influencing T-cell turnover in HIV-1-seropositive patients. *J. Clin. Invest.* **105**, R1–R8.
29. Kovacs JA, Lempicki RA, Sidorov IA et al. (2005) Induction of prolonged survival of CD4+ T lymphocytes by intermittent IL-2 therapy in HIV-infected patients. *J. Clin. Invest.* **115**, 2139–2148.
30. Read SW, Lempicki RA, Mascio MD et al. (2008) CD4 T cell survival after intermittent interleukin-2 therapy is predictive of an increase in the CD4 T cell count of HIV-infected patients. *J. Infect. Dis.* **198**, 843–850.
31. Macallan DC, Asquith B, Zhang Y et al. (2009) Measurement of proliferation and disappearance of rapid turnover cell populations in human studies using deuterium-labelled glucose. *Nat. Protoc.* **4**, 1313–1327.
32. Stevens RA, Lempicki RA, Natarajan V, Higgins J, Adelsberger JW, Metcalf JA. (2006) General immunologic evaluation of patients with human immunodeficiency virus infection. In: Detrick B, Hamilton RG, Folds JD, editors. Manual of molecular and clinical laboratory immunology. Washington, DC: ASM, 848–861.
33. Fox SD, Lempicki RA, Hosack DA et al. (2003) A comparison of microLC/electrospray ionization-MS and GC/MS for the measurement of stable isotope enrichment from a [2H2]-glucose metabolic probe in T-cell genomic DNA. *Anal. Chem.* **75**, 6517–6522.
34. Vrisekoop N, den Braber I, de Boer AB et al. (2008) Sparse production but preferential incorporation of recently produced naive T cells in the human peripheral pool. *Proc. Natl. Acad. Sci. USA* **105**, 6115–6120.
35. Asquith B, Zhang Y, Mosley AJ et al. (2007) In vivo T lymphocyte dynamics in humans and the impact of human T-lymphotropic virus 1 infection. *Proc. Natl. Acad. Sci. USA* **104**, 8035–8040.
36. Borghans JA, De Boer RJ. (2007) Quantification of T-cell dynamics: from telomeres to DNA labeling. *Immunol. Rev.* **216**, 35–47.

Chapter 17

Flow Cytometric Detection of Human Regulatory T Cells

Barbara Fazekas de St Groth, Erhua Zhu, Suzanne Asad and Loretta Lee

Abstract

Tregs are absolutely required for the maintenance of self tolerance in mouse and man. Major abnormalities in Treg number or function cause rare but fatal syndromes with autoimmune, allergic and inflammatory features. Whether subtle Treg abnormalities contribute to the pathogenesis of sporadic autoimmune, allergic and immunoinflammatory diseases in man remains controversial. Robust methods for identifying and isolating human Tregs in patients and healthy controls are essential if we are to understand their role in these increasingly common diseases. We have outlined below a flow cytometric technique to detect and isolate the entire human Treg population based on expression of CD4, CD25, and CD127. Use of a number of additional antibodies for defining subsets within the Treg compartment is described. For analysis, anti-Foxp3 can be added to the cocktail, but the necessity for fixation and permeabilisation may reduce the signal from other antibodies.

Key words: Treg, Human, Flow cytometry, CD127, Foxp3

1. Introduction

Tregs were originally identified as a distinct population of CD4$^+$ T cells expressing the alpha chain of the IL-2 receptor, CD25 (1). Laboratories attempting to replicate the murine studies in humans encountered an unexpected difficulty, namely that CD25 is also expressed by a significant proportion of CD45RO$^+$CD45RA$^-$ effector/memory CD4$^+$ T cells. A number of strategies were employed in an attempt to isolate a pure population of human Tregs. Using an in vitro suppression assay as readout, Baecher-Allen et al. first showed that when peripheral blood CD4$^+$ T cells were sorted according to CD25 expression, only the fraction expressing the highest level of CD25 could reliably suppress in vitro (2).

George Kassiotis and Adrian Liston (eds.), *Regulatory T Cells: Methods and Protocols*, Methods in Molecular Biology, vol. 707,
DOI 10.1007/978-1-61737-979-6_17, © Springer Science+Business Media, LLC 2011

The human CD25^{hi} phenotype was then defined as equivalent to the mouse CD25^{+}, even though only 1–3% of circulating human CD4^{+} T cells were identified as Tregs (2), compared with 6–7% in the mouse. The placement of the CD25 gate was difficult to standardize between laboratories, since human CD25 expression is essentially unimodal with a long tail (Fig. 1a), unlike the essentially bimodal mouse CD25 distribution. Indeed, in the 5 years after the initial publication describing the CD25^{hi} phenotype, the range for Treg in the blood of healthy controls varied between 1.4 and 20%, depending on where the CD25 gate was placed. This problem was particularly apparent in studies aimed at testing whether Treg numbers were altered in patients with autoimmune and inflammatory diseases, since "contamination" of the CD25^{hi} Treg gate by CD25-expressing effector/memory CD4^{+} T cells involved in disease pathogenesis could not be ruled out. Disease-related changes in Tregs in the early studies showed no consistent trend, no doubt in part due to these technical issues.

A number of labs, our own included, subsequently showed that CD25 was expressed bimodally within human CD45RA^{+}CD45RO^{-}

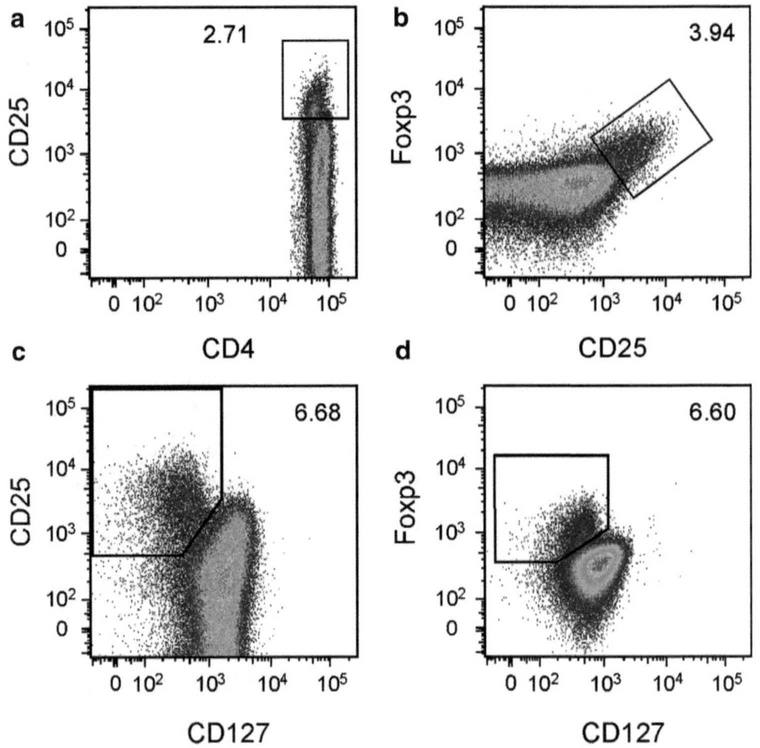

Fig. 1. Gating strategies for human Tregs. Plots are gated for peripheral blood CD4^{+} lymphocytes. (**a**) CD25^{hi} gate. (**b**) CD25/Foxp3 gate. (**c**) CD25/CD127 gate. (**d**) CD127/Foxp3 gate. The use of CD127 allows nonorthogonal gating for CD127^{lo} cells expressing intermediate levels of CD25 and/or Foxp3, identifying more Treg cells than can be detected with the CD25/Foxp3 gate.

naïve $CD4^+$ T cells, allowing identification of a well-defined subpopulation of naïve Tregs (3, 4). This population declined with age, as would be expected for a naïve, thymically derived T cell subset.

Within the $CD45RO^+RA^-$ subset, we subsequently discovered that costaining for CD25 and CD127, the alpha chain of the IL-7 receptor, separated a distinct $CD25^+CD127^{lo}$ population from the bulk of the $CD25^{lo\text{-}hi}CD127^{hi}$ cells (5, 6) (Fig. 1c). The $CD25^+CD127^{lo}$ population represented 6–7% of human peripheral blood $CD4^+$ T cells and expressed very high levels of *FOXP3* mRNA as assessed by qPCR of flow-sorted cells. When Foxp3 antibodies became available, we were able to show for the first time the correlation between Foxp3 and the subsets to which Treg function had been attributed (5).

Foxp3 is often regarded as the definitive marker for Treg cells, based on murine data. However, in humans, Foxp3 can also be expressed by activated cells (7). Currently no single marker can provide an unequivocal identification of all human Treg cells. In addition, Foxp3 detection requires an intracellular stain that is incompatible with the recovery of viable cells. Thus both the identification and sorting of human Tregs still requires the detection of small changes in the expression of multiple surface markers, one of the most challenging applications of flow cytometry.

2. Materials

2.1. Cell Preparation

1. Phosphate Buffered Saline (PBS), pH 7.2: Milli-Q filtered H_2O containing 8 g/L NaCl, 0.2 g/L KCL, 1.15 g/L Na_2HPO_4 prepared at the Centenary Institute, kept sterile and stored at room temperature.
2. Tissue Culture Medium (TCM): RPMI Medium 1640 (Gibco) with L-glutamine and HEPES buffer (Invitrogen) supplemented with 10% heat activated fetal calf serum (FCS) (JRH Bioscience), 2 mM L-Glutamine, 0.05 mM 2-Mercaptoethanol (Invitrogen), 1,000 units/L of Penicillin (Invitrogen) and 1,000 μg/L Streptomycin (Invitrogen).
3. 9 ml Lithium Heparin tubes (Greiner bio-one) for blood collection.
4. Ficoll-Paque Plus (GE Health).
5. Freezing medium: 10% DMSO (Sigma), 20% FCS (JRH Bioscience) in RPMI1640 (Gibco).

6. Tissue digestion enzymes: Collagenase Type III and DNase Type I (Sigma).
7. 14 ml round bottom tubes (BD) for tissue digestion.
8. 15 and 50 ml sterile conical tubes (Falcon) for cell processing.
9. 70 μm nylon cell strainer (BD Falcon).
10. 0.2% Eosin in phosphate-buffered saline (PBS), haemocytometer chamber.
11. 2 ml cryovials (Sardtedt).
12. Cryo-1°C/min Freezing Container (Nalgene).

2.2. Cell Staining: Basic Protocol

1. 96-well U bottom plate (Greiner bio-one).
2. Antibodies for basic staining cocktail.
   (a) Our-preferred 2-laser combination – chosen on the basis of comparison of a number of anti-CD25 antibodies and conjugates.

mAb/Stain	Clone	Fluorochrome	Source	Cat#
Mouse IgG1 κ, antihuman Foxp3	259D	AF488	BioLegend	320212
Mouse IgG1 κ, antihuman CD127	hIL-7R-M21	PE	BD Pharmingen	557938
Mouse IgG2b κ, antihuman CD4	OKT4 (alternative)	PerCP-Cy5.5	BD Pharmingen	341654
Mouse IgG2b κ, antihuman CD4	OKT4 (alternative)	PE-Cy7	BioLegend	317414
Mouse IgG1 κ, antihuman CD25	M-A251 (alternative)	APC	BD Pharmingen	555434
Mouse IgG1 κ, antihuman CD25	2A3 (alternative)	APC	BD	340939
LIVE-DEAD® Fixable Near-IR	–	–	Invitrogen	L10119

(b) Our-preferred 3-laser combination.

mAb/Stain	Clone	Fluorochrome	Source	Cat#
Mouse IgG1 κ, antihuman Foxp3	259D	AF488	BioLegend	320212
Mouse IgG1, antihuman CD127	eBioRDR5	Pacific Blue	eBioscience	57-1278-73
Mouse IgG2b κ, antihuman CD4	OKT4 (alternative)	PerCP-Cy5.5	BD Pharmingen	341654
Mouse IgG2b κ, antihuman CD4	OKT4 (alternative)	PE-Cy7	BioLegend	317414
Mouse IgG1 κ, antihuman CD25	M-A251 (alternative)	APC	BD Pharmingen	555434
Mouse IgG1 κ, antihuman CD25	2A3 (alternative)	APC	BD	340939
LIVE-DEAD® Fixable Near-IR	–	–	Invitrogen	L10119

3. Medium for cell staining (FACS wash): 5% FCS (JRH Bioscience) in PBS plus 0.05% Na Azide (Sigma), 0.22 μm filtered to remove particulates.
4. Fixative for biohazard reduction prior to running human samples on flow cytometer (FACS fix): 1% paraformaldehyde (BDH Laboratory Supplies) in PBS.
5. Foxp3 staining kit: our preferred kit contains anti-Foxp3 conjugated with AF 488 (Biolegend, 320212, 100 tests), plus Fix/Perm Buffer set (Biolegend, 421403, 100 tests). Many kits are marketed and we have not tested them all in parallel to determine the best kit.
6. 1.5 ml microcentrifuge tubes (Greiner bio-one).
7. 70 μm cell strainers (BD).
8. 5 ml round-bottom tubes (BD).

2.3. Data Collection, Analysis, and Sorting

1. Flow cytometer. Many flow cytometers are suitable for this procedure. We use a FACSCanto II from BD.
2. Analysis software. We use FlowJo from Treestar.

2.4. Variations of Basic Protocol

Additional antibodies

mAb/Stain	Clone	Fluorochrome	Source	Cat#
Antihuman CD45RO	UCHL1	Biotinylated	Conjugated in house	–
Streptavidin		Pacific Orange	Invitrogen	S32365
Mouse IgG1 κ, antihuman CD45RA	L48	PE-Cy7	BD Pharmingen	337167
Rat IgM, antihuman CLA	HECA-452	PE	Miltenyi Biotec	130-091-635
Rat IgG2a κ, antihuman β7	FIB504	PE	BD Pharmingen	555945
Mouse IgG2b κ, antihuman CCR6	TG7/CCR6	PerCP/Cy5.5	Biolegend	335505

3. Methods

This is a basic protocol for identifying Treg cells, using a minimal cocktail of antibodies to CD4, CD25, and CD127, with or without addition of anti-Foxp3. The use of anti-CD127 is important as it is the combination of 2 Treg markers, one expressed at a higher level than conventional cells (CD25 and/or Foxp3) and one at a lower level (CD127) that allows a relatively clean gate to be set even though neither marker is adequate on its own (Fig. 1c, d). In the commonly used combination of CD25 and Foxp3, cells generally co-express low levels of each marker and so the combination is only marginally more discriminating than use of either as a single marker (Fig. 1b).

3.1. Cell Preparation

3.1.1. Isolation of Mononuclear Cells from Peripheral Blood

This protocol is designed for a single venous blood sample collected in a 9 ml Lithium Heparin tube. All blood samples are handled in a Class 2 biosafety hood and under sterile conditions.

1. First take an aliquot (usually 50–100 μl is required) for determining the concentration of leucocytes and lymphocytes using an automated hematology analyser such as the Sysmex KX-21.
2. Dilute the remaining blood 1:2 in PBS and transfer to a 50-ml tube.
3. Layer 12 ml of Ficoll-Paque Plus solution under the blood.
4. Centrifuge for 30 min at 1,600 rpm, 22°C with the break disengaged.

5. Collect peripheral blood mononuclear cells (PBMCs) from the interface between the Ficoll-Paque and the diluted plasma and transfer into sterile 15 ml tubes.
6. Wash twice with PBS (10 min, 1,000 rpm, 22°C).
7. Resuspend cells in PBS and take an aliquot for counting.
8. Dilute the counting aliquot in 0.2% eosin and perform a viable cell count using a haemocytometer.
9. Centrifuge the remaining cells and resuspend in FACS wash for staining.

   OR
10. Resuspend in ice cold freezing medium at a concentration of $2–4 \times 10^6$ cells per ml for freezing. Freeze 1 ml aliquots in –70°C freezer at –1°C/min using a Cryo-1°C/min Freezing Container and then store in liquid NO_2 until use (see Note 4.1).

3.1.2. Isolation of Mononuclear Cells from Tissue Biopsies

This protocol is designed for a tissue biopsy up to 125 mm³.

1. Collect specimens in cold TCM and store on ice until processing.
2. Cut the tissue into <0.5 mm³ cubes.
3. Suspend in 1 ml TCM containing a final concentration of 100 U/ml Collagenase III and 0.1 mg/ml DNase I in 14 ml round bottom tubes. Incubate 60–90 mins at 37°C with shaking – for example at 200 rpm in an orbital mixer incubator (Ratek).
4. Pour the digested tissue through a 70 μm nylon cell strainer to remove debris. Rinse the strainer through with PBS to wash any trapped cells from the strainer.
5. Wash twice with PBS (1,500 rpm, 10 mins, 4°C).
6. Resuspend cells in PBS and take an aliquot for counting.
7. Dilute the aliquot in 0.2% eosin and perform a viable cell count using a haemocytometer.
8. Centrifuge the remaining cells and resuspend in FACS wash for staining.

   OR
9. Resuspend in ice cold freezing medium at a concentration of $2–4 \times 10^6$ cells per ml for freezing. Freeze 1 ml aliquots in –70°C freezer at –1°C/min using a Cryo-1°C/min Freezing Container and then store in liquid NO_2 until use.

3.2. Cell Staining

1. Thaw the frozen mononuclear cells rapidly in a 37°C water bath and transfer to tubes containing 9 ml of prewarmed TCM (37°C).
2. Centrifuge at 1,000 rpm, 10 min, 22°C.

3. Resuspend in 2 ml cold FACS wash.
4. Take an aliquot, dilute in 0.2% eosin and perform a viable cell count using a haemocytometer.
5. Transfer 1×10^6 viable cells from each sample to a 96-well U-bottom plate.
6. Centrifuge the plate at 1,500 rpm, 4°C for 3 min to pellet cells. Carefully remove the supernatant with a pipette.
7. Add 100 μl of prediluted antibody mixture to all samples and mix by pipetting. All antibodies used have been titrated to obtain the best profile for the staining (see Note 4.2).
8. Incubate on ice for 45 min.
9. Wash cells twice with FACS wash (centrifuge at 1,500 rpm for 3 min). Remove supernatant carefully and discard.
10. *For analysis without Foxp3 staining*: Resuspend cells in 150 μl FACS fix and incubate 15 mins on ice. Wash cells (centrifuge 1,500 rpm, 4°C, 5 mins), resuspend in 200 μl.
11. *For Foxp3 staining*:
    (a) Transfer cells to 1.5 ml microcentrifuge tubes.
    (b) Fix cells in 1 ml 1× Biolegend Foxp3 Fix/Perm solution for 20 min, 22°C in the dark.
    (c) Centrifuge cells at 1,100 rpm, 5 min, 22°C and remove supernatant.
    (d) Wash cells once with 200 μl FACS wash and discard supernatant.
    (e) Wash cells again with 1 ml of Foxp3 Perm buffer at 1,100 rpm, 5 min, 22°C and discard supernatant.
    (f) Resuspend cells in 1 ml Foxp3 Perm buffer and incubate for 15 min in the dark (22°C).
    (g) Spin cells at 1,500 rpm, 5 min, 22°C and remove supernatant.
    (h) Resuspend in 100 μl Foxp3 perm buffer and add 5 μl of anti-Foxp3 antibody.
    (i) Incubate for 30 min in the dark (22°C).
    (j) Wash cells twice with 200 μl FACS wash (1,500 rpm, 5 mins, 22°C). resuspend in 200 μl FW and keep at 4°C in the dark until data acquisition.
12. Filter cells with 50 μm nylon mesh into 5 ml FACS tubes before data acquisition.

3.3. Data Collection and Analysis

1. Prepare compensation controls and collect at least 10^4 events of each. We use antibody capture beads for convenience but PBMCs can also be used. For samples, collect at least $1–2 \times 10^5$ events (excluding dead cells and fragments with low FSC) on

flow cytometer. For tissue samples in which CD4⁺ T cells may comprise only 5–10% of total, it may be necessary to collect up to $1-2 \times 10^6$ events. We generally run samples uncompensated and perform compensation only during data analysis with FlowJo. This prevents errors due to overcompensation, which cannot be reversed after data collection.

2. Data analysis is a crucial step in Treg detection. Figure 2 outlines our gating strategy for standard CD4/25/127 staining of PBMCs. The first three gates are designed to remove artifacts due to fluidics instability and cell doublets, after which dead cells are excluded. After gating for $CD4^+$ T cells using a size parameter, the standard Treg CD25/127 gate is applied (5).

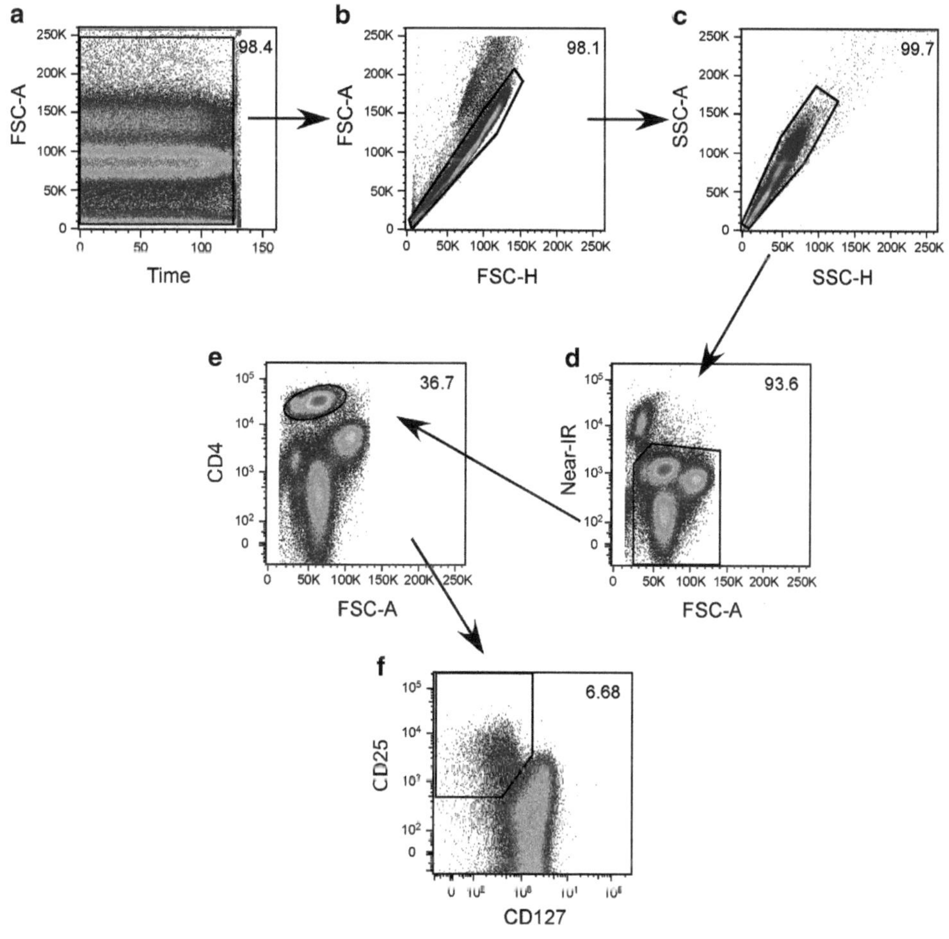

Fig. 2. Standard gating for Treg cells in human peripheral blood after staining for CD4, CD25, and CD127. (**a**) Cells are gated on the basis of time to exclude artifacts due to fluidics instability at the start and end of the sample. (**b**) and (**c**) Use of sequential pulse height vs. area gates for forward and side scatter allows doublets to be identified and excluded. (**d**) Use of the fixable dead cell exclusion dye LIVE-DEAD® Fixable Near-IR allows dead cells to be excluded from fixed human samples. (**e**) $CD4^{hi}$ cells with low forward scatter are lymphocytes. The cells with higher FSC-A and lower CD4 expression are monocytes. (**f**) Gating for Tregs using the nonorthogonal CD25/CD127gate.

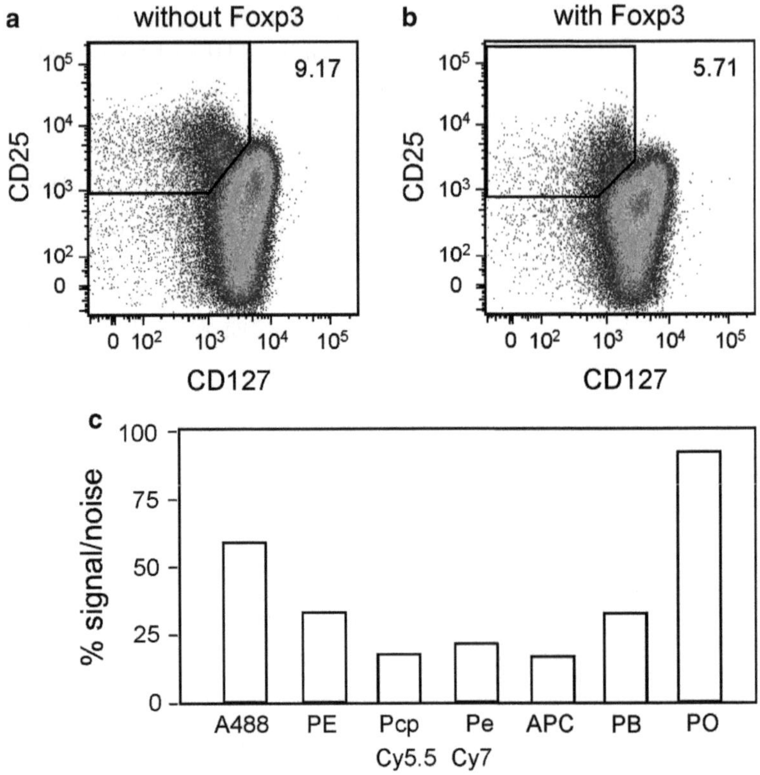

Fig. 3. Effect of Foxp3 fix/perm treatment on fluorescence of the conjugated antibodies used for Treg identification. (**a**) Expression of CD25 and CD127 by peripheral blood leukocytes gated for CD4+ T cells, after staining with CD4/CD25/CD127, Near-IR and formaldehyde fixation. (**b**) The same sample, with additional staining for Foxp3 according to the manufacturer's instructions. (**c**) Effect of Foxp3 fix/perm treatment on fluorescence. Single color compensation samples were stained and formaldehyde fixed. Half of each sample was then treated with fix/perm buffers as per the manufacturer's instructions for Foxp3 staining. This treatment reduced the signal and increased the background of unstained cells for each fluorochrome tested. Signal:noise ratio was calculated for all samples and the reduction in signal:noise after fix/perm was expressed as a percentage of signal:noise without fix/perm treatment. As expected, fluorescence of tandems such as PCP-Cy5.5 and PE-Cy7 was reduced, but the unexpected drop in the APC signal is problematical when anti-CD25-APC is part of the staining cocktail.

3. The addition of Foxp3 to staining cocktails may have unanticipated consequences, as illustrated in Fig. 3. In our hands, sensitive Foxp3 detection is inevitably accompanied by a reduction in sensitivity for other parameters. The fixation-dependent change in forward and side scatter prevents the use of FSC/SSC gating to exclude dead cells, but this problem can be overcome by including a fixable dead cell stain. However the reduction in fluorescence in the other channels is a major problem, particularly as it affects detection of CD25 via an APC conjugate and prevents the use of tandem fluorochrome conjugates to detect further markers on Treg subsets (Fig. 3).

4. To overcome the problems associated with Foxp3 staining, we have adopted the following strategy. We stain 2×10^6 cells from each sample with 200 μl CD4/25/127 cocktail. When staining is complete, we remove half the sample for additional Foxp3 staining. The matched samples are run in parallel and the number of $CD25^+CD127^{lo}$ cells is calculated from the first sample and the number of $Foxp3^+CD127^{lo}$ cells from the second sample (in which CD25 signal is too low for reliable $CD25^+CD127^{lo}$ gating, as shown in Fig. 3b). The estimates of Treg numbers from the 2 samples should be within 10% of each other.
5. Because the staining and gating procedure has such a profound effect on the detection sensitivity for Tregs, we believe that it is important to include full methodological details in publications. Flow data plots showing how gates have been placed, in addition to calculated values for each individual, are vital if data from different laboratories are to be usefully compared.
6. Treg numbers can be expressed either as concentration in venous blood or as a percentage of another parameter such as $CD4^+$ T cells. In our hands, lymphocyte concentrations are far more variable than percentages. In particular, the percentage of $CD4^+$ T cells within a lymphocyte FSC-A/SSC-A gate is stable in healthy subjects across a wide range of ages. We therefore routinely present Treg data expressed as a percentage of $CD4^+$ T cells. With essentially unlimited capacity to include supplementary data in most publications, it is also very useful to include concentrations in addition to percentages. For conditions such as HIV infection in which changes in $CD4^+$ T cell concentrations are a crucial and highly variable parameter, we suggest that both measures should be routinely presented.

3.4. Variations on the Basic Protocol

1. *Staining for CD45RA and/or CD45RO*: Include appropriate antibodies in the staining cocktail (see Subheading 3.2, step 7 above).
   (a) Gating for Treg subsets must take into account the lower expression of both CD25 and Foxp3 by $CD45RA^+RO^-$ Treg cells compared with $CD45RA^-RO^+$ Tregs. However the "background" expression of CD25 by $CD45RA^+RO^-$ non-Tregs is also lower. Hence the precise positioning of the Treg gate is often slightly different within the two subpopulations.
   (b) Figure 4 illustrates the use of CD45RO to subset Tregs. When the CD25/127 gate appropriate for total $CD4^+$ T cells (Fig. 4a) is applied to the $CD45RO^-$ subset, it underestimates Treg numbers (Fig. 4e) whereas for the

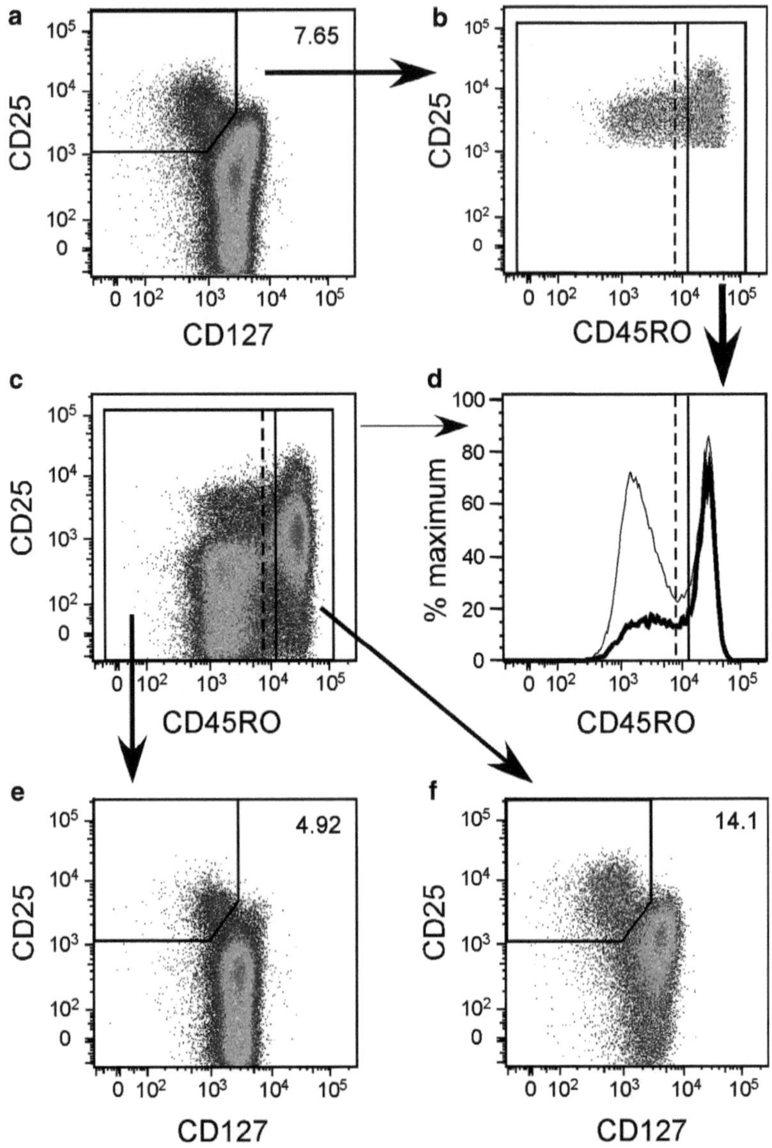

Fig. 4. Use of CD45RO staining to subset Tregs and total CD4+ T cells in human peripheral blood after staining for CD4, CD25, CD127, and CD45RO. (**a**) Standard gating of CD4+ T cells for Tregs, as shown in Fig. 2. (**b**) Expression of CD45RO by Tregs gated as in (**a**). Optimal gating of Tregs into CD45RO- and CD45RO+ subpopulations, based on expression of CD25 and CD45RO, is shown by the solid black lines. The *dotted line* is derived from optimal gating of conventional CD4+ T cells as in (**c**). (**c**) Expression of CD45RO by CD4+ T cells. Optimal gating into CD45RO- and CD45RO+ subpopulations is shown by the *dotted line*. (**d**) Comparison of CD45RO expression by Tregs (*bold line* with *solid black gate*) and total CD4+ T cells (fine line with *dotted black gate*). This overlay shows that the solid black gate suitable for CD45RO+ cells within total CD4+ T cells includes a subpopulation of Tregs that are phenotypically identical to CD45RO− Tregs in terms of CD25 expression. Thus the position of the CD45RO gate must be adjusted according to whether conventional or Treg cells are gated. (**e**) and (**f**) Gating for Tregs within CD45RO- (**e**) and CD45RO+ (**f**) CD4+ T cells, using the Treg gate shown in (**a**). These plots show that the gate is not optimal for either population, illustrating the necessity for modifying gates according to the differential expression of CD25 and CD127 by CD45RO- and CD45RO+ CD4+ T cells respectively.

CD45RO⁺ subset, it overestimates numbers (Fig. 4f). The most accurate analysis requires the percentage in each subset to be determined separately (5).

(c) Gating for CD45RO may also present a challenge. Comparison of CD45RO expression within Tregs (Fig. 4b) vs. total CD4⁺ T cells (Fig. 4c, overlays in Fig. 4d) shows that the optimal CD45RO⁺ gate for Tregs is placed at a significantly higher channel than for total CD4⁺ T cells. Thus it is often necessary to set a preliminary Treg CD25/127 gate, then set the CD45RO gate appropriate for Tregs, and then apply that CD45RO gate to total CD4⁺ T cells, after which the Treg CD25/127 gate can be adjusted to achieve the best accuracy within each subset.

2. *Staining for further Treg subsets*: Include appropriate antibodies in the staining cocktail (see 3.2.7 above). As noted above, use of Foxp3 staining is generally incompatible with cocktails containing multiple subset markers, especially if they rely on tandem dyes.

(a) A gating strategy for possible Treg subsets based on expression of tissue-homing integrins is illustrated in Fig. 5.

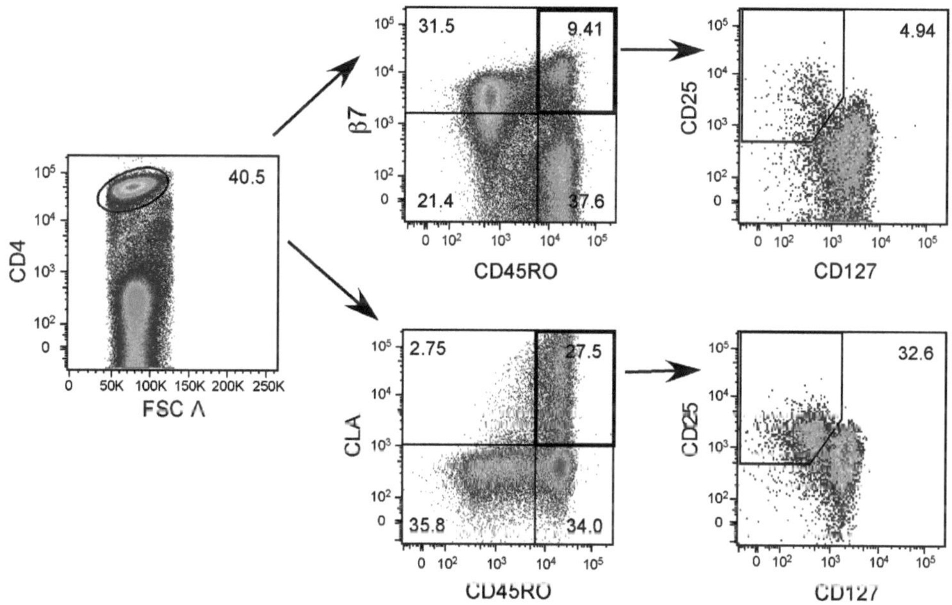

Fig. 5. Use of additional surface markers to subset human Tregs. Expression of integrin β7 and CLA by CD45RO⁺ CD4⁺ T cells is shown in the *bold rectangular gate*. Differential expression between conventional and Treg cells is indicated by the relative decrease in Treg cells within the integrin β7-expressing population and the marked increase in Treg cells within the CLA-expressing population. Use of an alternative gating strategy with sequential gates for Tregs, CD45RO and then integrin β7 or CLA should yield identical final populations, excluding unanticipated gating artifacts.

4. Notes

1. *Use of frozen cells*: Our comparison of results from fresh vs. frozen samples has indicated that virtually identical fluorescence values and population percentages are obtained for all the markers discussed here. Expression of a subset of chemokine receptors, including CCR5, may however be affected by freeze/thaw procedures.
2. *Titering antibodies and testing cocktails*: Reagents conjugated with different fluorochromes and/or from different companies can give vastly different signal levels.
   (a) The signal: noise ratio for detecting CD25, CD127 and Foxp3 expression by Tregs is in the mid to low range (unlike CD4, for which even a suboptimal signal is usually adequate), so it is vital that the antibody cocktail be optimised.
   (b) In general, highly expressed molecules or those for which very high affinity antibodies are available can be detected with the dimmer fluorochromes such as Pacific Blue and Pacific Orange (we detect CD4 and CD45RO respectively in those channels in our 6–8 color panels). Of the remaining fluorochromes, PE and APC are the brightest and so they can be reserved for CD25 and/or Foxp3. Substituting AF488 for FITC usually increases the signal in that channel, as the AF488 does not undergo intermolecular quenching, unlike FITC.
   (c) Avoid detecting highly expressed molecules in a fluorescence channel with significant cross-over into neighboring channels, especially when the molecule is expressed by only a subset of Tregs. For example, the signal from CD45RO-APC-Cy7 may interfere significantly with CD25-APC, leading to difficulty in detecting CD25 expression by $CD45RO^+$ but not $CD45RO^-$ T cells.
   (d) Most commercial suppliers market anti-human antibody conjugates as a number of tests of given volume, to be used in a final volume of 100 μl per sample. In many cases (for example, CD25-APC) these concentrations are not saturating, as indicated by the readily apparent drop in fluorescence with only twofold dilution (Fig. 6b, e). For an antibody such as CD25-APC, even a small decrease in fluorescence can cause gating problems, and the recommended amount must be used. For other antibodies such as CD4-PE-Cy7, the signal:noise ratio is so high that nonsaturating concentrations can be diluted even further without diminishing the accuracy of gating

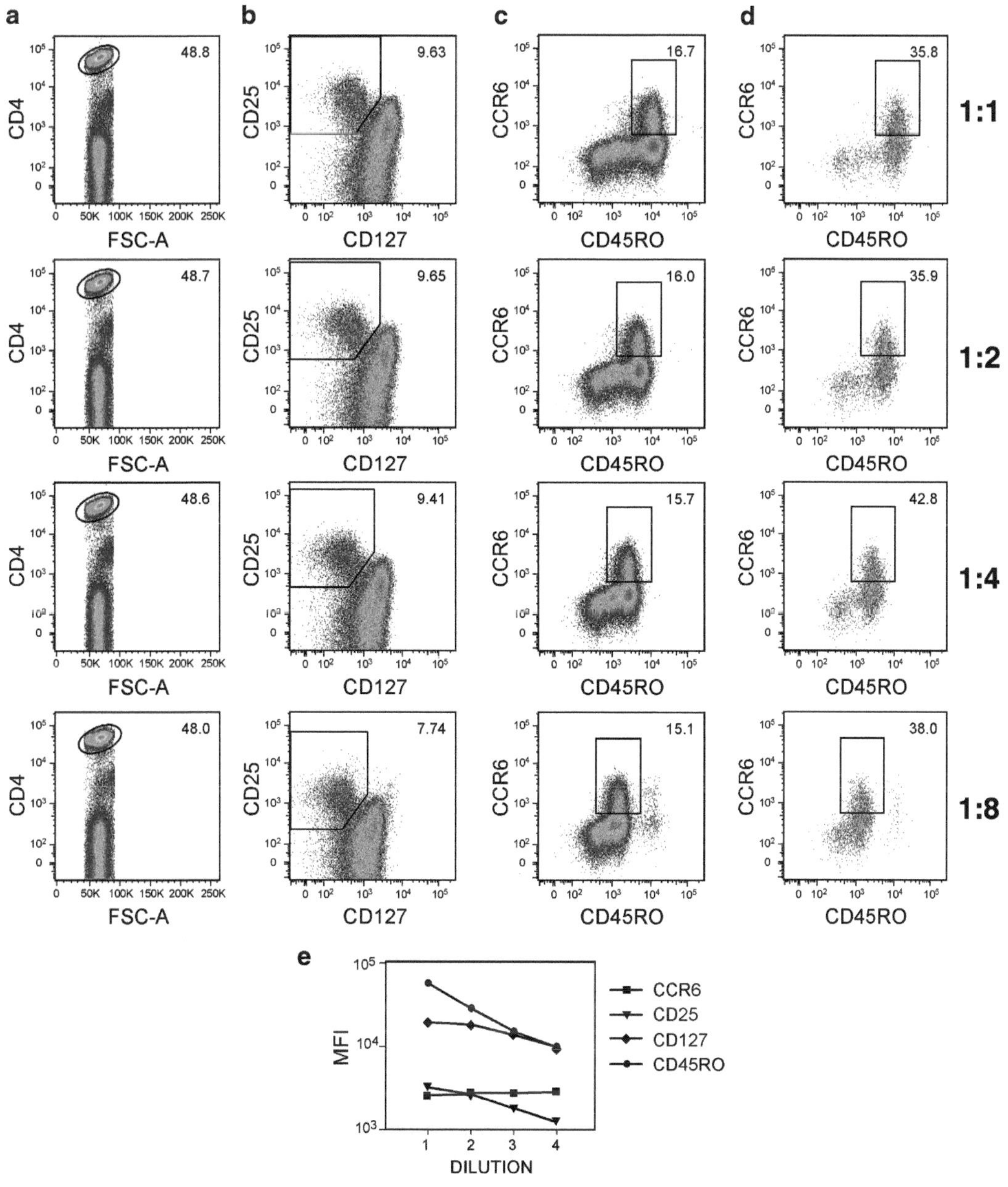

Fig. 6. Strategy for titering antibodies for Treg subsetting. In this example, a mixture of antibodies to CD4, CD25, CD127, CD45RO and CCR6 at the concentrations recommended by the manufacturer were diluted in a two-fold series for staining PBMCs. Each row is from a single sample stained at the indicated dilution and gated as indicated for CD4^{+} T cells (**a**), CD25^{+}CD127^{lo} Treg cells within CD4^{+} T cells (**b**), CCR6^{+}CD45RO^{+} cells within CD4^{+} T cells (**c**) and CCR6^{+}CD45RO^{+} cells within Tregs (**d**). (**e**) MFI of cells within the indicated gates is shown as a function of dilution. Although the signal from CD25, CD127 and CD45RO declines with each dilution, it is still possible to gate on Treg cells in each sample, in order to measure the signal from the test antibody. In this case, the recommended concentration of the CCR6 antibody can be reduced at least eightfold with no loss of signal.

(Fig. 6a). Lastly, some antibodies, such as the anti-CCR6 shown in Fig. 6c–e, can be diluted at least eightfold without any change in fluorescence intensity.

(e) Titering each antibody individually can be very expensive in terms of time and reagents, because each sample must also be stained for CD4/CD25/CD127 so that expression of the new marker by Tregs can be determined. We recently tested a new strategy in which the complete cocktail, containing CD4/CD25/CD127 as well as the test antibody or antibodies, is titered (Fig. 6). The cocktail is prepared using the recommended concentrations in a final volume of 200 μl, and then a twofold dilution series is performed to give final volumes of 100 μl. Although this reduces the signal from the CD4/25/127 cocktail simultaneously with that of the test Ab, the Treg population is clearly gateable out to a dilution of 1:8 (Fig. 6b). In our experience, the antibody volume calculated from titration can be as little as 1:20 of the manufacturer's recommended volume.

(f) Signal cross-over from channel to channel can vary widely, particularly with tandem dyes, so it is also important to pretest each of the cocktails to check for unanticipated interference in any of the channels.

3. *Validating gating strategies*: we have presented several of our "standard" gating strategies in the Figures, but we always check the multiple strategies with each new experiment to make sure that the same cells are included in (and excluded from) the final population.

4. *Validating staining reproducibility*: we always include control samples in each experiment. This requires freezing of multiple vials prepared from a single donor on a single occasion.

5. *Tissue-specific effects*: CD25 expression can vary widely in different tissues, presumably due to availability of IL-2 which, in the mouse at least, upregulates the expression of CD25. CD25 expression in synovium of some patients with rheumatoid arthritis is high enough to define a clear $CD25^{+}$ population without requiring either CD127 or Foxp3 staining (8). In bowel mucosa, on the other hand, our unpublished results show that a significant number of $Foxp3^{+}$ cells express undetectable levels of CD25. This, together with the significant number of $CD127^{lo}$ activated conventional $Foxp3^{-}CD4^{+}$ T cells in bowel mucosa, makes the use of anti-Foxp3 staining essential for identification of Treg cells.

References

1. Sakaguchi, S., N. Sakaguchi, M. Asano, M. Itoh, and M. Toda. 1995. Immunologic self-tolerance maintained by activated T cells expressing IL-2 receptor a-chains (CD25); breakdown of a single mechanism. *J Immunol* 155:1151–1164.
2. Baecher-Allan, C., J.A. Brown, G.J. Freeman, and D.A. Hafler. 2001. $CD4^{+}CD25^{high}$ regulatory cells in human peripheral blood. *J Immunol* 167:1245–1253.
3. Valmori, D., A. Merlo, N. Souleimania, C. Hesdorffer, and M. Ayyoub. 2005. A peripheral circulating compartment of natural naive CD4 Tregs. *J Clin Invest* 115:1953–1962.
4. Seddiki, N., B. Santner-Nanan, S.G. Tangye, S.I. Alexander, M. Solomon, S. Lee, R. Nanan, and B. Fazekas de St Groth. 2006. Persistence of naïve $CD45RA^{+}$ regulatory T cells in adult life. *Blood* 107:2830–2838.
5. Seddiki, N., B. Santner-Nanan, J. Martinson, J. Zaunders, S. Sasson, A. Landay, M. Solomon, W. Selby, S.I. Alexander, R. Nanan, A. Kelleher, and B. Fazekas de St Groth. 2006. Expression of IL-2 and IL-7 receptors discriminates between human regulatory and activated T cells. *J Exp Med* 203:1693–1700.
6. Liu, W., A.L. Putnam, Z. Xu-Yu, G.L. Szot, M.R. Lee, S. Zhu, P.A. Gottlieb, P. Kapranov, T.R. Gingeras, B. Fazekas de St Groth, C. Clayberger, D.M. Soper, S.F. Ziegler, and J.A. Bluestone. 2006. CD127 expression inversely correlates with FoxP3 and suppressive function of human $CD4^{+}$ T reg cells. *J Exp Med* 203:1701–1711.
7. Morgan, M.E., J.H. van Bilsen, A.M. Bakker, B. Heemskerk, M.W. Schilham, F.C. Hartgers, B.G. Elferink, L. van der Zanden, R.R. de Vries, T.W. Huizinga, T.H. Ottenhoff, and R.E. Toes. 2005. Expression of FOXP3 mRNA is not confined to $CD4^{+}CD25^{+}$ T regulatory cells in humans. *Hum Immunol* 66:13–20.
8. Cao, D., V. Malmstrom, C. Baecher-Allan, D. Hafler, L. Klareskog, and C. Trollmo. 2003. Isolation and functional characterization of regulatory $CD25^{bright}CD4^{+}$ T cells from the target organ of patients with rheumatoid arthritis. *Eur J Immunol* 33:215–223.

INDEX

George Kassiotis and Adrian Liston (eds.), *Regulatory T Cells: Methods and Protocols*, Methods in Molecular Biology, vol. 707, DOI 10.1007/978-1-61737-979-6, © Springer Science+Business Media, LLC 2011

MIX
Papier aus verantwortungsvollen Quellen
Paper from responsible sources
FSC® C105338

If you have any concerns about our products,
you can contact us on
ProductSafety@springernature.com

In case Publisher is established outside the EU,
the EU authorized representative is:
Springer Nature Customer Service Center GmbH
Europaplatz 3, 69115 Heidelberg, Germany

Printed by Libri Plureos GmbH
in Hamburg, Germany